Pipe Line Rules of Thumb Handbook

Gulf Publishing Company, Book Division, Houston, London, Paris, Tokyo

Pipe Line Rules of Thumb Handbook

Prepared by Gulf Publishing Company, Book Division, in association with the editors of
Pipe Line Industry

Pipe Line Rules of Thumb Handbook

First Printing, *January 1978*
Second Printing, *November 1979*

ISBN 0-87201-698-6
Library of Congress Catalog Number 76-52237

Contents

OIL & PRODUCTS—Pump Stations, 141

CORROSION, 149

ECONOMICS—Pipe Line and Station Operation, 162

CONVERSION FACTORS, 172

1. Construction—Cables, Ropes

Estimating strength of cable

Rule:

1. Change line diameter to eighths
2. Square the numerator
3. Divide by the denominator
4. Read the answer in tons

Example:

Estimate the strength of ½-inch steel cable:

$$\text{Diameter} = \frac{1}{2} = \frac{4}{8}$$

$$\frac{4^2}{8} = \frac{16}{8} = 2$$

The approximate strength of ½-inch steel cable is 2 tons.

Find the working strength of manila rope

The working strength of Manila rope is approximately 900 x (diameter)^2.

or

$$W = 900\ d^2$$

where d is expressed in inches.
W is given in pounds.

Example: What is the working strength of a ¾-inch Manila rope?

The maximum recommended pull is

$$W = 900 \times \frac{3 \times 3}{4 \times 4} = 506 \text{ lbs.}$$

Example: Find the maximum working pull for a 1½-inch Manila rope.

$$W = 900 \times \frac{3}{2} \times \frac{3}{2} = 2025 \text{ pounds.}$$

For rope diameters greater than 2-inches, a factor lower than 900 should be used: in working with heavier rigging it is well to refer to accepted handbooks to find safe working strengths.

How large should drums and sheaves be for various types of wire rope?

The diameter of sheaves or drums should preferably fall within the table* given below for most efficient utilization of the wire rope.

Type of Wire Rope	6 x 19	6 x 37	8 x 19	5 x 28	6 x 25	18 x 7	6 x 7
For best wear	45	27	31	36	45	51	72
Good practice	30	18	21	24	30	34	42
Critical	16	14	14	16	16	18	28

Example: What size should the hoisting drum on a dragline be, if the wire rope is 6 x 19 construction, ¾ inch in diameter?

From the table, good practice calls for 30 diameters which in this instance would be 22½ inches. Loads, speeds, bends and service conditions will also affect the life of wire rope so it is better to stay somewhere between the "good practice" and "best wear" factors in the table.

* *Construction Methods and Machinery,* by F. H. Kellogg, Prentice-Hall, Inc., 1954.

Find advantages of block and tackle, taking into account *pull out friction*

The efficiency of various sheaves differs. For one with roller bearings the efficiency has been estimated at 96.2 percent. For plain bearing sheaves a commonly used figure is 91.7 percent. The following formula will give close results.

$$MA = \frac{W}{w} = E\,\frac{1 - E^n}{1 - E}$$

MA = Mechanical Advantage

Where W = Total weight to be lifted by the assembly

w = Maximum line pull at the hoist

n = Number of working parts in the tackle

E = Efficiency of individual sheaves

It is assumed that the line leaving the upper block goes directly to the hoist without additional direction change (requiring a snatch block).

Example: Find the Mechanical Advantage of a 4-part block and tackle using upper and lower blocks having journal bearings, which have an efficiency of 91.7%.

$$MA = .917\,\frac{1 - .917^4}{1 - .917} = .917\left(\frac{1 - .707}{1 - .917}\right) = .917\left(\frac{.293}{.083}\right)$$

$$MA = 3.25$$

If the load weighed 3,250 pounds, what pull would be required on the lead line?

$$\frac{W}{w} = MA$$

$$\frac{3250}{w} = 3.25$$

$$w = 1000 \text{ pounds.}$$

Safe loads for wire rope

Safe Loading Factors For Wire Rope And Chains

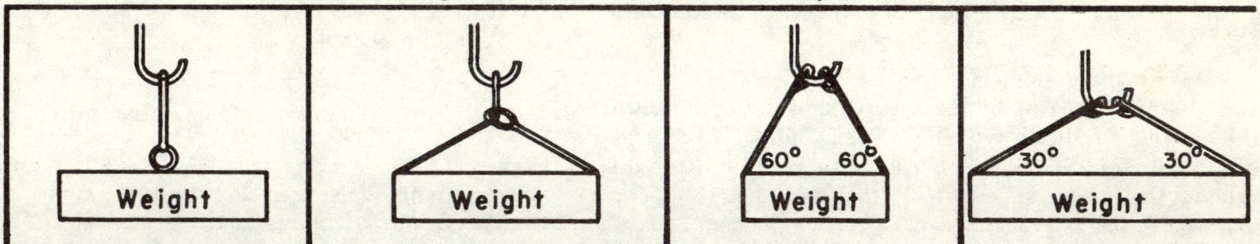

Dia. In.	SINGLE PART LINE			SINGLE PART-CHOKER HITCH			TWO PARTS			TWO PARTS		
	Wire Rope (Tons)	Alloy Chain (Tons)	Wrought Chain (Tons)	Wire Rope (Tons)	Alloy Chain (Tons)	Wrought Chain (Tons)	Wire Rope (Tons)	Alloy Chain (Tons)	Wrought Chain (Tons)	Wire Rope (Tons)	Alloy Chain (Tons)	Wrought Chain (Tons)
¼	.5	....	....	.35	....	....	.6	....	....	.35	....	....
9/32	.5	1.3	....	....	1	....	....	1.8	....	....	1	....
⅜	1	2.3	.8	.75	1.8	.63	1.4	3.1	1.1	.8	1.8	.64
7/16	1	3	1.1	....	2.4	.9	....	4.2	1.5	....	2.4	.9
½	1.8	4	1.4	1.4	3.2	1.1	2.4	5.6	1.9	1.4	3.2	1.1
9/16	1.8	5	....	....	4	....	....	7	....	....	4	....
⅝	2.8	6	2.2	2.1	4.8	1.8	3.7	8.2	3	2.2	4.8	1.8
¾	3.9	7.8	3.2	2.8	6.2	2.6	5.2	10.7	4.4	3	6.2	2.6
⅞	5.2	10.4	4.3	3.7	8.3	3.4	7	14.4	6	4	8.3	3.4
1	6.6	13.8	5.7	4.6	10.8	4.6	9	18.7	7.8	5.2	10.8	4.6
1⅛	6.6	17.8	6.7	....	14.2	5.4	....	24.7	9.3	....	14.2	5.4
1¼	9.7	23.	8.3	6.6	18.2	6.6	13	31.9	11.5	7.5	18.4	6.6
1⅜	9.7	....	10	....	....	8	....	....	13.9	....	....	8
1½	13.2	....	12	9.2	....	9.6	17.3	....	16.6	10	....	9.6
1⅝	13.2	....	14	....	....	11.2	....	....	19.4	....	....	11.2
1¾	17	....	16	12.1	....	12.8	21.5	....	22.2	12.4	....	12.8
2	20.6	....	21	15.4	....	16.8	26.7	....	29.1	15.4	....	16.8

There are various kinds of wire rope and chain available, and where it is possible strength factors should be reviewed before using any particular kind. Tables here may be used on construction jobs. They allow for a safety factor of four as based on mild plow steels, etc. Where the wire rope is made of improved plow steel, etc., it will have a greater safety factor.

Strength and weight of popular wire rope

The following tables give the breaking strength for wire rope of popular construction made of improved plow steel.

6 x 19

SIZE	Breaking Strength	Weight
¼	5,480	0.10
5/16	8,520	0.16
⅜	12,200	0.23
7/16	16,540	0.31
½	21,400	0.40
9/16	27,000	0.51
⅝	33,400	0.63
¾	47,600	0.90
⅞	64,400	1.23
1	83,600	1.60
1⅛	105,200	2.03
1¼	129,200	2.50
1⅜	155,400	3.03

6 x 19

Size	Breaking Strength	Weight
1½	184,000	3.60
1⅝	214,000	4.23
1¾	248,000	4.90
1⅞	282,000	5.63
2	320,000	6.40

Conversion Factors For Wire Rope of Other Construction:

To apply the above table to wire rope of other construction, multiply by the following factors:

Wire Rope Construction	6 x 19	6 x 29	6 x 37	18 x 7
Strength Factors	1.00	0.96	0.95	0.92
Weight Factors	1.00	0.97	0.97	1.08

For example: Find the breaking strength of 6 x 29 improved plow steel wire rope two inches in diameter.

Strength = 320,000 x 0.96 = 307,000 lbs.

The weight can be found the same way.

How to estimate length of material contained in roll

Where material of uniform thickness, like belting, is in a roll, the total length may be obtained by the following rule:

Measure the diameter of the hole in the center, and of the outside of the roll, both measurements in inches; count the number of turns; multiply the sum of the two measured diameters by the number of turns, and multiply this product by 0.13; the result is the total length of the material in feet.

Example: A roll of belting contains 24 turns. The diameter of the hole is two inches, and of the outside of the roll is 13 inches:

$$(2 + 13) \times 24 \times 0.13 = 46.8;$$

the roll contains 46.8 feet of belting.

Note: The rule can even be applied to materials as thin as pipe line felt; counting the turns is not as difficult as might appear without a trial.

Measuring the diameter of wire rope

Correct way

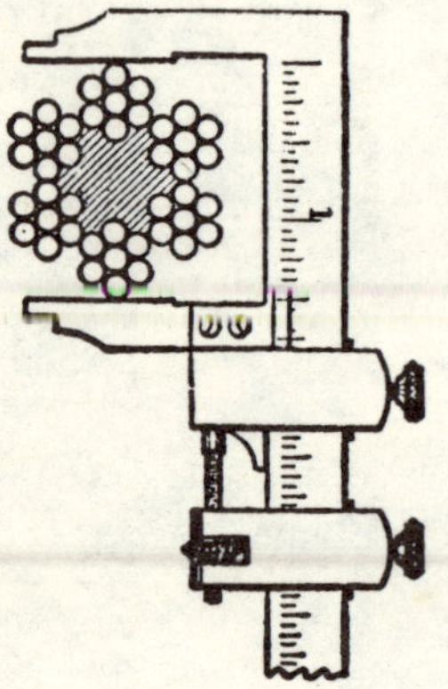

Incorrect way

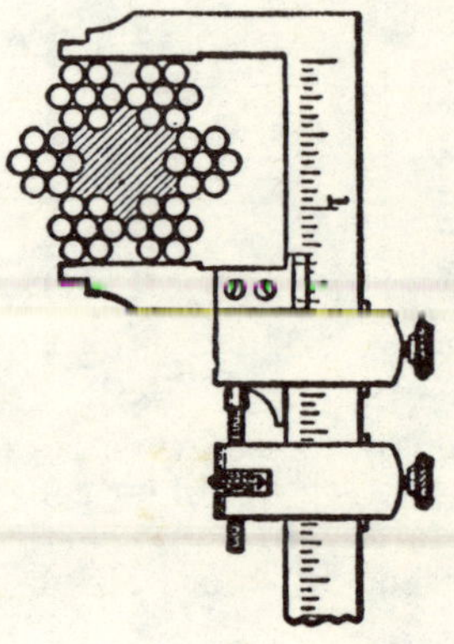

Wire rope: field troubles and their causes

All wire rope will eventually deteriorate in operation or have to be removed simply by virtue of the loads and reversals of load applied in normal service. There are, however, many conditions of service or inadvertent abuse which will materially shorten the normal life of a wire rope of proper construction although it is properly applied. The following field troubles and their causes give some of the field conditions and practices which result in the premature replacement of wire rope. It should be borne in mind that in all cases the contributory cause of removal may be one or more of these practices or conditions.

Wire-Rope Trouble	Cause
a. Rope broken (all strands).	Overload resulting from severe impact, kinking, damage, localized wear, weakening of one or more strands, or rust-bound condition and loss of elasticity.
b. One or more whole strands parted.	Overloading, kinking, divider interference, localized wear, or rust-bound condition. Fatigue, excessive speed, slipping, or running too loosely. Concentration of vibration at dead sheave or dead-end anchor.
c. Excessive corrosion.	Lack of lubrication. Exposure to salt spray, corrosive gases, alkaline water, acid water, mud, or dirt. Period of inactivity without adequate protection.
d. Rope damage by careless handling in hauling to the well or location.	Rolling reel over obstructions or dropping from car, truck, or platform. The use of chains for lashing, or the use of lever against rope instead of flange. Nailing through rope to flange.
e. Damage by improper socketing.	Improper seizing which allows slack from one or more strands to work back into rope; improper method of socketing or poor workmanship in socketing, frequently shown by rope being untwisted at socket, loose or drawn.
f. Kinks, dog legs, and other distorted places.	Kinking the rope and pulling out the loops such as in improper coiling or unreeling. Improper winding on the drum. Improper tie-down. Open-drum reels having longitudinal spokes too widely spaced. Divider interference. The addition of improperly spaced cleats to increase the drum diameter. Stressing while rope is over small sheave or obstacle.

g. Damage by hooking back slack too tightly to girt.	Operation of walking beam causing a bending action on wires at clamp and resulting in fatigue and cracking of wires, frequently before rope goes down into hole.
h. Damage or failure on a fishing job.	Rope improperly used on a fishing job, resulting in damage or failure as a result of the nature of the work.
i. Lengthening of lay and reduction of diameter.	Frequently produced by some type of overloading, such as an overload resulting in a collapse of the fiber core in swabbing lines. This may also occur in cable-tool lines as a result of concentrated pulsating or surging forces which may contribute to fiber-core collapse.
j. Premature breakage of wires.	Caused by frictional heat developed by pressure and slippage, regardless of drilling depth.
k. Excessive wear in spots.	Kinks or bends in rope due to improper handling during installation or service. Divider interference; also, wear against casing or hard shells or abrasive formations in a crooked hole. Too infrequent cut-offs on working end.
l. Spliced rope.	A splice is never as good as a continuous piece of rope, and slack is liable to work back and cause irregular wear.
m. Abrasion and broken wires in a straight line. Drawn or loosened strands. Rapid fatigue breaks.	Injury due to slipping rope through clamps.
n. Reduction in tensile strength or damage to rope.	Excessive heat due to careless exposure to fire or torch.
o. Distortion of wire rope.	Damage due to improperly attached clamps or wire-rope clips.
p. High strands.	Slipping through clamps, improper seizing, improper socketing or splicing, kinks, dog legs, and core popping.
q. Wear by abrasion.	Lack of lubrication. Slipping clamp unduly. Sandy or gritty working conditions. Rubbing against stationary object or abrasive surface. Faulty alignment. Undersized grooves and sheaves.
r. Fatigue breaks in wire.	Excessive vibration due to poor drilling conditions, i.e., high speed, rope slipping, concentration of vibration at dead sheave or dead-end anchor, undersized grooves and sheaves, and improper selection of rope construction. Prolonged bending action over spudder sheaves, such as that due to hard drilling.
s. Spiraling or curling.	Allowing rope to drag or rub over pipe, sill, or any object during installation or operation. It is recommended that a block with sheave diameter 16 times the nominal wire-rope diameter, or larger, be used during installation of the line.
t. Excessive flattening or crushing.	Heavy overload, loose winding on drum, or cross winding. Too infrequent cut-offs on working end of cable-tool lines. Improper cutoff and moving program for cable-tool lines.
u. Bird-caging or core-popping.	Sudden unloading of line such as hitting fluid with excessive speed. Improper drilling motion or jar action. Use of sheaves of too small diameter or passing line around sharp bend.
v. Whipping off of rope.	Running too loose.
w. Cutting in on drum.	Loose winding on drum. Improper cutoff and moving program for rotary drilling lines. Improper or worn drum grooving or line turn-back plate.

Capacity of Drums

The capacity of wire line drums may be figured from formula

$M = (A + B) \times A \times C \times K$

where (See figure below)

M = rope capacity of drum in feet

A = depth of flange—inches

B = diameter of drum—inches

C = width of the drum between flanges—inches

K = constant depending on rope size shown below

Rope Size	K	Rope Size	K
3/8	1.86	1	.262
7/16	1.37	1 1/8	.207
1/2	1.05	1 1/4	.167
9/16	.828	1 3/8	.138
5/8	.672	1 1/2	.116
3/4	.465	1 5/8	.099
7/8	.342	1 3/4	.085

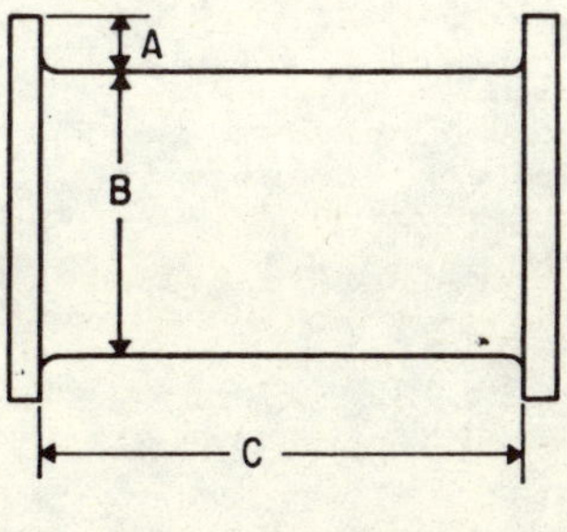

Capacity of wire line drums.

2. Construction—Coatings

NAPCA specification designations for enamel coating

SPECIFICATION SYMBOL												
*PART #1	**PART #2	SHOT BLAST	PRIMER	3/32″ MIN. ENAMEL	GLASS WRAP	15# FELT WRAP	1/32″ MIN. ENAMEL	SEAL COAT	GLASS WRAP	15# FELT WRAP	HEAVY KRAFT PAPER	ELECTRICAL INSPECTION
TF-		X	X	X		X					X	X
AF-		X	X	X		X					X	X
TG-		X	X	X	X						X	X
AG-		X	X	X	X						X	X
TGF-		X	X	X	X	X					X	X
AGF-		X	X	X	X	X					X	X
TGG-		X	X	X	X		X		X		X	X
AGG-		X	X	X	X		X		X		X	X
TGGF-		X	X	X	X		X		X	X	X	X
AGGF-		X	X	X	X		X		X	X	X	X
TFF-		X	X	X		X	X			X	X	X
AFF-		X	X	X		X	X			X	X	X
TFS-		X	X	X		X		X			X	X
AFS-		X	X	X		X		X			X	X
TGS-		X	X	X	X			X			X	X
AGS-		X	X	X	X			X			X	X
TGFS-		X	X	X	X	X		X			X	X
AGFS-		X	X	X	X	X		X			X	X

***PART #1 SYMBOLS — ENAMEL AND WRAPPERS**

T — Coal Tar
A — Asphalt
F — Felt
G — Glass
O — Outerwrap
S — Seal Coat

****PART #2 SYMBOLS — GRADES OF ENAMEL**

1 — Standard Coal Tar
2 — Semi-Plasticized Coal Tar
3 — Fully-Plasticized Coal Tar
4 — Hot Line Coal Tar
7 — Standard Asphalt
8 — Hot Line Asphalt
H — After Part #2 specifies 5/32″ total enamel thickness

To specify outerwrap, use letter O in place of F.
Whitewash may be substituted for kraft paper.

EXAMPLE: 3/32″ (MINIMUM) fully-plasticized coal tar enamel wrapped with glass, felt and kraft — the number would be TGF-3.

These are not all specifications approved by NAPCA. For other specifications consult your NAPCA Member.

How much primer for a mile of pipe?

Multiply pipe O.D. by 2¾ to get gallons of primer per mile.

Examples:

Pipe 24 inches in diameter will require 24 × 2¾ or 66 gallons of primer per mile;

Pipe 16 inches in diameter will require 16 × 2¾ or 44 gallons of primer per mile; and

Pipe 4½ inches in diameter will take 4½ × 2¾ or 12⅜ gallons of primer per mile (12 gallons in round figures).

The figure above is applicable only where the pipe is new and in good condition. For rough pipe the multiplying factor should be upped from 2¾ to 3.45. In other words, rough 20-inch pipe will require 20 × 3.45 or 69 gallons of primer per mile.

It is hard to estimate the amount of spillage and waste. Temperature and relative humidity have an effect on the way the primer goes on the pipe.

How much coal-tar enamel for a mile of pipe?

For an average cover of 3/32 inch, multiply outside diameter of pipe by 0.6. The answer is in tons of enamel per mile. (Add 15 percent to take care of gate valves, drips, bends, etc.)

Examples:

How many tons of coal-tar enamel will be required to apply a 3/32-inch coat to a 26-inch pipe line?

26 x 0.6 = 15.6 tons per mile.

Add 15 percent to take care of gate valves, bends etc.

15.6 + 2.3 = 17.9 or 18 tons per mile.

How many tons of coal-tar enamel will be required to apply a 3/32-inch coat to a 8.625-inch pipe line?

8.625 x 0.6 = 5.2 tons per mile

5.2 x 15% = 0.8

Total . . 6 tons per mile.

These factors strongly affect the amount of coating used per mile on pipe lines: 1 Temperature of the pipe, and ambient temperatures. (It is assumed that the temperature of the enamel can be controlled.) 2. The kind of dope machine. 3. Humidity. 4. The number of bends per mile. 5. The number of valves, crossovers, etc. 6. Condition of the pipe.

This latter point has a great effect on the amount of enamel required for a mile of pipe line. In fact, where reconditioned pipe is being coated, it is recommended that the amount of enamel be increased by about 20 percent.

How much wrapping for a mile of pipe?

For 14-inch to 48-inch pipe multiply diameter in inches by 15 to get number of squares (100 square feet). For pipe sizes of 12¾ inches and smaller add 1 (one) to nominal size of pipe and multiply by 15.

Examples:

How many squares of asbestos felt would be required to wrap one mile of 30-inch pipe line?

30 × 15 = 450 squares of felt.

How many squares of felt would be required to wrap one mile of four-inch pipe?

(4 + 1) × 15 = 75 squares of felt.

These figures are conservative and do not allow for wrapping drips, for patching holidays, and for short ends which are usually discarded. They allow for machine wrapping using ¾-inch lap on diameters up to 10¾-inches and using one-inch lap on larger diameters.

For smaller diameters of pipe this figure should be increased about 5 percent to take care of additional wastage and patching.

Estimating coating and wrapping materials required per mile of pipe

Nominal Pipe Size	Primer Gallons	3/32" Coal-tar Enamel Tons	Glass Wrap Squares	Kraft Paper Squares	15-Pound Felt Squares
4	11	3	75	75	75
6	17	4	105	105	105
8	22	6	135	135	135
10	28	7	165	165	165
12	33	8	195	195	195
14	39	10	210	210	210
16	44	11	240	240	240
18	50	13	270	270	270
20	55	14	300	300	300
24	66	17	360	360	360
30	83	21	450	450	450
36	99	25	540	540	540
42	116	29	630	630	630

Primer and Coal-Tar Enamel:

Gallons of primer and tons of coal-tar enamel are based on covering new pipe in good condition. For rough pipe, increase primer quantities by 25 percent and the coal-tar quantities by 20 percent. No allowance is included for spillage or waste.

Wrappings:

Squares are based on machine wrapping, allowing ¾-inch laps on pipe sizes through 10-inch, and 1-inch laps on pipe sizes greater than 10 inches. No allowance is included for wrapping drips, patching holidays or short ends which are usually discarded.

Coefficient of friction for pipe coating materials

Tests indicate that previous values appear to be valid for thinfilm epoxies and conservative for coal tars

J. B. Ligon, Assistant Professor, Mechanical Engineering-Engineering Mechanics Dept., Michigan Technological University, Houghton, Mich., and **G. R. Mayer,** Project Engineering Manager, Bechtel, Inc., San Francisco

A MAJOR FACTOR in the stress analysis of buried pipe lines is the movement that pipe undergoes in the presence of temperature and pressure differentials during its life. This movement is highly dependent upon friction resistance of the soil.

Although ample information is available on the static coefficient of friction for many materials, there is a lack of data on friction between soils and various coatings used in the pipe line industry. In the past, friction coefficient information was extrapolated from data in the literature that was believed to have a similarity to the external pipe coating to soil interface.

However, with the development of thinfilm epoxy resin coating systems and the increasing use of these systems in the pipe line industry, a change from a conventional coal tar felt coating to a thinfilm epoxy coating would indicate a significant change in the friction coefficient design criteria due to the extreme contrast in the surface texture of the two materials.

To evaluate the effect of the difference in surface texture on a pipe line system, test procedures were developed to determine the coefficient of friction for both coal tar felt and thinfilm epoxy to various soils and to obtain more reliable information for future pipe line designs using these coating materials.

Static friction tests were conducted to find the coefficient of friction between coal tar felt and thinfilm epoxy pipe coating and eight representative backfill soil samples from typical locations along a pipe line right of way. The results indicate that the friction coefficients are significantly larger than those previously extrapolated from literature and that coal tar has a higher friction resistance in respect to anchorage of a pipe line.

For the coal tar felt coating, the coefficient varies from 0.59 to 0.91 depending on the soil and moisture content. The thinfilm epoxy coating varies from 0.51 to 0.71 under the same conditions.

DEFINITIONS

Static coefficient of friction. The theoretical longtitudinal soil force acting on the pipe surface can be calculated from the relation:

$$F = \mu \int_A p \, dA$$

where

F = longitudinal soil friction force (lb.)

μ = coefficient of friction (dimensionless)

p = normal soil pressure acting on the pipe surface (psi)

dA = soil to pipe differential contact area (in.^{2})

$\int_A p \, dA$ = total normal soil force on pipe surface (lb.).

The above relation is independent of the pressure per unit area of the mating surfaces as long as very high contact pressures are not encountered, which is the case for a buried pipe line.

In terms of a buried pipe line supporting a soil burden up to three pipe diameters of depth, the soil force relation becomes:

$$F = \mu \left[2D\gamma \left(H - \frac{D}{2}\right) + Wp \right]$$

where

Wp = weight of pipe and contents (lb./in.)

D = pipe diameter (in.)

H = depth of the pipe centerline (in.)

γ = specific weight of the soil (lb./in.^{3}).

The sensitivity of the soil friction forces to the coefficient of friction is readily apparent from the above relation. Since this force is inversely proportional to the active length that a pipe line moves as a result of temperature and pressure expansion, the coefficient of friction becomes a major factor in pipe line stress design.

Test system. The theoretical soil friction force acting on the surface of a coated plate can be calculated from the relation:

$$F = \mu N$$

where

N = normal force acting on the plate surfaces (lb.)

W = total weight of plate and external load (lb.)

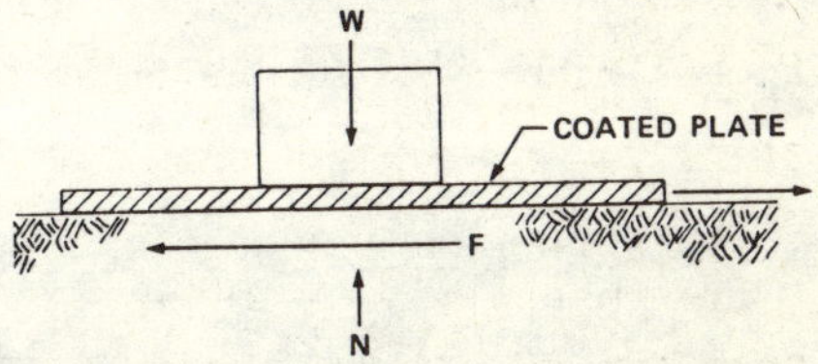

Since the coefficient of friction parameter is the same in all of the above expressions, the plate test system can be used to simulate the soil to pipe line interface.

Coal tar felt. The coal tar coating, an enamel and felt combination, was prepared to simulate an actual coated pipe line.

The surface of a steel plate was coal tar coated and then wrapped in a 15-lb. felt. Therefore, the soil surface contact was predominantly to the felt material.

Thinfilm epoxy. The epoxy coating simulated a thinfilm epoxy resin pipe line coating applied by the fusion method.

Plates used in the tests were coated by the fluidized bed method. However, the surface texture was the same as that expected by coating plant production equipment.

RESULTS

Coal tar to soils. Fig. 1 illustrates that the friction coefficient varies from 0.59 to 0.91 for the various soils tested at field moisture content.

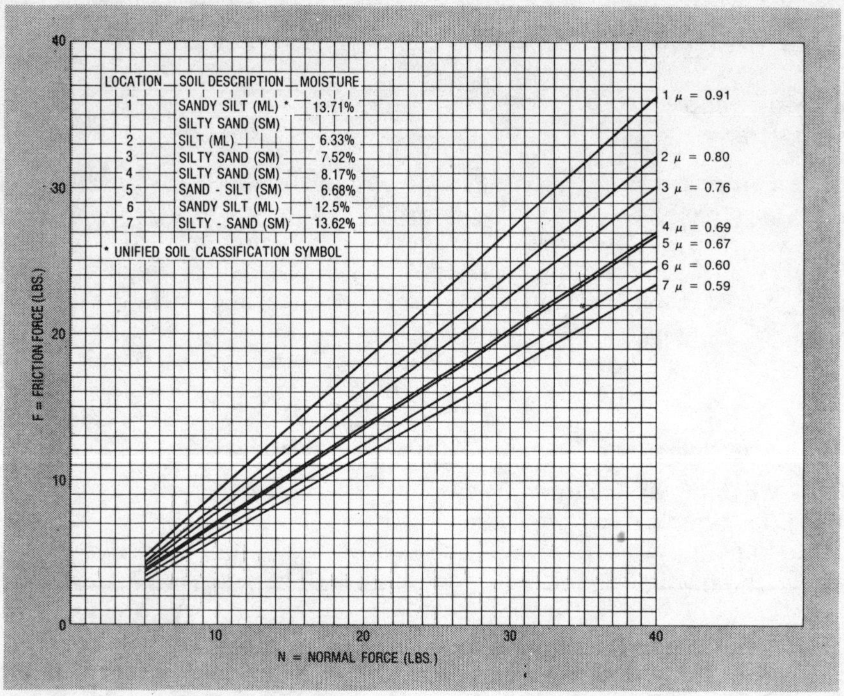

Fig. 1—Coefficient of friction for coal tar coating to soil.

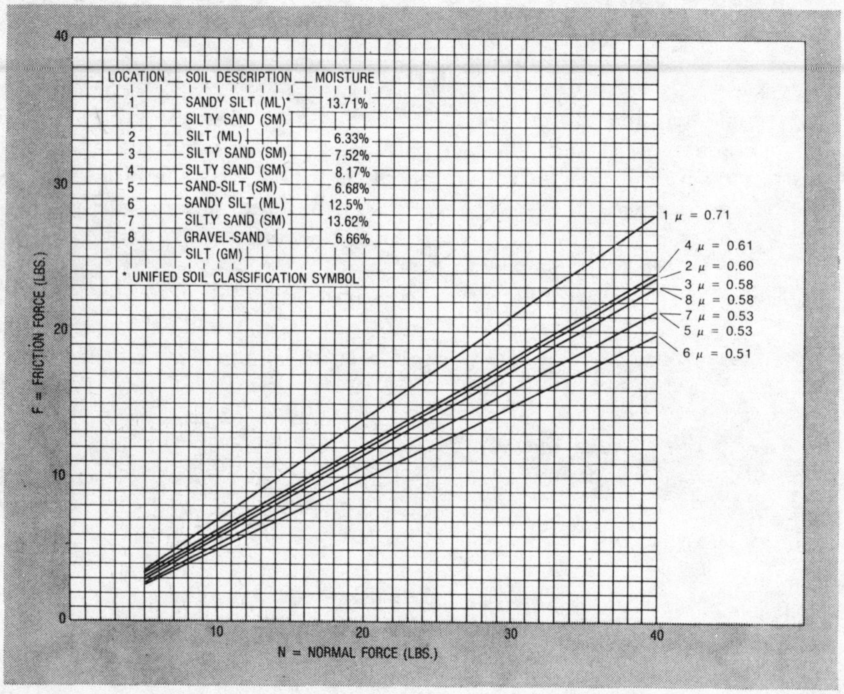

Fig. 2—Coefficient of friction for thin film epoxy to soil.

The mechanism by which the friction coefficient changes among the various soils is not apparent from the results of these tests since the soil samples used did not vary considerably in description. However, the range over which the factor changes is considerably greater than that previously extrapolated from the literature, i.e.:

Soil description	Coefficient of friction commonly used
Silt	0.3
Sand	0.4
Gravel	0.5

The tests also indicate that the moisture content alters the friction factor to some extent, as would be expected.

In order to investigate the influence of temperature, tests were conducted with coal tar felt wrapping heated to 120° F. Only a slight softening beneath the coating surface was observed, and it is believed that temperatures up to this range will not significantly affect the magnitude of the friction coefficient.

Thinfilm epoxy to soils. As expected, results shown in Fig. 2 indicate that the friction factor range of 0.51 to 0.71 for the epoxy is somewhat lower than that of the coal tar.

Again, the mechanism is not clear, but the results are fairly consistent with the friction coefficient increasing in most instances in the same order of magnitude as for coal tar. Since such information on this coating was not available in the literature, no comparison to similar tests could be made.

Temperatures in the 120° F range should have little or no effect on the coefficient of friction of epoxy to soil.

Conclusions. Although it is virtually impossible to precisely simulate the surface contact situation of a pipe line in a back-filled trench, the test procedure and apparatus reported here are a means of approximating it. The results indicate that coal tar coatings have a higher friction resistance than epoxy coatings as far as anchoring capabilities of the soil are concerned.

The selection of a coating, based on its soil friction resistance, could be of economic value in reducing those situations where extreme expansions call for reinforced fittings or elaborate culverts to overcome excessive stress levels in a pipe line system.

The test results show that previous values commonly used for the coefficient of friction were conservative for similar soils, and it is suggested that some conservatism still be incorporated in future analyses.

Since the tests indicated that the friction coefficient of the epoxies was similar to those previously used for coal tar coatings, their continued use for epoxy coated pipe lines should be valid. A more conservative approach to these values is recommended in soils where the presence of excessive moisture could change the friction factor substantially when interfaced with a smooth epoxy coating. ■

3. Construction — Right-of-Way, Ditching, Concrete Work

How to determine the crop acreage included in a right-of-way strip

Multiply the width of the strip in feet by the length in rods; divide this by 2640 to obtain the acreage. If the ends of the strip are not parallel, use the length of the center line of the right-of-way.

Example: A right-of-way 35 feet wide crosses a cultivated field for a length of 14 rods; how many acres of crop were destroyed?

$35 \times 14/2640 = 0.18$ acres, or almost one-fifth of an acre.

Example: A right-of-way 50 feet wide crosses a field for 330 rods. How many acres of crop were destroyed?

$50 \times 330/2640 = 6.25$ acres.

The rule is exact, not an approximation.

How many cubic yards of excavation in a mile of ditch?

Multiply the width in inches by the depth in inches by 1.36; the answer is cubic yards per mile.

Example: How many cubic yards of excavation in a mile of 12-inch ditch 30-inch deep? $12 \times 30 \times 1.36 = 490$ cubic yards per mile.

The rule is correct within about ⅕ of 1 percent; actually the errors in depth and width are much greater than this. To get the cubic yards per 1000 feet, as in computing rock ditching, use 0.257 instead of 1.36.

Shrinkage and expansion of excavated and compacted soil

Ever notice how the spoil from a ditch occupies a greater volume than the ditch itself? There's a reason. **Excavate sand and it expands about 10 percent. Ordinary soil expands about 25 percent and clay expands about 40 percent.**

Here's a summary of the bulk you can expect from excavated soil:

Type soil	Undisturbed	Excavated	Compacted
Sand	1.00	1.11	0.95
Ordinary earth	1.00	1.25	0.90
Clay	1.00	1.43	0.90

Example: Find the volume of loose spoil from a pipe line ditch 42 inches wide, 60 inches deep. The excavation is through clay.

Volume of undisturbed clay soil:
3.5 x 5.0 = 17.5 cu. ft. per lineal foot of ditch.

Volume of the spoil will be 143 percent of the undisturbed volume.

Volume of spoil = 1.43 x 17.5 = 25+ cu. ft. per lineal foot.

Through mechanical compaction this volume can be reduced to 15.75 cu. ft.

How to approximate sacks of cement needed to fill a form

To obtain a close estimate of the number of sacks of cement that will be required, first determine the volume (cubic feet) to be filled in the form. Divide the volume by 4.86 to approximate the number of sacks of cement needed.

Example: How many 94-pound sacks of cement will be required to fill a form for a concrete base 10 feet by 10 feet if it is to be 6 inches thick?

$$\text{number of sacks} = \frac{10 \times 10 \times 0.5}{4.86} = 10.3 \text{ or 11 sacks needed.}$$

What you should know about mixing and finishing concrete

To determine proper mix, divide the constant 44 by the sum of the parts of cement, the parts of sand and parts of gravel to determine the number of bags of Portland cement required. Multiply the number of bags of cement as determined above by parts of sand and the constant 0.035 to calculate the number of cubic yards of sand needed. To determine cubic yards of gravel needed, multiply bags of cement by parts of gravel x 0.035.

Example:

Calculate the quantities of cement, sand and gravel required for one cubic yard of 1:2:4 concrete.

$$C = \frac{44}{1+2+4}$$
$$= 6.28 \text{ bags of cement}$$
$$S = 6.28 \times 2 \times 0.035$$
$$= 0.44 \text{ cubic yard of sand}$$
$$G = 6.28 \times 4 \times 0.035$$
$$= 0.88 \text{ cubic yard of gravel}$$

To increase or decrease slump of concrete, add or subtract one gallon of water per cubic yard of mix and subtract or add 20 pounds of aggregate to maintain yield.

To adjust from no air to air-entrained concrete and maintain strength, reduce water ¼ gallon per sack of cement and reduce sand 10 pounds per sack of cement for each 1% of entrained air.

Clearing and grading right-of-way: *labor/equipment considerations*

To estimate labor crew and equipment spread for clearing and grading operations, the following items, as they may apply to a given project, should be given consideration:

1) Removal of trees, brush and stumps.
2) Grubbing and removal of stumps that are in the way of the ditch.
3) Disposal of all debris, including method of disposal and length of haul.
4) Clearing area spoil a sufficient distance from the ditch line so that the spoil-bank from the ditching operations will not fall in any foreign material that might become mixed with the excavated spoil.
5) Cutting of merchantable timber into standard lengths and stacked along the right-of-way for disposition by others if specifically required by the right-of-way agreement.
6) Providing temporary walks, passageways, fences or other structures so as not to interfere with traffic.
7) Providing sufficient and proper lighting where required.
8) Providing guards where required.
9) Preserving all trees, shrubs, hedges and lawns where required.
10) Grading irregularities where required.
11) Preserving topsoil for replacement, through all cultivated or improved fields and pastures, to its original position.
12) Proper grading of the terrain so as to allow passage of loaded trucks and equipment hauling materials and so ditching operations can be properly performed.
13) Protecting and preserving existing drainage facilities.
14) Protecting any existing structures or pipelines.
15) Protecting any telephone or utility lines and keeping them in service.
16) Cutting through fences and hedges where required and replacing these when necessary.
17) Installing gates and fencing where required.

Ditching and trenching: *labor/equipment considerations*

In determining the labor crew and equipment spread for ditching and trenching operations, the following should be given consideration should they apply to the particular project:

1) Ditching or trenching for buried pipelines should be in accordance with the following table of minimum width and coverage for all soil formations.

Nominal Pipe Size Inches	Minimum Width Inches	Normal Minimum Coverage Inches
4	22	30
6	26	30
8	26	30
10	26	30
12	30	30
14	32	30
16	36	30
18	38	30
20	40	30
24	44	30
30	50	30
36	52	36
42	58	36

2) In rock cut ditches at least six (6) inches wider.
3) If dirt-filled benches are used, ditch should be excavated deeper to obtain proper coverage.
4) Trench should be excavated to greater depth when required for proper installation of the pipe where the topography of the country warrants same.
5) Repair any damage to and maintain existing natural or other drainage facilities.
6) Do not open ditch too far in advance of pipelay crew.
7) Obtain permits for blasting.
8) When blasting, use extreme caution and protection.
9) Clean up of blasted rock to prevent damage to coated pipe.

Estimating manhours for removing trees

NET MANHOURS—EACH

Average Tree Diameter in Inches	Softwood Trees		Hardwood Trees	
	Open Area	Congested Area	Open Area	Congested Area
Cross-cut Saws				
4	1.49	1.86	1.88	2.35
6	2.26	2.83	2.82	3.53
8	3.24	4.02	4.00	4.96
10	4.10	5.08	5.00	6.20
12	4.98	6.18	6.00	7.44
14	6.39	7.86	7.70	9.47
16	7.39	9.09	8.80	10.82
18	8.32	10.23	9.90	12.18
20	10.58	12.91	12.60	15.37
24	12.71	15.51	15.12	18.45
30	17.09	20.85	20.10	24.52
36	20.50	25.01	24.12	29.43
Chain Saws				
4	0.37	0.46	0.47	0.59
6	0.57	0.71	0.71	0.89
8	0.81	1.00	1.00	1.24
10	1.03	1.28	1.25	1.55
12	1.25	1.55	1.50	1.86
14	1.60	1.97	1.93	2.37
16	1.85	2.28	2.20	2.71
18	2.08	2.56	2.48	3.05
20	2.65	3.23	3.15	3.84
24	3.18	3.88	3.78	4.61
30	4.28	5.22	5.03	6.14
36	5.13	6.26	6.03	7.36

Manhours include ax trimming, cutting down with cross-cut saws or chain saws and cutting into four feet lengths for the tree diameter sizes as listed above. Manhours are average for various heights of trees.

Manhours do not include hauling and piling and burning of trees, or branches, or the removal of stumps.

Estimating manhours for removing tree stumps

Manhours include excavating and removing by hand or blasting and removing with cables and tractors of stumps for the base of tree diameters as listed above.

Manhours do not include burning or removal from premises.

NET MANHOURS – EACH

Item	Laborer	Oper. Engr.	Powder Man	Total
Grub & Removal by Hand				
8″ to 12″ diameter	6.00	–	–	6.00
14″ to 18″ diameter	7.50	–	–	7.50
20″ to 24″ diameter	9.00	–	–	9.00
26″ to 36″ diameter	11.20	–	–	11.20
Blast & Pull with Tractor				
8″ to 12″ diameter	.83	.11	1.50	2.44
14″ to 18″ diameter	1.05	.23	2.33	3.61
20″ to 24″ diameter	1.50	.30	3.38	5.18
26″ to 36″ diameter	2.11	.42	4.76	7.29

Estimating manhours for clearing trees and brush

Item	Manhours		
	Laborer	Oper. Engr.	Total
Clear with Dozer, Pile & Burn			
Heavy timbered	27.00	11.25	38.25
Medium timbered	16.20	6.75	22.95
Light timbered	12.15	5.07	17.22
Clear Light Trees & Brush with:			
Tractors & Cables	.53	1.05	1.58

Clear with Dozer, Pile and Burn units include dozing into piles and burning.

Clear Light Trees and Brush units include removal with tractor and cable of saplings or light trees and brush, piling and burning.

Manhours do not include sawing, cutting or trimming of trees or the removal of stumps.

Overtime allowance for fire tender has not been considered above. If this is necessary, add for this operation.

4. Construction—Pipe & Fittings

How to calculate the contraction or expansion of a pipe line

Steel pipe contracts or expands approximately 0.8 inch per 100° F. temperature change for each 100 feet of pipe. For construction in the U. S., slack is not normally needed in welded lines for contraction or expansion unless abnormal variations are encountered. However, slack is often provided near road crossings where the pipe may have to be lowered at some future time.

$$\text{Contraction} = 0.8 \times \frac{\text{length of pipe (feet)}}{100} \times \frac{\text{temp. change (°F.)}}{100}$$

Example: Calculate the amount a 1000-foot section of pipe line would contract if laid at a temperature of 100° F. and cooled to a low of 0° F. during winter operation. (Assuming the line were free of soil or other resisting loads.)

$$\text{Contraction} = 0.8 \times \frac{1000}{100} \times \frac{100}{100}$$
$$= 0.8 \times 10 \times 1$$
$$= 8 \text{ inches}$$

Large temperature changes may arise on exposed line or in lines downstream from compressor stations.

When slack is desired in a line, the amount of sag or rise needed may be estimated quickly. To provide for 0.8 inch of movement longitudinally along the pipe, a sag of 21 inches would be needed in a 100-foot loop. For a 150-foot loop, 31 inches are necessary. Such slack put into the line at time of construction might allow for lowering the pipe under roadbeds without putting excessive stress upon the pipe.

Location of overbends, sags, and side bends should be considered when laying slack loops. When loops are placed, parts of the line should be backfilled to serve as a tie-down. Loops should be lowered during the coolest part of the day when the line is the shortest. After lowering-in, sags should rest on the bottom of the ditch; overbends should "ride high." Side bends should rest on the bottom of the ditch and against the outside wall. The line should be lowered so that all sections of the pipe are in compression.

Estimate weight of pipe in metric tons per kilometer

To estimate the weight of pipe in metric tons per kilometer, multiply the nominal diameter by the number of sixteenths of an inch in wall thickness.

Example: Find the weight of pipe in metric tons per kilometer for 20-inch diameter pipe, wall thickness ¼-inch.

$4 \times 20 = 80$ tons (metric) per kilometer

Actual answer from pipe tables is 79 metric tons per kilometer.

This rule of thumb is based on a density of 490 pounds per cubic foot for steel. For larger diameter thin wall pipe, this approximation gives an answer usually about one percent low. The accompanying table gives a comparison between actual weights in metric tons per kilometer, as compared to that calculated by this rule of thumb.

PIPE DIAMETER	WALL THICKNESS	ACTUAL TONS (METRIC)	R. O. T. TONS (METRIC)
10-inch	3/16-inch	31.5	30
12-inch	5/16-inch	62	60
16-inch	7/16-inch	109	112
20-inch	¼ -inch	79	80
20-inch	5/16-inch	99	100
20-inch	7/16-inch	137	140
24-inch	5/16-inch	118	120
24-inch	8/16-inch	189	192
28-inch	5/16-inch	139	140
30-inch	5/16-inch	148	150
32-inch	7/16-inch	221	224
36-inch	4/16-inch	143	144
36-inch	8/16-inch	285	288

How to find pipe weight from outside diameter and wall thickness

Weight (pounds per foot) = $(Dt - t^2) \times 10.68$, where D = outside diameter in inches, and t = wall thickness in inches.

Examples:

Outside diameter = 4.500", thickness = 0.250".

$W = (4.5 \times .25 - .25^2) \times 10.68$
$= (1.125 - .0625) \times 10.68$
$= 1.0625 \times 10.68 = 11.35$ pounds per foot.

The above formula is based on a density of 490 pounds per cubic foot for the steel. High-yield-point thin-wall pipe may run slightly heavier than indicated.

What is the maximum allowable lengths of unsupported line pipe?

For Schedule 40 pipe, the maximum span between supports is given by

$$S = 6.6\sqrt{P},$$

where S is the span length in feet and P is the actual outside diameter of the pipe in inches. For pipe smaller than 12-inch, the nominal size may be used, and

$$S = 7\sqrt{N}.$$

Example: How wide a ditch can be spanned with four-inch line pipe without the use of intermediate supports?

$$S = 6.6\sqrt{4.5} = 6.6 \times 2.12 = 13.99 \text{ (14 feet)}.$$

$$S = 7\sqrt{4} = 7 \times 2 = 14 \text{ feet}.$$

Note: The above rule is approximate, but conservative. It does not assume that the pipe is held fixed at either end; this condition can make longer spans safe. Also, it does not guarantee that the pipe will not be damaged by floating logs or debris! This hazard must be evaluated from the conditions encountered.

Identify the schedule number of pipe by direct measurement

Add three inches to the actual inside diameter; divide this by the wall thickness; the schedule number can then be identified by the following table:

Schedule 30	40 to 50
Schedule 40	29 to 39
Schedule 60	25 to 29
Schedule 80	20 to 23
Schedule 100	16 to 18
Schedule 120	13 to 15
Schedule 140	11 to 13
Schedule 160	9 to 11

Example: Pipe measures 12.5 inches inside diameter, wall thickness .75 inch; 15.5/.75 = 20.2; pipe is Schedule 80, 14 inches outside diameter.

The rule fails on small diameter pipe (under 6 inches), and also on schedules 10 and 20, because of the use of "standard" thicknesses in this range.

How to identify the series number of flanged fittings

Divide the flange thickness, expressed in sixteenths of an inch, by the nominal pipe size + 12. If the answer is less than 1.00, the flange is Series15; if between 1.15 and 1.33, it is Series 30; if between 1.37 and 1.54, it is Series 40; and if it is between 1.60 and 1.85, it is Series 60.

Example: a six-inch flange has a thickness of 1 7/16 inch = 23/16 inch.

23/18 = 1.28; the flange is series 30.

The rule is not applicable to fittings below four inches in size; it also fails for a four-inch Series 60,which comes out 1.50.

Determine buoyancy of bare steel pipe

Multiply the square of the outside diameter of the pipe, in inches, by 0.341, and subtract the pipe weight in pounds per foot; the answer is the buoyancy per foot in fresh water.

Example: 12" (12.75" O. D.) 33.4 pound pipe (¼" wall).

$$\text{Buoyancy} = 12.75^2 \times 0.341 - 33.4$$
$$= 162.56 \times 0.341 - 33.4$$
$$= 55.4 - 33.4 = 22 \text{ pounds per foot.}$$

Since all of the coating materials in common use are heavier than water (but not as much so as steel), coated pipe will have less buoyancy than indicated by the rule. For sea water, use 0.350 for the constant.

How to determine the degrees of bend in a pipe which must fit a ditch calling for a bend in both horizontal and vertical planes

Rule: To find the number of degrees in the combination bend, square the side bend and the sag or overbend; add them together and extract the square root. The answer will be the number of degrees necessary to make the pipe fit the ditch.

Example: Determine the bend to make in a pipe whose ditch has a 3° overbend and a 4° sidebend. Let X = the unknown angle:

$$X^2 = 3^2 + 4^2$$
$$X^2 = 9 + 16$$
$$X^2 = 25$$
$$X = 5°$$

Example: Determine the bend to make in a pipe whose ditch has a 9° sag and a 12° sidebend. Let X = the unknown angle:

$$X^2 = 9^2 + 12^2$$
$$X^2 = 225$$
$$X = 15°$$

How to bend pipe to fit ditch - sags, overbends and combination bends

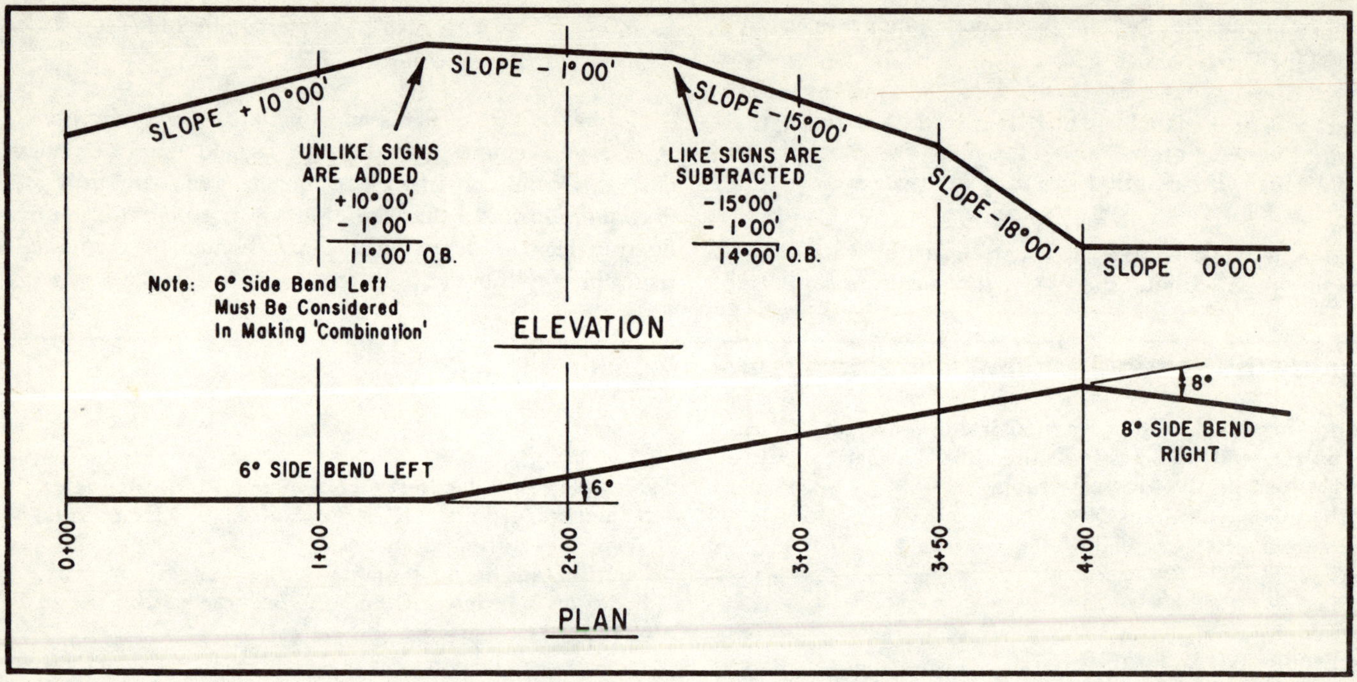

To make straight sags or overbends fit the ditch (see drawing) **add angles of unlike signs, subtract those of like signs.**

Example: (Sta. 2 + 40)
Slope — 15° 00′
Slope — 1° 00′
(sub.) 14° 00′ overbend

In cases involving either a sag or overbend, in addition to a side-bend, the rule becomes: **Make the combination bend equal to the largest angle plus ⅓ of the smallest.**

Example: (Sta. 1 + 40)
Slope + 10° 00′
Slope — 1° 00′
(add) 11° 00′ overbend (1)
6° 00′ side-bend left. (2)

Combination bend = 11° + (⅓ × 6°) = 13° 00′ overbend left.

Note: This rule gives an error of approximately 1° for a maximum bend of 18.5°

Dimensions of three-diameter ells with tangents

Here are face-to-face and center-to-face dimensions of the three-diameter radius weld ells being used in natural gas compressor station piping.

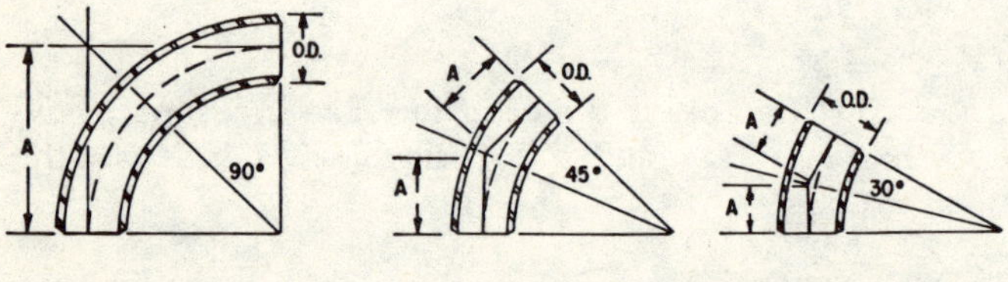

Degree Ell	Outside Diameter	A (Center To Face)	A With 4″ Tan	A With 6″ Tan
30°	16″	1′-0⅞″	1′- 4⅞″	1′- 6⅞″
30°	20″	1′-4$\frac{1}{16}$″	1′- 8$\frac{1}{16}$″	1′-10$\frac{1}{16}$″
30°	24″	1′-7$\frac{5}{16}$″	1′-11$\frac{5}{16}$″	2′- 1$\frac{5}{16}$″
30°	26″	1′-8⅞″	2′- 0⅞″	2′- 2⅞″
30°	30″	2′-0⅛″	2′- 4⅛″	2′- 6⅛″
45°	16″	1′-7⅞″	1′-11⅞″	2′- 1⅞″
45°	20″	2′-0⅞″	2′- 4⅞″	2′- 6⅞″
45°	24″	2′-5$\frac{13}{16}$″	2′- 9$\frac{13}{16}$″	2′-11$\frac{13}{16}$″
45°	26″	2′-8$\frac{5}{16}$″	3′- 0$\frac{5}{16}$″	3′- 2$\frac{5}{16}$″
45°	30″	3′-1¼″	3′- 5¼″	3′- 7¼″
90°	16″	4′-0″	4′- 4″	4′- 6″
90°	20″	5′-0″	5′- 4″	5′- 6″
90°	24″	6′-0″	6′- 4″	6′- 6″
90°	26″	6′-6″	6′-10″	7′- 0″
90°	30″	7′-6″	7′-10″	8′- 0″

Calculate maximum bend on cold pipe

How much of a bend can cold straight pipe take without over stressing? In other words, in laying straight pipe cross-country, what is the smallest radius of curvature possible within the allowable stress?

The following is an analytical solution to these questions.

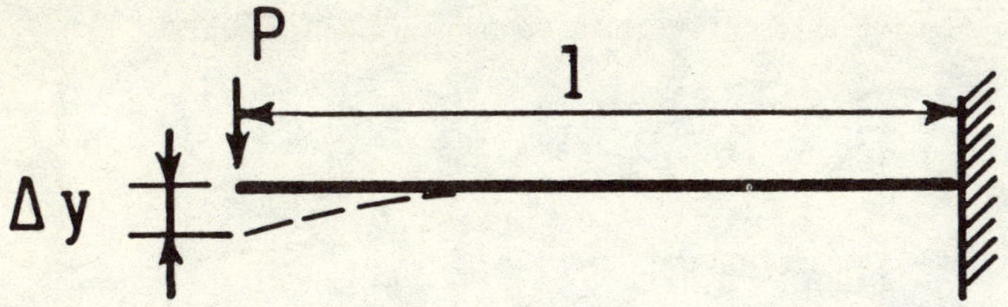

The deflection equation for a length of straight pipe used as a cantilever beam with a concentrated end load is:

(1) $$\Delta y = \frac{Pl^3}{3EI}$$

The maximum stress in the beam is:

(2) $$s = \frac{M}{Z}$$

The maximum bending moment is:

(3) $$M = Pl$$

If the allowable stress is S_A:

Rewriting (2)

(4) $$S_A = \frac{M}{Z}$$

(5) $$S_A = \frac{Pl}{Z}$$

(6) $$Pl = Z\,S_A$$

Rewriting (1)

(7) $$\Delta y = \frac{(Pl)\,l^2}{3EI} = \frac{Z\,S_A\,l^2}{3EI}$$

Taking the first and second derivatives of (7)

(8) $$\frac{d\,\Delta y}{dl} = \frac{2\,l\,ZS_A}{3\,EI}$$

(9) $$\frac{d^2\,\Delta y}{d\,l^2} = \frac{2\,Z\,S_A}{3\,EI}$$

Radius of curvature is given by the equation

(10) $$R = \frac{\left[1+\left(\frac{d\,\Delta y}{d\,l}\right)^2\right]^{3/2}}{\frac{d^2\,\Delta y}{d\,l^2}}$$

Substituting in (10)

(11) $$R = \frac{\left[1+\left(\frac{2\,l\,ZS_A}{3\,EI}\right)^2\right]^{3/2}}{\frac{2\,Z\,S_A}{3\,EI}}$$

(12) $$\lim_{l\to 0} R = \frac{1^{3/2}}{\frac{2\,Z\,S_A}{3\,EI}} = \frac{3\,EI}{2\,Z\,S_A}$$

NOTATION:

Δy — deflection, inches
l — length, inches
P — load, lb.
E — modulus of elasticity, lbs./sq. in.
I — cross-sectional moment of inertia, in^4
S — Stress, lb./sq. in.
S_A — allowable stress, lb./sq. in. from Code for Pressure Piping
M — bending moment, inch pounds
Z — section modulus, in^3.
R — Radius of curvature, inches

EXAMPLE:
What is the smallest radius of curvature that can be used without overstressing straight 24 inch Schedule 20 seamless steel pipe ASTM 53? The pipe is at 70°F, and is to be used for oil outside of refinery limits.

$$R = \frac{3\,EI}{2\,Z\,S_A}$$

$$E = 29 \times 10^6 \text{ lb./sq. in.}$$

$$I = 1943 \text{ in}^4$$

$$Z = 162 \text{ in}^3$$

$S_A = 25{,}500$ psi from Code for Pressure Piping

$$R = \frac{3 \times 29 \times 10^6 \times 1943}{2 \times 162 \times 25{,}500} = 20{,}460 \text{ inches or } 1{,}705 \text{ feet}$$

Determine length of a pipe bend

For smooth bends (other than wrinkle bends) use the following formula:

L = R x D x .017 + T

Where L = length of bend
R = radius of bend (center line of pipe)
D = number of degrees in the bend
T = total length of the two tangents at either end.

Example: Find the length of a piece of pipe required to make a 90° bend with a radius of 60 inches. There's to be a four-inch tangent at either end.

L = ?
R = 60
D = 90
T = 8
L = 60 x 90 x .01745 + 8
= 94.23 + 8
= 102.23 inches

Determine buoyancy of bare and concrete coated steel pipe in water and mud

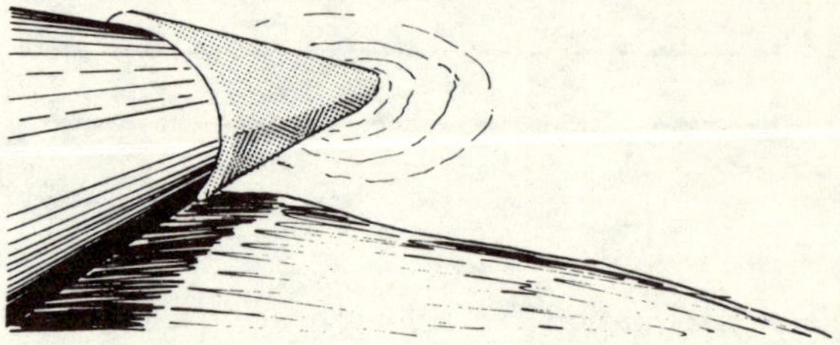

To find the buoyancy of bare steel and concrete coated pipe in water in pounds per foot:

$$\text{Buoyancy (B)} = \frac{D}{3}(D - 32t) + 11t^2 \text{ for bare pipe;}$$

$$\text{and } B = \frac{D}{3}(D - 32t) + t_1 D \frac{63 - W_c}{48}$$

for coated pipe;

where D is outside diameter of pipe in inches
t is wall thickness of pipe in inches
t_1 is thickness of concrete coating in inches
W_c is weight of concrete in pounds per cubic foot

To find the buoyancy of bare steel and coated pipe in mud in pounds per foot:

$$\text{Buoyancy (B)} = 10.7\left(\frac{DW_m}{2000} - t\right)$$

$$+ 11\,t^2 \text{ for bare pipe;}$$

$$\text{and } B = 10.7\left(\frac{DW_m}{2000} - t\right) + t_1 D\left(\frac{W_m - W_c}{48}\right)$$

for coated pipe;
where W_m is weight of mud in pounds per cubic foot.

Example 1: Find the buoyancy of steel pipe with 20-inch OD and ½-inch wall thickness in water. What thickness of concrete coating would be required to give this pipe a negative buoyancy of 100 pounds per foot ($W_c = 143$)?

$$(B) = \frac{20}{3}[20 - 32(½)] + 11(½)^2$$

$$= 26.7 + 2.8 = 29.5 \text{ pounds per foot positive buoyancy.}$$

To give this pipe a negative buoyancy of 100 pounds per foot:

$$-100 = \frac{20}{3}[20 - 32(½)] + t_1(20)\left(\frac{63 - 143}{48}\right)$$

$$-100 = 26.7 - 33.3\,t_1$$

$$t_1 = 3.8 \text{ inches of concrete coating.}$$

The error introduced by this method is about 15% but would be less as the thickness of coating decreased.

Example 2: Bottom conditions at a certain crossing that require the pipe to have a negative buoyancy of at least 50 pounds per foot. Let's check to see if a 10-inch pipe having a 2-inch coating of concrete ($W_c = 143$) has sufficient negative buoyancy?

$$B = \frac{10}{3}\left(10 - 32(½)\right) + 2(10)\left(\frac{63 - 143}{48}\right)$$

$B = -20 - 33.3 = -53.3$ pounds per foot; so coated pipe meets design requirements.

Formulae and constants of value in solving problems relating to tubular goods

Area of Metal in Tubular Section

Area in square inches = 3.1416 (D—t) t
where D = Outside diameter in inches
t = Wall thickness in inches.

Capacity

$$B_{100} = .1237\,A = .0972\,D^2$$

where
B_{100} = Barrels per 100 feet
A = Internal area of pipe in square inches
D = Internal diameter in inches

and

$$F_B = \frac{808.5}{A} = \frac{1029.4}{D^2}$$

where
F_B = Number of feet filled by one barrel
A = Internal area of pipe in square inches
D = Internal diameter in inches.

Complete tables giving capacities of different types of tubular goods are given in chapters dealing with these types.

Steel constants* (applicable to tubular goods)

One cubic inch = 0.2833 pounds
One cubic foot = 489.542 pounds
Specific gravity = 7.851

Approximate coefficient of expansion for commercial casing, drill pipe, and tubing is
Coefficient = 6.9×10^{-6} per degree Fahrenheit, over range from 0 to 400° F.

Formula for expansion of pipe due to change of temperature is as follows:

$$L_t = L_o(1 + .0000069\,t)$$

where
L_o = Length at atmospheric temperature
L_t = Length after change of temperature t.

Modulus:

$$\text{Section Modulus } \frac{\pi}{64}(D^4 - D_1^4)$$

$$\text{Polar Modulus } \frac{\pi}{32}(D^4 - D_1^4)$$

Where:
D = Outside diameter.
D_1 = Inside diameter.
$\pi = 3.1416$.

Relation Between Twisting Effort, Torsional Strain and Stress of Cylindrical Shaft or Tube

$$\frac{q}{r} = \frac{T}{J} = \frac{C\theta}{L}$$

* Engineering data Spang-Chalfant Division of the National Supply Co.

Length of pipe in arc subtended by any angle

DEGREES						MINUTES			
Degrees	Length Of Arc	Degrees	Length Of Arc	Degrees	Length Of Arc	Minutes	Length Of Arc	Minutes	Length Of Arc
1	.0175	31	.5411	61	1.0647	1	.000291	31	.009018
2	.0349	32	.5585	62	1.0821	2	.000582	32	.009308
3	.0524	33	.5760	63	1.0996	3	.000873	33	.009599
4	.0698	34	.5934	64	1.1170	4	.001164	34	.009890
5	.0873	35	.6109	65	1.1345	5	.001454	35	.010181
6	.1047	36	.6283	66	1.1519	6	.001745	36	.010472
7	.1222	37	.6458	67	1.1694	7	.002036	37	.010763
8	.1396	38	.6632	68	1.1868	8	.002327	38	.011054
9	.1571	39	.6807	69	1.2043	9	.002618	39	.011345
10	.1745	40	.6981	70	1.2217	10	.002909	40	.011636
11	.1920	41	.7156	71	1.2392	11	.003200	41	.011926
12	.2094	42	.7330	72	1.2566	12	.003491	42	.012217
13	.2269	43	.7505	73	1.2741	13	.003782	43	.012508
14	.2443	44	.7679	74	1.2915	14	.004072	44	.012799
15	.2618	45	.7854	75	1.3090	15	.004363	45	.013090
16	.2793	46	.8029	76	1.3265	16	.004654	46	.013381
17	.2967	47	.8203	77	1.3439	17	.004945	47	.013672
18	.3142	48	.8378	78	1.3614	18	.005236	48	.013963
19	.3316	49	.8552	79	1.3788	19	.005527	49	.014254
20	.3491	50	.8727	80	1.3963	20	.005818	50	.014544
21	.3665	51	.8901	81	1.4137	21	.006109	51	.014835
22	.3840	52	.9076	82	1.4312	22	.006400	52	.015126
23	.4014	53	.9250	83	1.4486	23	.006690	53	.015417
24	.4189	54	.9425	84	1.4661	24	.006981	54	.015708
25	.4363	55	.9599	85	1.4835	25	.007272	55	.015999
26	.4538	56	.9774	86	1.5010	26	.007563	56	.016290
27	.4712	57	.9948	87	1.5184	27	.007854	57	.016581
28	.4887	58	1.0123	88	1.5359	28	.008145	58	.016872
29	.5061	59	1.0297	89	1.5533	29	.008436	59	.017162
30	.5236	60	1.0472	90	1.5708	30	.008727	60	.017453

Application:

To find the length of pipe subtended by an angle of 42°-35 feet with a radius of 24 inches.

Assuming R = 1 inch,

angle of 42° subtends an arc = 0.7330″

angle of 35′ subtends an arc = 0.01018″

0.7432″

The arc subtended for any angle is directly proportional to radius:

$$\frac{1 \text{ inch}}{24 \text{ inches}} = \frac{0.7432 \text{ inches}}{\text{Arc}}$$

$$\text{Arc} = 0.7432 \times 24 = 17\tfrac{13}{16}''$$

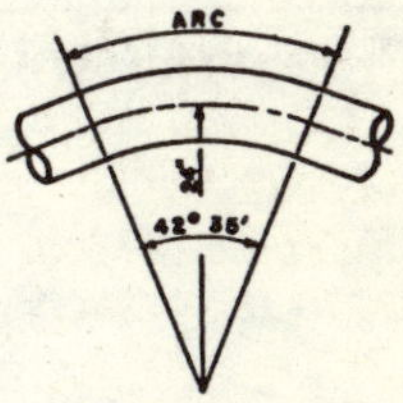

Average pipelay table—Underground

Linear feet of pipelay per ten (10) hour day in trench in level, rock-free terrain

Nominal Pipe Size	PIPE WALL THICKNESS IN INCHES					
	0.000 through 0.250	0.251 through 0.375	0.376 through 0.500	0.501 through 0.625	0.626 through 0.750	0.756 through 1.000
4	10,800	10,500	10,300	—	—	—
6	9,600	9,400	9,210	—	—	—
8	9,120	8,900	8,720	8,550	—	—
10	8,900	8,760	8,580	8,410	—	—
12	8,600	8,400	8,230	8,070	7,910	—
14	8,400	8,230	8,160	8,000	7,840	—
16	8,000	7,840	7,680	7,530	7,380	7,230
18	7,800	7,640	7,500	7,350	7,200	7,050
20	7,400	7,250	7,100	6,960	6,820	6,680
24	—	6,600	6,470	6,340	6,210	6,080
30	—	6,000	5,880	5,760	5,640	5,530
36	—	—	5,400	5,300	5,200	5,090
42	—	—	4,800	4,700	4,600	4,500

Productivity will vary where different types of terrain/or rock are encountered.

Above footage is based on installing double random joints of pipe.

Average pipelay table—On supports

Linear feet of pipelay per ten (10) hour day on waist-high supports on level ground

Nominal Pipe Size	PIPE WALL THICKNESS IN INCHES 0.000 through 0.250	0.251 through 0.375	0.376 through 0.500	0.501 through 0.625	0.626 through 0.750	0.751 through 1.000
4	11,320	11,100	10,880	—	—	—
6	10,080	9,880	9,680	—	—	—
8	9,560	9,370	9,190	9,010	—	—
10	9,390	9,200	9,020	8,840	—	—
12	8,800	8,620	8,500	8,330	8,160	—
14	8,700	8,570	8,400	8,250	8,090	—
16	8,040	7,880	7,730	7,580	7,430	7,280
18	7,800	7,650	7,500	7,350	7,200	7,060
20	7,550	7,400	7,250	7,100	6,960	6,820
24	—	6,720	6,590	6,460	6,330	6,200
30	—	6,120	6,000	5,880	5,770	5,660
36	—	—	5,440	5,350	5,240	5,140
42	—	—	4,840	4,740	4,650	4,560

Productivity will vary with contour of terrain and accessibility.

Above footage is based on installing double random joints of pipe.

Pipe wall thickness and weight

Nominal sizes 2-12"

POUNDS PER LINEAR FOOT

Schedule Number	Nominal Size	2"	2½"	3"	3½"	4"	5"	6"	8"	10"	12"
	O. D.	2.375	2.875	3.5	4.0	4.5	5.563	6.625	8.625	10.75	12.75
10	Wall Thickness	—	—	—	—	—	—	—	—	—	—
	Weight/Foot	—	—	—	—	—	—	—	—	—	—
20	Wall Thickness	—	—	—	—	—	—	—	.250	.250	.250
	Weight/Foot	—	—	—	—	—	—	—	22.36	28.04	33.38
30	Wall Thickness	—	—	—	—	—	—	—	.277	.307	.330
	Weight/Foot	—	—	—	—	—	—	—	24.70	34.24	43.77
40	Wall Thickness	.154	.203	.216	.226	.237	.258	.280	.322	.365	.406
	Weight/Foot	3.65	5.79	7.59	9.11	10.79	14.62	18.97	28.55	40.48	53.52
60	Wall Thickness	—	—	—	—	—	—	—	.406	.500	.562
	Weight/Foot	—	—	—	—	—	—	—	35.64	54.75	73.15
80	Wall Thickness	.218	.276	.300	.318	.337	.375	4.32	.500	.594	.688
	Weight/Foot	5.02	7.66	10.25	12.51	14.98	20.78	28.57	43.39	64.43	88.63
100	Wall Thickness	—	—	—	—	—	—	—	.594	.719	.844
	Weight/Foot	—	—	—	—	—	—	—	50.95	77.03	107.32
120	Wall Thickness	—	—	—	—	.438	.500	.562	.719	.844	1.000
	Weight/Foot	—	—	—	—	19.00	27.04	36.39	60.71	89.29	125.49
140	Wall Thickness	—	—	—	—	—	—	—	.812	1.000	1.125
	Weight/Foot	—	—	—	—	—	—	—	67.76	104.13	139.67
160	Wall Thickness	.344	.375	.438	—	.531	.625	.719	.906	1.125	1.312
	Weight/Foot	7.46	10.01	14.32	—	22.51	32.96	45.35	74.69	115.64	160.27
Double Extra Heavy	Wall Thickness	.436	.552	.600	.636	.674	.750	.864	.875	—	—
	Weight/Foot	9.03	13.70	18.58	22.85	27.54	38.55	53.16	72.42	—	—
Non-Scheduled	Wall Thickness	—	—	.156	—	.156	.188	.188	.188	.188	.188
	Weight/Foot	—	—	5.57	—	7.24	10.79	12.92	16.94	21.21	25.22
Non-Scheduled	Wall Thickness	—	—	.188	—	.188	—	.250	.219	.203	.375
	Weight/Foot	—	—	6.65	—	8.66	—	17.02	19.66	22.87	49.56
Non-Scheduled	Wall Thickness	—	—	—	—	—	—	.375	.375	.279	.500
	Weight/Foot	—	—	—	—	—	—	25.03	33.04	31.20	65.42

Weights of piping materials

The weight per foot of steel pipe is subject to the following tolerances:

SPECIFICATION	TOLERANCE
A.S.T.M. A-53 A.S.T.M. A-120	STD WT + 5%, − 5% XS WT + 5%, − 5% XXS WT +10%, −10%
A.S.T.M. A-106	SCH 10–120 +6.5%, −3.5% SCH 140–160 +10%, −3.5%
A.S.T.M. A-158 A.S.T.M. A-206 A.S.T.M. A-280	12″ and under +6.5%, −3.5% over 12″ +10%, −5%
API 5L	All sizes +6.5%, −3.5%

Weight of Tube $= F \times 10.6802 \times T \times (D - T)$ pounds/foot

T = wall thickness in inches

D = outside diameter in inches

F = relative weight factor

The weight of tube furnished in this piping data is based on low carbon steel weighing 0.2833 pounds per cubic inch.

Relative Weight Factor F of various metals

Aluminum	= 0.35
Brass	= 1.12
Cast Iron	= 0.91
Copper	= 1.14
Lead	= 1.44
Ferritic Stainless Steel	= 0.95
Austenitic Stainless Steel	= 1.02
Steel	= 1.00
Tin	= 0.93
Wrought Iron	= 0.98

(Weight of Contents of a Tube $= G \times 0.3405 \times (D - 2T)^2$ pounds per foot)

G = Specific Gravity of Contents

T = Tube Wall Thickness, inches

D = Tube Outside Diameter, inches

The weight of Welding Tees and Laterals is for full size fittings. The weight of reducing fittings is approximately the same as for full size fittings.

The weight of Welding Reducers is for one size reduction, and is approximately correct for other reductions.

Pipe Covering temperature ranges are intended as a guide only and do not constitute a recommendation for specific thickness of material.

Pipe Covering thicknesses and weights indicate average conditions and include all allowance for wire, cement, canvas, bands, and paint. The listed thicknesses of combination covering is the sum of the inner and the outer layer thickness. When specific inner and outer layer thicknesses are known, add them, and use the weight for the nearest tabulated thickness.

To find the weight of covering on Fittings, Valves, or Flanges, multiply the weight factor (light faced subscript) by the weight per foot of covering used on straight pipe. All Flange weights include the proportional weight of bolts or studs required to make up all joints.

Lap Joint Flange weights include the weight of the lap.

Welding Neck Flange weights are compensated to allow for the weight of pipe displaced by the flange. Pipe should be measured from the face of the flange.

All Flanged Fitting weights include the proportional weight of bolts or studs required to make up all joints.

To find the approximate weight of Reducing Flanged Fittings, subtract the weight of a full size Slip-On Flange and add the weight of reduced size Slip-On Flange.

Weights of valves of the same type may vary because of individual Manufacturer's design. Listed valve weights are approximate only. When it is possible to obtain specific weights from the Manufacturer, such weights should be used.

To obtain the approximate weight of Flanged End Steel Valves, add the weight of two Slip-On Flanges of the same size and series to the weight of the corresponding Welding End Valves.

Tubular size, wall thickness and weight

Wall thicknesses .250-2.125″

POUNDS PER LINEAR FOOT

Nominal Size Inches	WALL THICKNESS IN INCHES															
	.250	.375	.500	.625	.750	.875	1.000	1.125	1.250	1.375	1.500	1.625	1.750	1.875	2.000	2.125
24	63.41	94.62	125.49	156.05	186.23	216.13	245.67	274.88	303.75	332.29	360.50	388.37	415.91	443.11	469.98	496.52
25	66.09	98.64	130.85	162.73	194.27	225.48	256.35	286.90	317.11	346.98	376.52	405.73	434.60	463.14	491.35	519.22
26	68.75	102.63	136.17	169.38	202.28	234.82	267.04	298.91	330.46	361.67	392.54	423.09	453.09	483.17	512.71	541.92
27	71.43	106.65	141.53	176.08	210.29	244.17	277.72	310.93	343.81	376.35	408.56	440.44	471.99	503.20	534.07	564.61
28	74.10	110.65	146.87	182.75	218.30	253.52	288.40	322.95	357.16	391.04	424.59	457.80	490.68	523.22	555.43	587.31
29	76.77	114.66	152.21	189.43	226.31	262.86	299.08	334.96	370.51	405.73	440.61	475.16	509.37	543.25	576.30	610.01
30	79.43	118.65	157.53	196.08	234.29	272.21	309.76	346.98	383.86	420.41	456.63	492.51	528.06	563.28	598.16	632.71
31	82.11	122.67	162.89	202.78	242.34	281.56	320.44	359.00	397.22	435.10	472.65	509.87	546.76	583.31	619.52	655.41
32	84.78	126.68	168.23	209.46	250.35	290.90	331.12	371.01	410.57	449.79	488.67	527.23	565.45	603.33	640.89	678.10
33	87.45	130.68	173.57	216.13	258.36	300.25	341.81	383.03	423.92	464.48	504.70	544.59	584.14	623.36	662.25	700.80
34	90.12	134.69	178.91	222.81	266.37	309.59	352.49	395.05	437.27	479.16	520.72	561.94	602.83	643.39	683.61	723.50
35	92.80	138.69	184.25	229.48	274.38	318.94	303.17	407.06	450.62	493.85	536.75	579.30	621.53	663.42	704.98	746.20
36	95.45	142.68	189.57	236.13	282.37	328.29	373.85	419.08	463.98	508.54	552.76	596.66	640.22	683.45	726.34	768.90
37	98.14	146.70	194.94	242.84	290.40	337.63	384.53	431.10	477.33	523.22	568.78	614.02	658.91	703.47	747.70	791.59
38	100.81	150.71	200.28	249.51	298.41	346.98	395.21	443.11	490.68	537.91	584.81	631.37	677.60	725.50	769.06	814.29
39	103.48	154.71	205.62	256.19	306.42	356.33	405.89	455.13	504.03	552.60	600.83	648.73	696.30	743.53	790.43	838.99
40	106.15	158.72	210.96	262.86	314.43	365.67	416.58	467.15	517.38	567.28	616.85	666.09	714.99	763.56	811.79	859.16
41	108.82	162.73	216.30	269.54	322.45	375.02	427.26	479.16	530.73	581.97	632.88	683.45	733.68	783.58	833.15	882.39
42	111.49	166.73	221.64	276.22	330.46	384.36	437.94	491.18	544.09	596.66	648.90	700.80	752.37	803.61	854.52	905.09
43	114.16	170.74	226.98	282.89	338.47	393.71	448.62	503.20	557.44	611.35	664.92	718.16	771.07	823.64	875.88	927.78
44	116.83	174.74	232.32	289.57	346.48	403.06	459.30	515.21	570.79	626.03	680.94	735.52	789.76	843.67	897.24	950.48
45	119.50	178.75	237.66	296.24	354.49	412.40	469.98	527.23	584.14	640.72	696.96	752.87	808.95	863.69	918.60	973.18
46	122.17	182.75	243.00	302.92	362.50	421.75	480.66	539.25	597.49	655.41	712.88	770.23	827.14	883.72	939.97	995.88
47	124.84	186.76	248.34	309.59	370.51	431.10	491.35	551.26	610.84	670.09	729.01	787.59	845.84	903.75	961.33	1018.58
48	127.51	190.76	253.68	316.27	378.52	440.44	502.03	563.28	624.20	684.78	745.03	804.95	864.53	923.78	982.69	1041.27
49	130.18	194.77	259.02	322.95	386.53	449.79	512.71	575.30	637.55	699.47	761.05	822.30	883.22	948.81	1004.06	1063.97
50	132.85	198.77	264.37	329.62	394.55	459.14	523.39	587.31	650.90	714.15	777.07	839.66	901.91	963.83	1025.42	1086.67
51	135.52	202.78	269.71	336.30	402.56	468.48	534.07	599.33	664.25	728.84	793.10	857.02	920.61	983.86	1046.78	1109.37
52	138.19	206.79	275.05	342.97	410.57	477.83	544.75	611.35	677.60	743.53	809.12	874.38	939.30	1003.89	1068.14	1132.07
53	140.86	210.79	280.39	349.65	418.58	487.17	555.43	623.36	690.96	758.22	825.14	891.73	957.99	1023.92	1089.51	1154.76
54	143.53	214.80	285.73	356.33	426.59	496.52	566.12	635.38	704.31	772.90	841.16	909.09	976.68	1043.94	1110.87	1177.46
55	146.20	218.80	291.07	363.00	434.60	505.87	576.80	647.40	717.66	787.59	857.19	926.45	995.38	1063.97	1132.23	1200.16
56	148.87	222.81	296.41	369.68	442.61	515.21	587.48	659.41	731.01	802.28	873.21	943.81	1014.07	1084.00	1153.60	1222.86
57	151.54	226.81	301.75	376.35	450.62	524.56	598.16	671.43	744.36	816.96	889.23	961.16	1032.76	1104.03	1174.96	1245.56
58	154.21	230.82	307.09	383.03	458.63	533.91	608.84	683.45	757.71	831.65	905.25	978.52	1051.45	1124.05	1196.32	1268.25
59	156.88	234.82	312.43	389.71	466.65	543.25	619.52	695.46	771.07	846.34	921.27	995.88	1070.15	1144.08	1217.68	1290.95
60	159.55	238.83	317.77	396.38	474.66	552.60	630.20	707.48	784,42	861.02	937.30	1013.23	1088.84	1164.11	1239.05	1313.65
61	162.22	242.84	323.11	403.06	482.67	561.94	640.89	719.50	797.77	875.71	953.32	1030.59	1107.53	1184.14	1260.41	1336.35
62	164.89	246.84	328.45	409.73	490.68	571.29	651.57	731.51	811.12	890.40	963.34	1047.95	1126.22	1204.17	1281.77	1359.05

Allowable pipe span between supports

When it is desired to run pipelines on supports above the ground, the engineer must often calculate the allowable pipe span between supports from a structural viewpoint. The formula for the span length can be obtained by equating the maximum bending moment caused by the loading to the resisting moment of the pipe. This formula is:

$$L = \left(\frac{SI}{wc}\right)^{0.5}$$

where:

L, span length, ft.
c, radius to outside of pipe, in.
I, moment of inertia of pipe, $in.^4$
S, maximum allowable fiber stress of material, $lb./in.^2$
w, uniform loading per unit length of pipe, lb./ft.

The above formula applies to pipes above two inches in diameter. Reference 1 gives calculated pipe span data as well as estimated pipe loading for different pipe diameters taking into account the fluid load, dead load, and the horizontal component of the wind loading.

The design of the pipe span can be speeded up with Figure 1.

Example

Find the allowable span length for a 6-in. continuous steel pipe strung between supports. The estimated loading (sum of pipe, fluid, dead and wind loads) is 36 lb./ft. The allowable fiber stress for steel can be taken at 20,000 $lb./in.^2$

From any suitable handbook, c for a 6-in. pipe is 6.625/2 = 3.3175 in., I is 28.14 $in.^4$ Now calculate I/c = 28.14/3.3175 = 8.5 $in.^3$ Connect S = 20,000 $lb./in.^2$ on the S scale with 8.5 on the I/c scale to the intersection with the reference line. With the reference line as a pivot, connect the pivot point with W = 36. At the intersection with the L scale, obtain a pipe span of 68 feet.

Source

Kuong, J. F., *Petroleum Refiner 39,* No. 10, 196 (1960).

Reference

1. Sweeney, R. J., *Chem. Eng., 63,* No. 3 199 (1956).

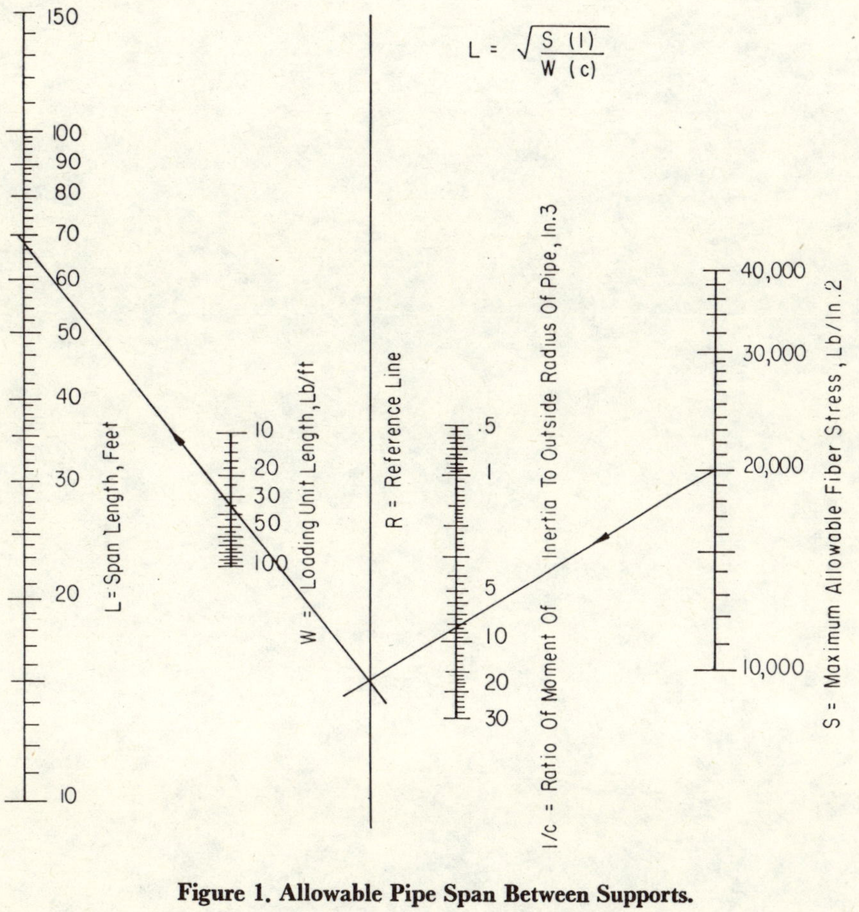

Figure 1. Allowable Pipe Span Between Supports.

5. Construction—Welding

When should steel be pre-heated before welding

From the chemistry of the steel determine the carbon equivalent which is—

$$\text{Carbon Equivalent} = C + \frac{Mn}{4}$$

If it exceeds 0.58 the steel may be crack sensitive and should be preheated before welding in ambient temperatures below 40° F.

Example: If steel pipe having a carbon content of 0.25 and manganese content of 0.70 is to be welded in springtime temperatures, ranging from 40° F to 80° F, is preheat necessary?

$$\text{Carbon Equivalent} = .25 + \frac{.70}{4} = .25 + .175 = 0.425$$

It is not necessary to preheat this particular steel before welding it.

But for another example:

$$\text{Carbon} = 0.20$$

$$\text{Manganese} = 1.60$$

$$\text{Carbon Equivalent} = .20 + \frac{1.60}{4} = 0.60$$

This steel should be preheated, particularly for early morning welding.

Why does preheating prevent cracking? It slows the cooling rate and reduces the amount of austenite retained as the weld cools. This prevents microcracking. Other alloying elements and pipe wall thickness may also influence when joints of high strength pipe should be preheated.

How many welds will the average welder make per hour?

Where stringer beads have been run, the average welder can complete about 140 inches of weld per hour on ordinary one-fourth-inch wall line pipe. To find the average number of welds per hour, divide the circumference of the pipe into 140.

$N = \frac{140}{C}$ where circumference (C) = pipe diameter in inches times 3.14

Example: How many welds per hour will a welder complete on quarter-inch wall 10¾-inch line pipe, if the stringer beads have been run?

$$N = \frac{140}{(10.75)\ (3.14)}$$

$$N = \frac{140}{33.76}$$

N = 4.15 welds per hour or about 42 welds per 10-hour day.

On large diameter pipe where wall thickness has been increased, welders average about 100 inches of completed weld per hour.

Example: How many welds per hour will a welder complete on 30-inch line pipe if the stringer beads have been run?

$$N = \frac{100}{C}$$

$$N = \frac{100}{(30)\ (3.14)} = \frac{100}{94.2} = 1.1 \text{ welds per hour}$$

or 11 welds per 10 hour day

How much welding rod is required for a mile of schedule 40 pipe line?

Rule: For 40-foot joints of schedule 40 pipe, multiply the nominal pipe diameter by 22½; the answer is pounds of electrode per mile of pipe line.

Example: 6-inch pipe; 6 × 22½ = 135 pounds per mile.

Example: 10-inch pipe, 30-foot joints; 10 × 22½ = 225 pounds per mile, for 40-foot joints; $225 \times \frac{40}{30} = 300$ pounds per mile for 30-foot joints.

The rule is based on standard weight (schedule 40) pipe, and the usual number of passes, which varies from two for small pipe up to four for 16 inches; larger pipe is usually thinner wall; the rule includes an allowance for wastage.

How many pounds of electrodes are required per weld on line pipe

Divide the nominal pipe size by two and multiply the result by one-fourth pound.

$$\text{Pounds of electrode} = \frac{N}{2} \times .25$$

Example: How many pounds of electrode will be used per weld on 10-inch line pipe?

$$\text{Pounds of electrode} = \frac{10}{2} \times .25 = 1.25 \text{ pounds per weld.}$$

Example: How many pounds of electrode will be used per weld on 24-inch line pipe?

Solution:

Pounds of electrode = 12 x .25 or 3 pounds per weld.

Circumferences of pipe for computing welding material

CIRCUMFERENCE OF PIPE IN INCHES

Nominal Pipe Size	Schedule Numbers									
	10	20	30	40	60	80	100	120	140	160
1	--	--	--	3.98	--	4.27	--	--	--	4.71
1-1/4	--	--	--	4.81	--	5.13	--	--	--	5.50
1-1/2	--	--	--	5.62	--	5.97	--	--	--	6.47
2	--	--	--	7.25	--	7.65	--	--	--	8.44
3	--	--	--	10.78	--	11.31	--	--	--	12.17
4	--	--	--	14.06	--	14.68	--	15.31	--	15.90
5	--	--	--	17.33	--	18.06	--	18.85	--	19.64
6	--	--	--	20.61	--	21.56	--	22.38	--	22.44
8	--	26.70	26.87	27.16	27.68	28.27	28.86	29.64	30.23	30.83
10	--	32.99	33.34	33.71	34.56	35.14	35.93	36.71	37.70	38.48
12	--	39.27	39.77	40.25	41.23	42.02	43.00	43.98	44.77	45.94
14	45.55	45.94	46.34	46.73	47.71	48.69	49.87	50.85	51.84	52.82
16	51.81	52.23	52.62	53.11	53.82	55.56	56.74	57.92	59.29	60.27
18	58.12	58.51	59.29	60.09	61.26	62.44	63.81	65.19	66.36	67.74
20	64.40	65.19	65.97	66.56	67.93	69.31	70.88	72.26	73.83	75.20
24	76.97	77.75	78.54	79.71	81.48	83.05	85.02	86.78	88.35	90.12

Circumferences of heavy wall pipe for computing welding material

CIRCUMFERENCE OF PIPE IN INCHES

Nominal Pipe Size	Wall Thickness in Inches							
	.500	.750	1.00	1.25	1.50	1.75	2.00	2.25
3	12.57	14.14	15.71	17.28	18.85	20.42	21.99	23.56
4	15.71	17.28	18.85	20.42	21.99	23.56	25.13	26.70
5	18.85	20.42	21.99	23.56	25.13	26.70	28.27	29.85
6	21.99	23.56	25.13	26.70	28.27	29.85	31.42	32.99
8	28.27	29.85	31.42	32.99	34.56	36.13	37.70	39.27
10	34.56	36.13	37.70	39.27	40.84	42.41	43.98	45.55
12	40.84	42.41	43.98	45.55	47.12	48.69	50.27	51.84
14	47.12	48.69	50.27	51.84	53.41	54.98	56.55	58.12
16	53.41	54.98	56.55	58.12	59.69	61.26	62.83	64.40
18	59.69	61.26	62.83	64.40	65.97	67.54	69.12	70.69
20	65.97	67.54	69.12	70.69	72.26	73.83	75.40	76.37
22	72.26	73.83	75.40	76.97	78.54	80.11	81.68	83.25
24	78.54	80.11	81.68	83.25	84.82	86.39	87.96	89.54
	2.50	2.75	3.00	3.25	3.50	3.75	4.00	4.25
10	47.12	48.69	50.27	51.84	53.41	54.98	56.55	58.12
12	53.41	54.98	56.55	58.12	59.69	61.26	62.83	64.40
14	59.69	61.26	62.83	64.40	65.97	67.54	69.12	70.69
16	65.97	67.54	69.12	70.69	72.26	73.83	75.40	76.97
18	72.26	73.83	75.40	76.97	78.54	80.11	81.68	83.25
20	78.54	80.11	81.68	83.25	84.82	86.39	87.96	89.54
22	84.82	86.39	87.96	89.54	91.11	92.68	94.25	95.82
24	91.11	92.68	94.25	95.82	97.39	98.96	100.53	102.10
	4.50	4.75	5.00	5.25	5.50	5.75	6.00	
20	91.11	92.68	94.25	95.82	97.39	98.96	100.53	
22	97.39	98.96	100.53	102.10	103.67	105.24	106.81	
24	103.67	105.24	106.81	108.39	109.96	111.53	113.10	

Circumferences of large O.D. pipe for computing welding material

CIRCUMFERENCE OF PIPE IN INCHES

Nominal Pipe Size	WALL THICKNESS IN INCHES							
	.375	.500	.750	1.00	1.25	1.50	1.75	2.00
26	84.04	84.82	86.39	87.96	89.54	91.11	92.68	94.25
28	90.32	91.11	92.68	94.25	95.82	97.39	98.96	100.53
30	96.60	97.39	98.96	100.53	102.10	103.67	105.24	106.81
32	102.89	103.67	105.24	106.81	108.39	109.96	111.53	113.10
34	109.17	109.96	111.53	113.10	114.67	116.24	117.81	119.38
36	115.45	116.24	117.81	119.38	120.95	122.52	124.09	125.66
38	121.74	122.52	124.09	125.66	127.23	128.81	130.38	131.95
40	128.02	128.81	130.38	131.94	133.52	135.09	136.66	138.23
42	134.30	135.09	136.66	138.23	139.80	141.37	142.94	144.51
44	140.59	141.37	142.94	144.51	146.08	147.66	149.23	150.80
46	146.87	147.66	149.23	150.80	152.36	153.94	155.51	157.08
48	153.15	153.94	155.51	157.08	158.65	160.22	161.79	163.36
54	172.00	172.79	174.36	175.93	177.50	179.07	180.64	182.21
60	190.82	191.64	193.21	194.78	196.35	197.92	199.49	201.06
	2.25	2.50	2.75	3.00	3.25	3.50	3.75	4.00
26	95.82	97.39	98.96	100.53	102.10	103.67	105.24	106.81
28	102.10	103.67	105.24	106.81	108.39	109.96	111.53	113.10
30	108.39	109.96	111.53	113.10	114.67	116.24	117.81	119.38
32	114.67	116.24	117.81	119.38	120.95	122.52	124.09	125.66
34	120.95	122.52	124.09	125.66	127.23	128.81	130.38	131.95
36	127.23	128.81	130.38	131.95	133.52	135.09	136.66	138.23
	4.25	4.50	4.75	5.00	5.25	5.50	5.75	6.00
26	108.39	109.96	111.53	113.10	114.67	116.24	117.81	119.38
28	114.67	116.24	117.81	119.38	120.95	122.52	124.09	125.66
30	120.95	122.52	124.09	125.64	127.23	128.81	130.38	131.95
32	127.23	128.81	130.38	131.95	133.52	135.09	136.66	138.23
34	133.52	135.09	136.66	138.23	139.80	141.37	142.94	144.51
36	139.80	141.37	142.94	144.51	146.08	147.66	149.23	150.80

Surface area and circumferences of pipe for computing painting and welding

Nominal Size Inches	Surface Area of Pipe Per Lin. Ft. Sq. Ft./Lin. Ft. (Painting)	Circumferences of Pipe In Inches (Welding)	Nominal Size Inches	Surface Area of Pipe Per Lin. Ft. Sq. Ft./Lin. Ft. (Painting)	Circumferences of Pipe In Inches (Welding)
1	0.344	4.24	12	3.34	37.7
1½	0.497	6.28	14	3.67	44.0
2	0.622	8.0	16	4.19	50.27
2½	0.753	9.0	18	4.71	56.55
3	0.916	11.0	20	5.24	62.83
3½	1.047	12.6	24	6.28	75.40
4	1.178	14.2	30	7.85	94.25
5	1.456	17.48	36	9.43	113.10
6	1.734	21.0	42	11.00	131.95
8	2.258	28.0	48	12.57	150.80
10	2.81	33.77	60	15.71	188.50

Welding rod requirements

WALL THICKNESS AND POUNDS OF WELD ROD PER JOINT

PIPE SCHEDULE NUMBERS

Nominal Pipe Size Inches	10		20		30		STD.		40		60	
	A	B	A	B	A	B	A	B	A	B	A	B
2	.109	.019	—	—	—	—	.154	.058	.154	.058	—	—
3	.120	.025	—	—	—	—	.216	.126	.216	.126	—	—
4	.120	.037	—	—	—	—	.237	.190	.237	.190	—	—
6	.134	.168	—	—	—	—	.280	.669	.280	.669	—	—
8	.148	.216	.250	.356	.227	.860	.322	.860	.322	.860	.406	1.580
10	.165	.250	.250	.411	.307	.613	.365	.993	.365	.993	.500	1.830
12	.180	.320	.250	.526	.330	1.270	.375	1.270	.406	2.340	.562	3.440
14	.250	.580	.312	.140	.375	1.400	.375	1.400	.438	2.580	.593	3.790
16	.250	.662	.312	.160	.375	1.600	.375	1.600	.500	2.950	.653	6.470
18	.250	.744	.312	.178	.438	3.310	.375	1.780	.562	3.310	.750	7.230
20	.250	.828	.375	2.000	.500	3.680	.375	2.000	.593	6.290	.812	8.910
22	.250	.910	.375	2.200	.500	4.050	.375	2.200	—	—	.875	11.460
24	.250	.992	.375	2.400	—	—	.375	2.400	.687	7.540	.968	15.860
26	.312	1.610	.500	4.790	—	—	.375	2.600	—	—	.625	7.030
28	.312	2.800	.500	5.150	.625	7.560	.375	2.800	—	—	1.000	18.520
30	.312	3.000	.500	5.520	.625	8.100	.375	3.000	—	—	1.000	19.830
32	.312	3.200	.500	5.890	.625	8.650	.375	3.200	.688	10.060	1.000	21.170
34	.312	3.400	.500	6.260	.625	9.180	.375	3.400	.688	10.680	1.000	22.480
36	.312	3.600	.500	6.620	.625	9.720	.375	3.600	.750	14.450	1.000	23.790
38	.250	1.570	.375	3.800	.500	6.990	.625	10.270	.750	15.260	1.000	25.130
40	.250	1.650	.375	4.000	.500	7.360	.625	10.810	.750	16.060	1.000	26.450
42	.250	1.750	.625	11.340	.750	16.860	.375	4.200	1.000	27.760	1.250	41.500

WALL THICKNESS AND POUNDS OF WELD ROD PER JOINT

PIPE SCHEDULE NUMBERS

Nominal Pipe Size Inches	XH		80		120		160		XXH	
	A	B	A	B	A	B	A	B	A	B
2	.218	.095	.218	.095	—	—	.343	.229	.436	.470
3	.300	.189	.300	.189	—	—	.438	.562	.600	.826
4	.337	.458	.337	.458	.438	.844	.531	1.240	.674	1.840
6	.432	1.230	.432	1.230	.562	1.810	.718	2.680	.864	3.480
8	.500	1.580	.500	1.580	.718	3.450	.906	5.680	.875	4.480
10	.500	1.830	.593	2.680	.843	5.170	1.125	8.100	—	—
12	.500	2.340	.843	6.630	1.000	8.410	1.312	14.940	—	—
14	.500	2.580	.750	5.630	1.093	11.440	1.406	19.340	—	—
16	.500	2.950	.843	8.340	1.218	13.060	1.593	25.560	—	—
18	.500	3.310	.937	11.900	1.375	21.130	1.781	37.490	—	—
20	.500	3.680	1.031	21.480	1.500	27.610	1.968	43.230	—	—
22	.500	4.050	1.125	23.640	1.625	35.140	2.125	54.140	—	—
24	.500	4.410	1.218	23.710	1.872	47.230	2.343	69.080	—	—
26	.500	4.790	.750	10.450	1.875	54.210	2.000	61.150	—	—
28	.500	5.150	.750	11.240	1.875	58.350	2.000	65.820	—	—
30	.500	5.520	.750	12.040	1.875	62.490	2.000	70.490	—	—
32	.500	5.890	.750	12.850	1.875	66.700	2.000	75.250	—	—
34	.500	6.260	.750	13.650	1.875	70.840	2.000	79.920	—	—
36	.500	6.620	.750	14.454	1.875	74.980	2.000	84.590	—	—
38	1.250	37.070	1.500	52.430	1.875	79.200	2.000	89.350	—	—
40	1.250	39.530	1.500	55.450	1.875	83.340	2.000	94.020	—	—
42	.500	7.730	1.500	57.910	1.875	87.480	2.000	98.690	—	—

Code:

A Wall thickness of pipe.

B Pounds of weld metal required per joint.

Above pounds of weld metal required per joint is based on 60° total bevel.

With the exception of standard weight and extra heavy wall 42-inch pipe, the wall thicknesses shown below the ruled line are not regularly scheduled pipe.

6. Construction & Design—General Estimating

Determine the most economical size for electric power conductors

To calculate quickly the most economical size copper wire for carrying a specified current load, use the formula:

$$A = (59)(I)\sqrt{\frac{C_e \times t}{C_c \times F}}$$

Where: A = Conductor cross-sectional area (circular mils)
I = Current load (amperes)
C_e = Cost of electrical power (cents per KWH)
C_c = Cost of copper (cents per pound)
t = Hours of service per year
F = Factor for fixed charges (amortization, insurance, taxes, etc.)

Selection of the proper size electric conductor will depend on the load to be carried and the mechanical strength required, as well as economics. Once these are considered, the most economical size conductor is that for which the annual energy cost equals the copper cost. This method can be used for other metallic conductors.

Example: Determine the most economical size copper conductor for an installation operating 365 days a year, 8 hours per day, with a 100 ampere load. Energy costs are $0.00875 per KWH. Fixed charges are 23 percent, considering a five-year amortization with 3 percent for insurance and taxes. Copper costs $0.46 per pound.

$$A = (59)(100)\sqrt{\frac{0.00875 \times 365 \times 8}{0.46 \times 0.23}}$$

$$= 92{,}500 \text{ circular mils}$$

How to find the resistance and weight of copper wires

It is easy to calculate mentally the approximate resistance and weight of standard sizes of copper wires, by remembering the following rules:

Rule 1: Number 10 wire has a resistance of 1 ohm per thousand feet. Number 2 wire weighs 200 pounds per thousand feet.

Rule 2: An increase of ten numbers in the size is an increase of ten times in the resistance, and a decrease to one-tenth of the weight.

Rule 3: An increase of three numbers in the size doubles the resistance and halves the weight.

Rule 4: An increase of one number in the size multiplies the resistance by 5/4 and the weight by 4/5.

Always proceed from the known value (No. 10 for resistance, No. 2 for the weight) in the smallest number of jumps; that is, go by tens, then by threes, and finally by ones; the jumps do not all have to be in the same direction.

Example: What is the weight and resistance of 200 feet of No. 18? Apply the rules in order, as follows; resistance first:

Number *10* wire has a resistance of *1 ohm* per *thousand feet.* (Rule 1)
Number *20* wire has a resistance of *10 ohms* per *thousand feet.* (Rule 2)
Number *17* wire has a resistance of *5 ohms* per *thousand feet.* (Rule 3)
Number *18* wire has a resistance of *6.25 ohms* per *thousand feet.* (Rule 4)
Number *18* wire has a resistance of *1.25 ohms* per *200 feet.* Answer.
Number *2* wire weighs *200 pounds* per ***thousand feet.*** (Rule 1)
Number *12* wire weighs *20 pounds* per ***thousand feet.*** (Rule 2)
Number *22* wire weighs *2 pounds* per ***thousand feet.*** (Rule 2)
Number *19* wire weighs *4 pounds* per ***thousand feet.*** (Rule 3)
Number *18* wire weighs *5 pounds* per ***thousand feet.*** (Rule 4)
Number *18* wire weighs *1 pound* per *200 feet.* (Answer)

The weight determined by the above rules is that of bare copper wire. Weight of insulated wire varies widely according to the type of insulation. Resistance values are correct for bare or insulated wire, and for solid or stranded. Errors resulting from the use of the rules will rarely exceed 2 percent.

Determine length of a V-belt

Rule: To find the nominal length of a V-belt; lay the belt on a table, and place within it a pair of circular objects of the same diameter (flanges, tin cans, or whatever may be at hand). Pull them apart until the belt is fully extended without stretching. Then measure the shortest distance between the two circles, in inches. The belt size is then twice this figure, plus 5.14 times the diameter of the circles. This value, for any standard belt, should be a whole number of inches, which is the belt size.

V-belts are made in four standard sections, classified as A, B, C, and D; the widths (at the widest part) are one-half, five-eighths, seven-eighths and 1⅛ inches respectively. The complete designation of the belt is the letter showing the width, followed by the length in inches; thus, an A26 belt is one-half inch wide, and 26 inches long on the inside edge. The *pitch length* of the belt is measured along a median section, and corresponds to the length which runs on the pully diameter which determines the actual speed ratio—about halfway of the depth of the groove. Pitch lengths for A, B, C, and D belts are greater than their nominal lengths by 1.3, 1.8, 2.9, and 3.3 inches respectively.

What you should remember about electrical formulas

Ohm's Law:

Easily Remember It, or $E = IR$

Power:

DC Power Is Easy, or $P = IE$

It Really Is, or $P = IRI = I^2R$

Other Facts:

The number of **watts in a hp** is equal to the year Columbus discovered America divided by two, or $1492/2 = 746$ watts/hp.

The horsepower of a reciprocating engine necessary to drive a three phase, 60 cycle electric generator can be determined by multiplying KW by 1.5.

How to calculate microwave hops on level ground

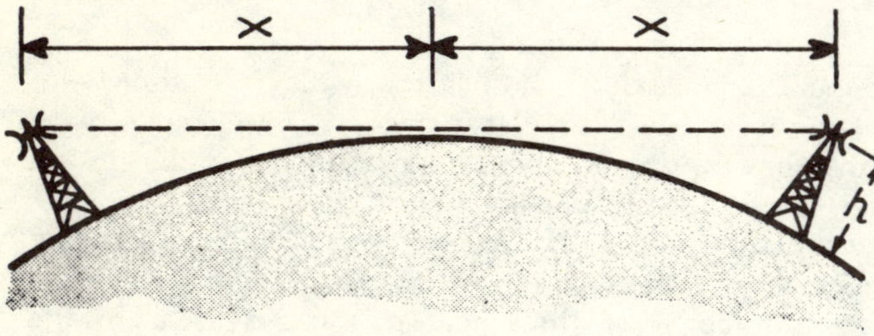

RULE: Excluding obstructions such as hills etc., the microwave (line-of-sight) distances between two towers will be twice the distance from the transmitter to the horizon. The line-of-sight distance in miles from the top of the tower to the horizon can be found by adding the square root of the tower height to the square root of the first square root (the latter of course is the fourth root).

$$x = \sqrt{h} + \sqrt{\sqrt{h}} \text{ or } \sqrt{h} + \sqrt[4]{h}$$

Example: How far apart can 100-foot-high towers be spaced to provide line-of-sight transmission if there are no obstructions between them.?

$$X = \sqrt{100} + \sqrt{10}$$
$$X = 10 + 3 + \text{ or } 13 \text{ miles}$$

Distance between towers can be 2X or 26 miles.

This same formula can be used to estimate distances in an airplane.

Example: How far is the line of sight distance to the horizon in an airplane flying at an altitude of 4900 feet.

$$(\sqrt{4900} = 70)$$
$$X = \sqrt{4900} + \sqrt{70}$$
$$X = 70 + 8+$$

Distance to horizon is 78 miles.

Calculate V-belt length using simple equation

Rule:

$$L = 1.57\,(D + d) + 2C$$

Where:

L = Belt length, in.
D = Diameter of larger sheave, in.
d = Diameter of smaller sheave, in.
C = Distance between sheave centers, in.

For quick determination of the horsepower per ampere for induction motors (3 phase) at different voltages

Voltage	HP per ampere
480	1
2400	5
4160	8

Predict subsurface temperature ranges

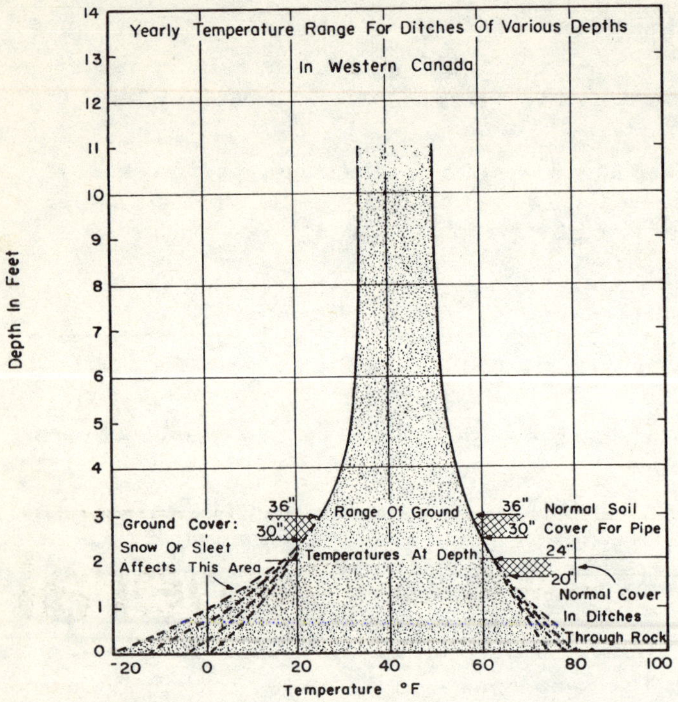

Ground temperatures are vital to pipe line design, for two reasons. First, the heat exchange between the contents of the line and the ground are greatly affected by the differential temperatures of the pipe and the soil. Second, the temperature of the pipe at time of installation will have a considerable effect on stresses, particularly where the subsequent ground temperatures will vary considerably from the installation temperatures.

For example, in the recent construction of a 34-inch line that is to operate at temperatures well over 100° F, care was exercised to limit bends when it was installed at temperatures ranging from 50 to 70° F and the underground temperature will range from about 50 to 80° F over a year's time.

The accompanying curve and table shows the range of underground temperatures you can expect in the prairie regions of Canada. At shallow depths, there is some variation and uncertainty as to temperature because of snow or sleet covering. But from three feet down, the temperature stabilizes.

Generally speaking, most areas will have somewhat similar curves from three feet on down. Of course, with depth, they approach the temperature of underground water, and the summer-winter variation ceases.

To adapt for U.S. work. As a comparison the following temperature ranges are encountered in winter and summer at depths of three feet in various areas in the United States.

California	51 F to 81 F
Colorado plains	35 F to 65 F
Texas (Panhandle)	42 F to 75 F
Texas (Coast)	60 F to 80 F
Mississippi	57 F to 80 F
Idaho	37 F to 56 F
Illinois	41 F to 56 F
Montana	33 F to 54 F
Michigan	38 F to 60 F
Wyoming	26 F to 65 F

Generally, a curve plotted for any one of the above will closely parallel the curve for Canada.

7. Gas—General

Know the gas laws

The symbols which will be used in the mathematical expression of the various gas laws are:

V_1 = volume of the gas in cubic feet with original conditions
V_2 = volume of gas in cubic feet under the second set of conditions
T_1 = initial temperature of the gas in °R (°F + 460)
T_2 = temperature of gas under second conditions °R
P_1 = pressure of gas in psia under original conditions
P_2 = pressure of gas in psia under second conditions

Any other symbols used in mathematically expressing the gas laws will be explained at the time of their introduction.

Boyle's Law may be stated—If the temperature remains the same, the volume of a given quantity of gas will vary inversely as the absolute pressure. This may be expressed mathematically as:

$$\frac{V_1}{V_2} = \frac{P_2}{P_1}$$

Inasmuch as in the application of Boyle's Law we are generally interested in the volume at a second set of pressure conditions, a rearrangement of the formula more readily used is:

$$V_2 = V_1 \times \frac{P_1}{P_2}$$

Example: A quantity of gas at a pressure of 42 psig has a volume of 1,000 cubic feet. If the gas is compressed to 100 psig, what volume would it occupy? Assume the barometric pressure to be 14.2 psia, and the temperature to remain the same. Substituting in the second arrangement of the formula for Boyle's Law above would give:

$$V_2 = 1000 \times \frac{42 + 14.2}{100 + 14.2}$$

$$= 492.1 \text{ cubic feet}$$

Charles' Law (sometimes called Gay-Lussac's Law)—If the pressure remains the same, the volume of a given quantity of gas will vary directly as the absolute temperature.

This may be expressed mathematically as:

$$\frac{V_1}{V_2} = \frac{T_1}{T_2}$$

Again since we are usually more interested in the volume at a second set of temperature conditions than any other information, an arrangement of the formula which would be handy is:

$$V_2 = V_1 \times \frac{T_2}{T_1}$$

A second part of Charles' Law is—If the volume of a quantity of gas does not change, the absolute pressure will vary directly as the absolute temperature.

Expressed mathematically this is:

$$\frac{P_1}{P_2} = \frac{T_1}{T_2}$$

In this instance, we would probably be more interested in the pressure at a second temperature condition which could be expressed as:

$$P_2 = P_1 \times \frac{T_2}{T_1}$$

Example: A given weight of gas has a volume of 450 cubic feet when the temperature is 45° F, and the pressure is 10 psig. If the pressure remains the same, but the temperature is changed to 90° F, what will be the volume of the gas?

Substitution in the formula for the first part of the Charles' Law gives:

$$V_2 = 450 \times \frac{(90 + 460)}{(45 + 460)}$$

$$= 490 \text{ cubic feet.}$$

It is desired to determine what the new pressure would be for the gas in the above example if the volume remains the same and the temperature changes from 45° F to 90° F as indicated. (Atmospheric pressure is 14.4 psia.)

Substitution in the formula gives:

$$P_2 = (10 + 14.4) \times \frac{(90 + 460)}{(45 + 460)}$$

$$= 26.6 \text{ psia or } 12.2 \text{ psig.}$$

A convenient arrangement of a combination of Boyle's and Charles' laws which is easy to remember and use can be expressed mathematically as:

$$\frac{P_1V_1}{T_1} = \frac{P_2V_2}{T_2}$$

One can substitute known values in the combination formula and solve for any one unknown value. In cases where one of the parameters, such as temperature, is not to be considered, it may be treated as having the same value on both sides of the formula and consequently it can be cancelled out.

TABLE 1—Molecular Weights of Hydrocarbons and Other Compounds Associated with Natural Gas

Compound	Atomic Formula	Molecular Weight
Methane	CH_4	16.043
Ethane	C_2H_6	30.070
Propane	C_3H_8	44.097
Butane	C_4H_{10}	58.124
Pentane	C_5H_{12}	72.151
Hexane	C_6H_{14}	86.178
Heptane	C_7H_{16}	100.205
Carbon Dioxide	CO_2	44.011
Nitrogen	N_2	28.016
Oxygen	O_2	32.00
Water	H_20	18.016
Air		28.967

Note: Molecular weights based on following values of atomic weights: Hydrogen 1.008, Carbon 12.011, Nitrogen 14.008, Oxygen 16.000, and Argon 39.944. Air was assumed to consist of 78.09% nitrogen, 20.95% oxygen, 0.93% argon, and 0.03% carbon dioxide.

Know the gas laws

Avogadro's Law states—Equal volumes of all gases at the same pressure and temperature conditions contain the same number of molecules.

From this it may be seen that the weight of a given volume of gas is a function of the weights the molecules and that there is some volume at which the gas would weigh, in pounds, the numerical value of its molecular weight.

The volume at which the weight of the gas in pounds is equal to the numerical value of its molecular weight (known as the "mol-volume") is 378.9 cubic feet for gases at a temperature of 60° F and a pressure of 14.73 psia. Table 1 gives the atomic formula and molecular weights for hydrocarbons and other compounds frequently associated with natural gas. Reference to the table reveals that the molecular weight for methane is 16.043. Going back to the mol-volume explanation, shows that 378.9 cubic feet of methane at 60° F and a pressure of 14.73 psia would weigh 16.043 pounds.

Avogadro's law ties in closely with what is usually known as the ideal gas law.

The Ideal Gas Law, although expressed in many slightly different arrangements, is most frequently expressed:

$$pV = nRT$$

In which:

p = pressure of the gas
V = volume of the gas
n = number of pound-mols of gas
R = the universal gas constant which varies depending upon the units of pressure, volume, and temperature employed.

Since the number of pound-mols of a gas would be equal to the weight of the gas divided by the molecular weight of the gas, we can express the ideal gas law as:

$$pV = 10.722 \times \frac{W}{M} \times T$$

In which:

p = pressure of the gas in psia
V = volume of the gas in cubic feet
W = weight of the gas in pounds
M = molecular weight of the gas
T = temperature of the gas °R

The constant 10.722 is based upon the generally used value for the universal gas constant of 1,544 when the pressure is expressed in pounds per square foot absolute.

This formula can be used in many arrangements. An arrangement which may be used to determine the weight of a quantity of gas is:

$$W = 0.0933 \times \frac{MVp}{T}$$

when the symbols and units are as above.

Example: It is desired to find the weight of a gas in a 1,000-cubic-foot container if the gas is at a pressure of 150 psig and a temperature of 90°F. The molecular weight of the gas is known to be 16.816, and the barometric pressure is 14.3 psia.

Substitution in the formula gives:

$$W = \frac{.0933 \times 16.816 \times 1000 \times (150 + 14.3)}{(90 + 460)}$$

$$= 468.7 \text{ pounds}$$

The formula above may be used when the molecular weight of a gas is known, however, at times it is desirable to determine the weight of a given volume of gas when the molecular weight is not known or cannot be readily determined. This may be accomplished if the specific gravity of the gas is known by using the formula:

$$W = 2.698 \times \frac{GVp}{T}$$

in which:

G = the specific gravity of the gas (air equals 1.000) and the other symbols and units are as previously given.

Example: If the molecular weight of the gas in the preceding example was unknown, but the specific gravity was known to be 0.581, substitution in the formula would give:

$$W = 2.698 \times \frac{0.581 \times 1000 \times (150 + 14.3)}{(90 + 460)}$$

$$= 468.3 \text{ pounds.}$$

Gaseous Mixtures. If an analysis giving the percent by volume of the various components of a gaseous mixture is available, the average molecular weight of the gaseous mixture may be readily determined. In order to keep the terminology straight we will use the term "mol fraction" in explaining the procedure.

However, the mol fraction of a gas is equal to the percent by volume of the constituent. This is perhaps better explained by use of an illustration.. Suppose it is desirable to find the average molecular weight of a natural gas with the following analysis:

Methane	94.6%
Ethane	4.6%
Nitrogen	0.6%
Carbon dioxide	0.2%

Calculation of the molecular weight of the natural gas would be (molecular weights for the components in this example are obtained from Table 1):

Molecular Weight of Component		Mole Fraction (% by volume)	
16.043	×	0.946	= 15.177
30.070	×	0.046	= 1.383
28.016	×	0.006	= 0.168
44.011	×	0.002	= 0.088
Molecular weight of mixture			= 16.816

Facts about methane and its behavior

A study of the physical, transport and thermodynamic properties of methane

Dr. Robert O. Parker and **Dr. William H. Kapfer**
New York University, New York, N.Y.

INTEREST IN LIQUEFIED natural gas, (LNG), requires that the physical, transport, and thermodynamic properties of methane, the principal constituent of natural gas, be readily available. This article will present values most frequently used in process calculations and in other studies associated with the storage and processing of natural gas. In addition, sources of more detailed information are listed.

Methane, CH_4, is the hydrocarbon of lowest molecular weight (16.042). The molecule is best visualized as a regular tetrahedron with the carbon atom at the geometric center and the hydrogen atoms at each of the four apices. Methane is an end product of the anaerobic (without air) decay of plants, that is, of the breakdown of certain very complicated molecules. As such, it is the major constituent (up to 97%), of natural gas. It is the firedamp of the coal mines, and can be seen as marsh gas bubbling to the surface of swamps.

The physical, transport, and thermodynamic properties required are:

Physical
(1) critical constants
(2) equation of state, (PVT relationships)
(3) vapor pressure, including b.p. and f.p.
(4) vapor and liquid density

Transport
(1) viscosity of liquid and gas
(2) thermal conductivity of liquid and gas
(3) diffusivity

Thermodynamic
(1) heat capacity
(2) latent heat
(3) enthalpy
(4) entropy

They are required for process calculations: material and energy balances; heat transfer, mass-transfer and pressure loss calculations.

Critical Constants

$T_c = -115.8°$ F ($-82.1°$ C)[1]
$P_c =$ 673 psia (45.8 atm.)[2]
$d_c =$ 10.17 lb/ft.^{3} (0.1630 g/cc)[2]

The main utility of critical constants to the engineer is related to the fact that correlations of many of the physical, transport, and thermodynamic properties of hydrocarbons are best expressed in terms of reduced temperature, T_R, and pressure, $P_R \left(T_R = \frac{T}{T_c}, P_R = \frac{P}{P_c} \right)$. This is simply an application of the "principle of corresponding states" and is recommended only in the absence of sufficient reliable experimental data.

Equations of State—(Pressure - Volume - Temperature Behavior)

For ordinary conditions, low to moderate pressures and temperatures not too close to the normal boiling point, i.e., when inspection of a compressibility factor chart shows it to be nearly unity, the ideal gas law

$$PV = \frac{W}{M} RT \tag{1}$$

where

P = pressure, psia
V = volume, ft.^{3}
W = weight of gas, lb.
M = molecular weight (here 16.042)
R = gas constant $= 10.73 \frac{\text{psia (ft}^3)}{\text{lb. mole (°R)}}$
T = temperature, °R

is sufficient. For most other conditions Equation (1) is modified by inclusion of the compressibility factor, Z, in the right-hand member, viz.

$$PV = \frac{W}{M} ZRT \tag{1a}$$

The writers recommend compressibility factors of Nelson and Obert.[3] If an analytical equation of state is preferred to Equation (1a) the writers recommend the 8 constant equation of Benedict, Webb and Rubin.[4]

$$P = RTd + (B_oRT - A_o - C_o/T^2)\, d^2 + (bRT - a)\, d^3 + a\,\alpha\, d^6 + \frac{cd^3 (1 + \gamma d^2)\, e^{-\gamma d^2}}{T^2} \tag{2}$$

where d = density, g-moles/l
in which the constants[4] for methane are:

$B_o = 0.042600$
$A_o = 1.85500$
$C_o = 22{,}570.0$
$b = 0.00338004$
$a = 0.0494000$
$c = 2{,}545.00$
$\gamma = 0.0060000$
$\alpha = 0.000124359$

With these constants, $R = 0.0820 \dfrac{\text{liter-atm}}{(°\text{K})\ (\text{gram mole})}$

P is in atmospheres, T is in °K, and e is the base of natural logarithms. Constants in this equation are available for ethane, ethylene, propane, propylene, *n*-butane, *i*-butane, *i*-butene, *n*-pentane, *i*-pentane, *n*-hexane, and *n*-heptane.[4] Using Equation (2) to predict the pressure corresponding to 89 data points[5,6] it was found that the average absolute and maximum deviations for methane were 0.376% and + 2.25% respectively. Ref. 6 discusses a generalization of the Benedict-Webb-Rubin equation.

Vapor Pressure[2]

Freezing point in air at 1 atm. = − 296.46 ° F (182.48 ° C)

Boiling point at 29.921″ Hg = − 258.68 ° F (− 161.49 ° C)

Vapor pressures are tabulated on p. 121 of Ref. 2 from 10 mm to 1,500 mm Hg and are represented by the Antoine equation.

$$\log_{10} P = A - B/(C + t) \qquad (3)$$

or

$$t = B/(A - \log_{10} P) - C \qquad (3a)$$

with P in mm Hg and t in °C.

$A = 6.61184$
$B = 389.93$
$C = 266.00$

Transport Properties

Viscosity (liquid)[2]

t °C	μ, centipoises
−185	0.226 (subcooled)
−180	0.188
−175	0.161
−170	0.142
−165	0.127
−160	0.115 (at sat. pressure)

Viscosity (vapor)[7] at 1 atm. pressure

t °C	μ, centipoises
−181.6	0.00348
− 78.5	0.00760
0	0.01026
20	0.01087
100	0.01331
200.5	0.01605
284.	0.01813
380	0.02026
499	0.02264

Additional data[8] are available.

Thermal Conductivity (liquid)

	t °F	K, Btu/hr ft. °F
	−280	0.0841
n.b.p.	−258.68	0.0790
	−200	0.0568
crit. temp.	−115.8	0.029 (reference[10])

Values for liquid thermal conductivity represent estimates by the authors using the method of Viswanath[9] which seems to be quite reliable for organic liquids (halogenated compounds excepted).

Thermal Conductivity (vapor)[7]

t °F	K, Btu/hr ft. °F
−300	0.00541
−200	0.00895
−100	0.01265
− 40	0.0149
− 20	0.0157
0	0.01645
20	0.01725
40	0.01805
60	0.0190
80	0.01985
100	0.0208
120	0.0216
200	0.0258

Diffusivity

The transport property, mass diffusivity (D), is of importance in calculations usually entering computations in combination with viscosity and density as the Schmidt No, $\mu/\rho D$. Reliable estimates of D_{12}, the diffusivity of species 1 through species 2 are best made by the method of Hirschfelder, Bird, and Spotz.[11]

$$D_{12} = \frac{0.001858 T^{3/2} [(M_1 + M_2)/M_1 M_2]^{1/2}}{P \sigma_{12}^2 \Omega_D} \quad (4)$$

D = diffusivity, cm^2/sec

T = temperature, $°K$

M = molecular wgt, M_1 of species 1, M_2 of species 2

P = pressure, atm.

$\sigma_{12} = (\sigma_1 + \sigma_2)/2$ = collision diameter, A°.

Ω_D = collision integral = function of $\frac{KT}{\epsilon_{12}}$

K = Boltzmann's constant

ϵ = a force constant

$\epsilon_{12} = \sqrt{\epsilon_1 \epsilon_2}$

Force constants and collision diameters for air and methane are:

	ϵ/K, °K	σ, A°
Air	97	3.617
Methane	136.5	3.822

These are known as the Lennard-Jones force constants and an extensive table is given in.[11] Ω_D, which depends upon the temperature through KT/ε, is tabulated in the reference.[12] A few values are:

KT/ϵ	Ω_D
0.30	2.662
0.50	2.066
1.00	1.439
1.50	1.198
2.00	1.075
4.00	0.8836
10	0.7424
50	0.5756
100	0.5130
400	0.4170

If the calculation requiring diffusivities is a crucial one, we recommend that the engineer consult the principal literature reference.[11]

Thermodynamic Properties

The most extensive, reliable compilation of the thermodynamic properties is the one by Matthews and Hurd.[13] In addition to a pressure-enthalpy diagram, tabulations are provided for saturated methane from −280° F to −115.8° F, the critical temperature, and for superheated methane to 500° F and 1,500 psia. This tabulated information is reproduced in the Chemical Engineers' Handbook 3rd Ed.[14] In the 4th Ed. (1963) superheat tables are omitted.

Corresponding to the pressure-temperature pair, values are given for:

Cold box where liquefaction of natural gas is accomplished.

(a) ratio fugacity (where fugacity[13] refers to "escaping tendency") to pressure

(b) specific volume of liquid and vapor

(c) enthalpy of liquid, vaporization, and vapor

(d) entropy of liquid and vapor

A few numbers of immediate interest to the process engineer are:

(a) latent heat of vaporization at normal boiling point = 219.22 Btu/lb.

(b) liquid density at normal boiling point = 26.6 lb/ft.^{3}

(c) vapor density at normal boiling point = 0.113 lb/ft.^{3}

The liquid density is required to compute the volume of a diked area to contain a spill. Latent heat is useful in estimating the time required for a spill to be vaporized. The vapor density points up the fact that, although methane has a specific gravity referred to air of about 0.55, vapors generated at a spill are about 1.5 times as heavy as ambient air; a matter of importance in various studies.

Heat Capacities—

If heat capacities are required for process calculations, they can be extracted from the thermodynamic tables mentioned above by the following procedure:

$$C_p = \frac{H_1 - H_2}{T_1 - T_2}, \text{ Btu/lb. °F} \quad (5)$$

Example: Find the heat capacity for superheated methane at 15 psia and 100° F.

From the Matthews and Hurd superheat table:[13 14]

at 15 psia

$T_1 = 110 \qquad H_1 = 426.6$

$T_2 = 90 \qquad H_2 = 415.8$

Then

$$C_p = \frac{426.6 - 415.8}{110 - 90} = 0.54 \frac{\text{Btu}}{\text{lb. °F.}}$$

The procedure given above is adequate for liquid heat capacities when enthalpy values are taken from the table for saturated methane. This works because the effect of pressure on liquid heat capacity is essentially negligible.

Alternatively, instantaneous values of heat capacity may be determined.[15]

$$C_p° = 0.213 + 1.11 \times 10^{-3}\, T - 2.59 \times 10^{-7}\, T^2, \frac{\text{Btu}}{\text{lb. °F}} \quad (6)$$

Note carefully that equation (6) yields the heat capacity at low pressure, 1 atm.; at elevated pressures a correction may be required. If equation (5) above is used, no pressure correction is necessary.

Heat Capacity Ratio

The ratio $Cp/Cv = \gamma$ useful in process calculations (adapted from reference 14) is, for atmospheric pressure.

t, °F	$\gamma = C_p/C_v$
−150	1.39
−100	1.36
− 50	1.33
0	1.31
50	1.28
100	1.27
150	1.25

Flammability Characteristics

Since methane is a combustible gas, consideration of any process involving its use or storage must include studies of the associated fire hazard. Basic data are flammable limits and the autoignition temperature. For methane in air these are [16, 17] at atmospheric pressure:

(a) Lower Flammable, Limit @ 25° C about 5% by vol.
(b) Upper Flammable, Limit @ 25° C about 15% by vol.
(c) Antoignition Temperature 1004 °F.

These figures mean that mixtures of methane and air having less than 5% methane by volume are too lean to burn, those with more than 15% are too rich. The autoignition temperature is the minimum temperature at which ignition can occur. Figures on flammable limits refer to "homogeneous" mixtures. Industrially, few mixtures are completely homogeneous. In safety computations, therefore, the engineer must establish that the "highest local concentration" is below the lower flammable limit by a suitable margin before saying there is no fire hazard. A method dealing with LNG spills has been proposed by Parker and Spata.[18]

LITERATURE CITED

1 Kobe, K. A., and R. E. Lynn, Chem. Revs. 52, 117, (1953).
2 American Petroleum Institute. "Selected Values of Physical and Thermodynamic Properties of Hydrocarbons and Related Compounds." Project 44, Carnegie Press, Pittsburgh, Pa. 1953, (NBS-C-461).
3 Nelson, L. C., and E. F. Obert, Trans. A.S.M.E. 76, 1057, (1954).
4 Benedict, M., G. B. Webb, and L. C. Rubin and L. Friend, Chem. Eng. Prog. 47, 419, 449, 571, 609 (1951).
5 Kvalnes, H. M. and V. L. Gaddy, J. Am. Chem. Soc. 53, 394 (1931).
6 Edmister, W. C., J. Vairogs, and A. J. Klekers, A.I.Ch.E. Journal 14, No. 3, p. 479 (1968).
7 Handbook of Chemistry and Physics, 47th Edition, 1966-67.
8 Huang, E. T. S., G. W. Swift and F. Kurata, A.I.Ch.E. Jour. 12; p. 932, (1966).
9 Viswanath, D.S., A.I.Ch.E. Jour. 13, 850 (1967).
10 Gamson, B. W., Chem. Eng. Prog. 45, 154 (1949).
11 Hirschfelder, J. O., Bird, R. B., and Spotz, E. L., Trans. ASME 71, 921, (1949).
12 Hirschfelder, J. O., C. F. Curtiss, and R. B. Bird, "Molecular Theory of Gases and Liquids," John Wiley and Sons, Inc. New York, 1954.
13 Matthews, C. S. and C. O. Hurd, Trans. A.I.Ch.E. 42, p. 55 (1946); ibid., pp. 78 1009.
14 Perry, J. H. (ed.) "Chemical Engineers Handbook, 3rd ed., McGraw-Hill, New York, 1950.
15 Reid, R. C. and T. K. Sherwood, "The Properties of Gases and Liquids," McGraw-Hill, New York (1958).
16 Coward, H. F. and G. W. Jones, U.S. Bureau of Mines, Bulletin 503 (1952).
17 Zabetakis, M. G., U.S. Bureau of Mines, Bulletin 627 (1965).
18 Parker, R. O. and J. K. Spata, Proc. 1st Intern. Conf. on L.N.G. Chicago, April 7-12, 1968, Paper No. 24.

Conversion table for pure methane

		Long tons liquid	ft^3 liquid	ft^3 gas	m^3 liquid	m^3 gas	U.S. bbl liquid	Therms	Thermies
long ton liquid	=	1	84.56	52,886	2.394	1,419.7	15.06	535.2	13,491
ft^3 liquid	=	0.011826	1	625.43	0.028311	16.789	0.1781	6.329	159.54
ft^3 gas x 10^6	=	18.91	1,599	1 x 10^6	45.27	26,847	284.785	10,121	255,115
m^3 liquid	=	0.4177	35.315	22,090	1	593	6.29	223.55	5,635.2
m^3 gas x 10^9	=	704,374	59.562 x 10^6	37.252 x 10^9	1.686 x 10^6	1 x 10^9	10.608 x 10^6	0.377 x 10^9	9.503 x 10^9
U.S. bbl liquid	=	0.0664	5.615	3,512	0.15896	94.268	1	35.54	895.8
Therm x 10^6	=	1,868	157,958	98.791 x 10^6	4,472	2.652 x 10^6	28,132	1 x 10^6	25.201 x 10^6
Thermie x 10^6	=	74.12	6,268	3.92 x 10^6	177.44	105,228	1,116.2	39,669	1 x 10^6

Notes

1. ft^3 measured at 60 °F, 14.696 lbf/in^2, dry
2. m^3 measured at 0 °C, 760 mmHg, dry
3. Gross heat content of methane is 1012 Btu/ft^3 in gaseous form: Specific gravity 0.554 (Air = 1)
   Boiling point −258.7 °F (14.696 lbf/in^2)
4. 1 x 10^6 ft^3 gas = 24.42 long tons (0.4417 b/d) fuel oil equivalent
5. The above factors are based on the properties of pure methane; LNG as shipped in practice will vary in composition according to the presence of heavier hydrocarbons (C_2, C_3, C_4) in the gas stream.

Categories of natural gas and reserves terminology

Associated and Non-Associated Natural Gas

Natural gas is found in underground structures similar to those containing crude oil. There are three types of natural gas reservoirs:

1. Structures from which only gas can be produced economically—called **non-associated gas** (or unassociated gas).
2. Condensate reservoirs which yield relatively large amounts of gas per barrel of light liquid hydrocarbons. Although many condensate reservoirs are produced primarily for gas, there are cases where gas is re-injected or "re-cycled" to improve liquid recovery, particularly if no gas market is yet available. This gas also is termed **non-associated.**
3. Reservoirs where gas is found dissolved in crude oil (**solution gas**) and in some cases also in contact with underlying gas-saturated crude (**gas-cap gas**). Both are called **associated gas.** (Gas-cap gas is almost never produced until most of the economically recoverable oil has been yielded.)

In such fields, gas production rates will depend on oil output, with the oil usually representing the major part in terms of energy equivalents.

Natural Gas Reserves

There are no generally accepted definitions of natural gas reserves, and wide variations both in terms and principles may be met. The following definitions are the most commonly used:

Expected Ultimate Recovery is the total volume of hydrocarbons, gas and/or oil, which may be expected to be recoverable commercially from a given area at current or anticipated prices and costs, under existing or anticipated regulatory practices and with known methods and equipment. This total can be expressed as the sum of the "Ultimate Recovery from Existing Fields" and the "Expectation from Future Discoveries."

Ultimate Recovery from Existing Fields is the sum of the "Cumulative Production" from such fields and the "Reserves from Existing Fields."

Reserves from Existing Fields are the sum of "Proven Reserves," "Discounted (i.e. 50% of) Probable Reserves" and "Discounted (i.e. 25% of) Possible Reserves."

Proven Reserves represent the quantities of crude oil and/or natural gas and natural gas liquids which geological and engineering data demonstrate with reasonable certainty to be recoverable in the future from known oil and/or gas reservoirs. They represent strictly technical judgements, and are not knowingly influenced by attitudes of conservatism or optimism.

Discounted (i.e. 50% of) Probable Reserves are those quantities of crude oil and/or natural gas and natural gas liquids for which there exists a probability of 50% that they will materialize. Such reserves are usually allocated to some conjectural part of a field or reservoir as yet undrilled or incompletely evaluated where, while characteristics of the productive zone and fluid content are reasonably favorable, other uncertain features may have equal chances of being favorable or unfavorable for commercial production.

Discounted (i.e. 25% of) Possible Reserves are those quantities of crude oil and/or natural gas, and natural gas liquids thought to be potentially producible but where the chance of their being realized is thought to be on the order of 25%. Reserves in this category are usually allocated to possible extensions of existing fields where, although geological data may be favorable, uncertainties as to the characteristics of the productive zone and/or fluid content are such as to preclude a probability greater than 25%.

It should be noted that the terms "gas-in-place" and "oil-in-place" represent estimates of the total hydrocarbons present which, in conjunction with various engineering and economic criteria, provide the basis for the "reserve" estimates. They should never be confused with "reserves," which word implies that physical recovery is possible.

Terminology of Reserves

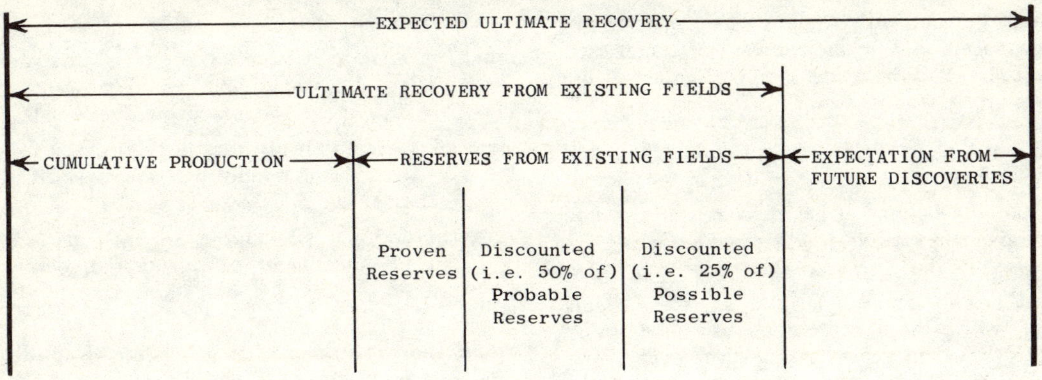

Glossary of common gas industry terms

(italics denote cross reference)

absolute pressure *gauge pressure* plus *atmospheric pressure.*

associated gas see preceding page.

atmospheric pressure the pressure of the weight of air and water vapor on the surface of the earth. Approximately 14.7 lbf/in^2 at sea level.

blue water gas also called "water gas." Made in a cyclic process in which an incandescent bed of coke or coal is alternately subjected to blasts of air and steam. The gas consists mainly of equal proportions of carbon monoxide and hydrogen and has a gross heat content of about 300 Btu/ft^3 (see also *carbureted water gas*).

British thermal unit the heat required to raise the temperature of one pound of water one degree Fahrenheit.

burner capacity the maximum Btu per hour that can be released by a burner while burning with a stable flame.

calorie the quantity of heat required to raise the temperature of one gram of water one degree Celsius (centigrade).

calorific value the quantity of heat released by the combustion, in a calora meter at constant pressure of one atmosphere, of a unit quantity of fuel measured under given conditions. In the case of gas, the conditions of temperature, pressure, and water vapor content under which the unit volume was measured affect the calorific value.

calorimeter an instrument for determining the heating value of a fuel.

carbureted water gas (**carbureted blue gas**) the gas resulting from the enrichment of *blue water gas* during its manufacture by a simultaneous process of light distillate, gas oil or fuel oil gasification. The gas has a gross heat content of about 500 to 550 Btu/ft^3.

city gate a measuring station, which may also include pressure regulation, at which a distributing gas utility receives gas from a natural gas pipeline company or the transmission system.

coal gas or **coke oven gas** a *manufactured gas* made by destructive distillation ("carbonization") of bituminous coal in a gas retort or by-product coke oven. Its chief components are methane (20 to 30%) and hydrogen (about 50%). This gas generally has a gross heating value of 500 to 550 Btu/ft^3. When the process takes place in a closed oven (with gas as a by-product of coke production) it is generally designated as "Coke Oven Gas," and when produced in retorts it is called "Coal Gas."

commodity charge that portion of the charge or rate based upon the total volume of gas consumed or billed.

condensate a liquid/hydrocarbon mixture of 45 to 65 degrees A.P.I. gravity, which may be recovered at the surface from some *non-associated gas* reservoirs.

connection charge an amount to be paid by the customer in a lump sum, or in installments, for connecting the customer's facilities to the supplier's facilities.

contract demand the volume that the supplier of gas service agrees to deliver and, in general, the amount that the customer agrees to take or pay for.

cubic foot and **cubic meter (gas)** common units of measurement of gas volume. They are the amounts of gas required to fill a volume of one cubic foot or one cubic meter under stated conditions of temperature, pressure and water vapor content.

customer charge a fixed amount to be paid periodically by the customer without regard to *demand* or *commodity* consumption.

cycling or **recycling** the process by which a *non-associated gas* reservoir is produced only for the recovery of condensate. The gas, after removal of the condensate, is re-injected into the formation.

daily average send-out the total volume of gas delivered during a period of time divided by number of days in the period.

daily peak the maximum volume of gas delivered in any one day during a given period, usually a calendar year.

degree day a measure of the extent to which the mean daily temperature falls below an assumed base, say 65° F. Thus each degree by which the mean temperature for any day is less than 65° F would represent one degree day. (In Continental Europe, °C are used instead of °F, and the assumed base temperature is generally taken as 16° C equivalent to 60.8° F).

demand (load) rate of flow of gas in a specified time interval, usually expressed in cubic feet (or cubic meters) per hour, per day or per year.

demand charge that portion of the charge or rate for gas service based upon the customer's *demand* characteristics.

distribution company or **gas utility** a company which obtains the major portion of its gas operating revenues from the operation of a retail gas *distribution system* and which operates no transportation system other than incidental connections to a transportation system of another company. For purposes of American Gas Association (A.G.A.) statistics, a distribution company obtains at least 95% of its gas operating revenues from the operation of its retail gas distribution system (see also *transmission company*).

distribution system feeders, mains, services, and equipment which carry or control the supply of gas from the point or points of local supply (usually the *city gate* station) to and including the consumer *meters.*

diversity factor the ratio of the sum of the non-coincident maximum demands of two or more loads to their coincident maximum demand for the same period.

dyne the standard centimeter-gram-second unit of force, equal to the force that produces an acceleration on one centimeter per second on a mass of one gram.

erg the centimeter-gram-second unit of work or energy, equal to the work done by a force of one dyne when its point of application moves through a distance of one centimeter in the direction of the force.

flat rate schedule a rate schedule that provides for a specified charge irrespective of the quantity of gas used or the *demand.*

gauge pressure the pressure generally shown by measuring devices. This is the pressure in excess of that exerted by the atmosphere (see *absolute pressure*).

gross calorific (heating) value the gross calorific value at constant pressure of a gaseous fuel is the number of heat units produced when unit volume of the fuel, measured under standard conditions, is burned in excess air in such a way that the materials after combustion consist of the gases carbon dioxide, sulfur dioxide, oxygen and nitrogen, water vapor equal in quantity to that in the gaseous fuel and the air before combustion, and liquid water equal in quantity to that produced during combustion, and that the pressure and temperature of the gaseous fuel, the air, and the materials after combustion are one standard atmosphere and 25° C.

inch of mercury a unit of pressure (equivalent to 0.491154 lbf/in^2) by which one pound per square inch equals 2.036009 inches of mercury column at 0° C.

inch of water a unit of pressure (equivalent to 0.03606 lbf/in^2) by which one pound per square inch equals 27.68049 inches of water column at 4° C.

interruptible gas gas made available under agreements which permit curtailment or cessation of delivery by the supplier.

joule the work done when the point of application of a force of one newton is displaced through a distance of one meter in the direction of the force. It is equal to 10^7 *ergs,* and in electrical units is the energy dissipated by one watt in a second.

kilocalorie one thousand *calories.*

kilowatt-hour a unit of energy equivalent to the energy transferred or expanded in one hour by one kilowatt of power.

line pack a method of *peak-shaving* by withdrawing gas from a section of a pipeline system in excess of the input into that section, i.e. normally the difference between the actual volume of gas in the pipeline at low flow (increased pressure) and that at normal flow.

liquefied natural gas (LNG) natural gas that has been liquefied by cooling to minus 258° F (−161° C) at atmospheric pressure.

liquefied petroleum gas (LPG) any hydrocarbon mixture, in either the liquid or gaseous state, the chief components of which consist of propane, propylene, butane, iso-butane, butylene or mixtures thereof in any ratio.

load curve a graph in which the *send-out* of a gas system, or segment of a system, is plotted against intervals of time.

load density the concentration of gas load for a given area, expressed as gas volume per unit of time and per unit of area.

load duration curve a curve of loads, plotted in descending order of magnitude, against time intervals for a specified period. The coordinates may be absolute quantities or percentages.

load factor the ratio of the average load over a designated period to the *peak load* occurring in that period. Usually expressed as a percentage.

low pressure gas distribution system a gas *distribution system,* or the mains of a segment of a distribution system, operated at low pressures of less than 15 *inches water column.*

LPG-air mixtures mixtures of *liquefied petroleum gas* and air to obtain a desired heating value and capable of being distributed through a *distribution system* also used for stand-by and *peak-shaving* purposes by gas utilities.

manufactured gas combustible gases derived from primary energy sources by processes involving chemical reaction. For instance, gas produced from coal, coke or liquid hy-hydrocarbons (see also *town gas*).

meter (gas) a mechanical device for automatically measuring quantities of gas. Some of the more important types are:

Diaphragm Meter a gas meter in which the passage of gas through two or more chambers moves diaphragms which are geared to a volume-indicating dial. Sizes up to 20,000 cubic feet per hour and 1,000 lbf/in^2 working pressure are available.

Mass Flowmeter a type of meter in which the gas flow is measured in terms of mass by utilizing the velocity and density of the gas stream to calculate the mass of a gas flowing in unit time.

Orifice Meter meter for measuring the flow of fluid through a pipe or duct by measurement of the pressure differential across a plate having a precision machined hole in its center. Ratings up to very high throughputs and pressure are available.

Rotary Displacement Meter the positive-pressure blower principle in reverse is used in this meter in which gas pressure turns twin matching impellers and the quantity of gas passing through is proportional to the number of revolutions. These meters are made in the range of 3,000 to 1,000,000 cubic feet per hour and for pressures up to 1,200 lbf/in^2.

Turbine Flowmeter a type of meter in which there is a turbine wheel or rotor made to rotate by the flowing gas. The rate or speed at which the turbine wheel revolves is a measure of the velocity of the gas.

Venturi Meter a meter in which the flow is determined by measuring the pressure drop caused by the flow through a Venturi throat. The flow rate is proportional to the square root of the pressure drop across the throat.

minimum charge a clause which provides that the charge for a prescribed period shall not be less than a specified amount.

natural gasoline those liquid hydrocarbon mixtures containing essentially pentanes and heavier hydrocarbons which have been extracted from natural gas.

natural gas liquids a term used for those mixtures of hydrocarbon fractions that can be extracted in liquid form from natural gas. In the widest sense, they can comprise any combination of the following: ethane, propylene, propane, butylene, isobutane, n-butane, pentanes, hexanes, heptanes, and octanes.

net calorific (heating) value the net calorific value at constant pressure of a gaseous fuel is the number of heat units produced when unit volume of the fuel, measured under standard conditions, is burned in excess air in such a way that the materials after combustion consist of the gases carbon dioxide, sulfur dioxide, oxygen, nitrogen and water vapor, and that the pressure and temperature of the gaseous fuel, the air and the materials after combustion are one standard atmosphere and 25°C. This value is derived from the measured gross figure.

non-associated gas see preceding section on "reserves terminology."

off-peak the period during a day, week, month or year when the load being delivered by a gas system is not at or near the maximum volume delivered by that system.

peak or **peak load** the maximum load consumed or produced by a unit or group of units in a stated period of time.

peak shaving the practice of augmenting the normal supply of gas during peak or emergency periods from another source where gas may have either been stored during periods of low demand, or manufactured specifically to meet the peak demand.

pressure maintenance (repressing). A process in which natural gas is injected into a formation capable of producing crude petroleum to aid in maintaining pressure in an underground reservoir for the purpose of assisting in the recovery of crude.

producer gas a gas manufactured by burning coal or coke with a regulated deficiency of air, normally saturated with steam. The principal combustible component is carbon monoxide (about 30%) and the gross heat content is between 120 and 160 Btu/ft^3.

recycling see *cycling.*

reforming the process of thermal or catalytic cracking of natural gas, *liquefied petroleum gas, refinery gas* or gas from oil, resulting in the production of a gas having a different chemical composition.

refinery gas a gas resulting from oil refinery operations consisting mainly of hydrogen, methane, ethylene, propylene and the butylenes. Other gases such as nitrogen and carbon dioxide may also be present. The composition can be highly variable and the heat content can range from 1,000 to 2,000 Btu/ft^3.

reserves see preceding section.

send-out the quantity of gas delivered by a plant or system during a specified period of time.

specific gravity of gas the ratio of the density of gas to the density of dry air at the same temperature and pressure.

standard metering base standard conditions, plus agreed corrections, to which all gas volumes are corrected for purposes of comparison and payment.

summer valley the decrease which occurs in the summer months in the volume of the daily load of a gas *distribution system.*

therm 100,000 Btu's.

thermie the heat required to raise the temperature of one metric ton of water one degree Celsius (centigrade).

town gas gas piped to consumers from a gas plant. The gas can comprise both *manufactured gas* (secondary energy) and natural gas (primary energy) used for enrichment.

transmission company a company which operates a natural gas transmission system and which either operates no retail distribution system, or, as defined by the American Gas Association (A.G.A.) receives less than 5% of its gas operating revenues from such retail distribution system (see also *distribution company or gas utility*).

unaccounted for gas the difference between the total gas available from all sources and the total gas accounted for as sales, net interchange and company use. The difference includes leakage or other actual losses, discrepancies due to *meter* inaccuracies, variations of temperature, and pressure and other variants.

unassociated gas see preceding section.

utilization factor the ratio of the maximum *demand* on a system or part of a system to the rated capacity of the system or part of the system under consideration.

water gas see *blue water gas.*

watt the meter-kilogram-second unit of power, equivalent to one joule per second and equal to the power in a circuit in which a current on one ampier flows across a potential difference on one volt.

wobbe index the gross calorific value of the gas divided by the square root of the density of the gas as compared with air.

8. Gas—Compressors

Estimate hp required to compress natural gas

To estimate the horsepower to compress a million cubic feet of gas per day, use the following formula:

$$BHP/MMCfd = \frac{R}{R + RJ}\left(\frac{5.16 + 124 \text{ Log } R}{0.97 - 0.03\,R}\right)$$

where R = compression ratio. Absolute discharge pressure divided by absolute suction pressure

J = Supercompressibility factor — assumed 0.022 per 100 psia. suction pressure.

Here is an example: How much horsepower should be installed to raise the pressure of 10 million cubic feet of gas per day from 185.3 psi to 985.3 psi?

This gives absolute pressures of 200 and 1000.

$$\text{Then } R = \frac{1000}{200} = 5.0$$

Substituting in the formula:

$$BHP/MMcfd = \frac{5.0}{5.0 + 5 \times 0.044} \times \frac{5.16 + 124 \times .699}{.97 - .03 \times 5}$$

$$= 106.5 \text{ hp} = \text{BHP for 10 MMcfd} = 1065 \text{ hp}$$

Where the suction pressure is about 400 psia, the brake horsepower per MMcfd can be read from the chart.

The above formula may be used to calculate horsepower requirements for various suction pressures and gas physical properties to plot a family of curves.

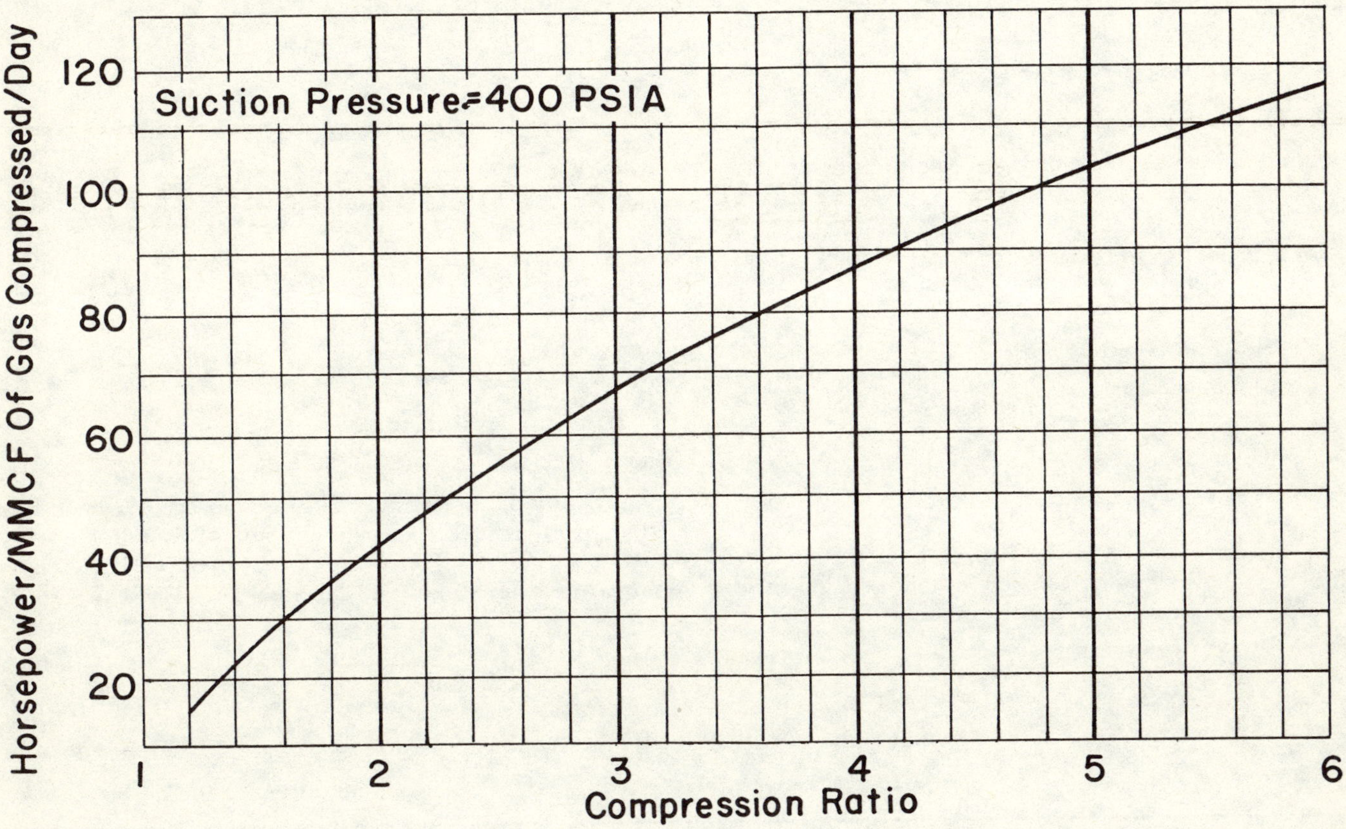

Estimate compressor hp where discharge pressure is 1,000 psi

Rule: Use the following formula for reciprocating compressors where the compression ratio is between 1.20 and 1.40. Brake horsepower per MMCF per day = 11 + (42.5) (R — 1.20) R is compression ratio.

Example: How much horsepower would have to be installed in a field compressor station where the discharge pressure would be 1055 psia and suction would be 800 psia. The station must handle 10,000,000 cubic feet per day.

Solution: The compression ratio is $\frac{1055}{800} = 1.32$ substituting in the formula: Brake horsepower

$$\text{per MMCF} = 11 + (42.5)(1.32 - 1.20)$$

$$= 11 + 5.1$$

$$= 16.1 \text{ hp.}$$

The station would require 161 hp.

For a centrifugal compressor the formula would be: Brake horsepower per MMCF per day = 10 + 49 (R — 1.20)

Example: How much horsepower should be installed in centrifugal compressor station handling 100,000,000 cubic ft. per day if the suction and discharge are the same as above?

Solution: Brake horsepower per MMCF =

$$10 + (49)(0.12) = 15.9$$

For 100 MMCF the horsepower would be 1590.

How to find the size fuel gas line for a compressor station

Suppose that an 8000 hp compressor building is added to an existing compressor plant, what size fuel line should be run from the meter building 1600 feet distant? Take-off pressure at the meter building will be 50 pounds per square inch, and it is desirable to maintain 25 psi pressure on the large engine fuel header in the building. Assume that the gas is heated prior to expansion so that outlet temperature is 60° F.

The first step is to estimate the ultimate compressor building horsepower. Assume in the foreseeable future the building will be expanded to 12,000 horsepower.

Next, estimate the fuel needs.

$Q = 12{,}000 \times 240$ cubic feet per day.

$Q = 2{,}880{,}000$ cu. ft./day.

To estimate the size line which must be run for fuel use this formula:

$$Q = \frac{(36{,}300)\ d^3 \sqrt{P_1^2 - P_2^2}}{\sqrt{l}}$$

Q = cubic feet of gas per 24 hrs.

d = pipe I. D. in inches

l = length of pipe in feet

P_1 = psi (abs.) at starting point

P_2 = psi (abs.) at ending point

Using this formula try several pipe diameters to determine the approximate size. For instance, if six-inch I. D. pipe were used, the approximate flow could be found as follows:

$$Q = \frac{36{,}300 \times 6^3 \sqrt{(50+15)^2 - (25+15)^2}}{\sqrt{1600}}$$

$$Q = \frac{36{,}300 \times 216 \sqrt{65^2 - 40^2}}{40}$$

Q = approximately 10,036,000 cubic feet per 24 hours.

From this it is obvious that six-inch pipe would be considerably oversized, especially if the fuel gas header in the compressor building is of large diameter.

Try substituting four-inch diameter pipe in this formula to find an approximate size pipe.

$$Q = \frac{36{,}300 \times 4^3 \sqrt{(50+15)^2 - (25+15)^2}}{\sqrt{1600}}$$

$$Q = \frac{36{,}300 \times 64 \sqrt{65^2 - 40^2}}{40}$$

Q = 2,973,700 cu. ft. per 24 hrs.

Apparently four-inch pipe will be correct size. To check it, use this more accurate formula which is based on the Weymouth:

$$Q = \frac{63{,}320\ d^{8/3} \sqrt{P_1^2 - P_2^2}}{\sqrt{l}}$$

$$Q = \frac{63{,}320 \times 41 \sqrt{65^2 - 40^2}}{\sqrt{1600}}$$

Four-inch pipe is adequate.

Q = 3,322,000 cubic feet per day.

Note: This latter formula is based on the Weymouth formula, for gas with a specific gravity of 0.60 and a temperature of 60° F. For every 10 degrees variation in temperature, the answer will be approximately 1 percent in error. For every 0.01 variation in specific gravity, the answer will be three-fourth percent in error.

Estimate engine cooling water requirements

This formula can be used for calculating engine jacket water requirements as well as lube oil cooling water requirements:

$$GPM = \frac{H \times BHP}{500\ \Delta t}$$

where:

H = Heat dissipation in Btu's per BHP/Hr. This will vary for different engines; where they are available, the manufacturers' values should be used. Otherwise, you will be safe in substituting the following values in the formula: For engines with water-cooled exhaust manifolds: Engine jacket water = 2200 Btu's per BHP/Hr. Lube oil cooling water = 600 Btu's per BHP/Hr.

For engines with dry type manifolds (so far as cooling water is concerned) use 1500 Btu's/BHP/Hr. for the engine jackets and 650 Btu's/BHP/Hr. for lube oil cooling water requirements.

BHP = Brake Horsepower Hour

Δt = Temperature differential across engine. Usually manufacturers recommend this not exceed 15°F; 10°F is preferable.

Example: Find the jacket water requirements for a 2,000 hp gas engine which has no water jacket around the exhaust manifold.

Solution:

$$GPM = \frac{1500 \times 2000}{500 \times 10}$$

$$GPM = \frac{3{,}000{,}000}{5000} = 600 \text{ gallons per minute}$$

The lube oil cooling water requirements could be calculated in like manner.

Calculate brake horsepower required to compress gas

The approximate brake horsepower per million cubic feet of gas to be compressed may be obtained from this table. The total brake horsepower required is calculated by multiplying the total million cubic feet of gas compressed. In this estimation, gas volumes are expressed at 14.7 psia and 60° F, specific gravity as 0.6 and K (the ratio of specific heats) as 1.27. The suction temperature is assumed as 80° F and interstage temperature as 130° F. The relative humidity of the air equals 100 percent.

(Table by the Cooper-Bessemer Corp.)

Example: What is the brake horsepower required to compress 5.0 million cubic feet of gas from 200 psig suction pressure to 1,000 psig discharge pressure?

On the chart, read opposite a suction pressure of 200 and under a discharge pressure of 1000 psig, that 100 brake horsepower is required per million cubic feet of gas compressed. Multiply 5.0 times 100 = 500 bhp.

HORSEPOWER SELECTION CHART
NATURAL GAS COMPRESSION

SUCTION PRESSURE	DISCHARGE PRESSURE									
	250 psig	375 psig	500 psig	750 psig	1000 psig	1250 psig	1500 psig	2000 psig	3500 psig	6000 psig
0 psig	205	235	260	280	300	320	330	360	390	—
10 psig	165	195	215	250	260	275	290	310	345	390
25 psig	130	155	175	205	230	250	255	275	320	370
50 psig	95	125	140	170	190	205	215	245	280	325
75 psig	70	95	120	145	165	180	195	215	250	300
100 psig	50	75	95	120	145	160	175	195	230	275
200 psig	—	—	55	80	100	120	130	150	190	240
300 psig	—	—	30	55	75	90	105	120	160	205
500 psig	—	—	—	25	40	55	65	85	125	165
750 psig	—	—	—	—	15	30	40	60	100	140
1000 psig	—	—	—	—	—	—	25	40	75	125
1500 psig	—	—	—	—	—	—	—	—	50	90
2000 psig	—	—	—	—	—	—	—	—	—	70
2500 psig	—	—	—	—	—	—	—	—	—	55

Convenient anti-freeze chart for winterizing cooling systems

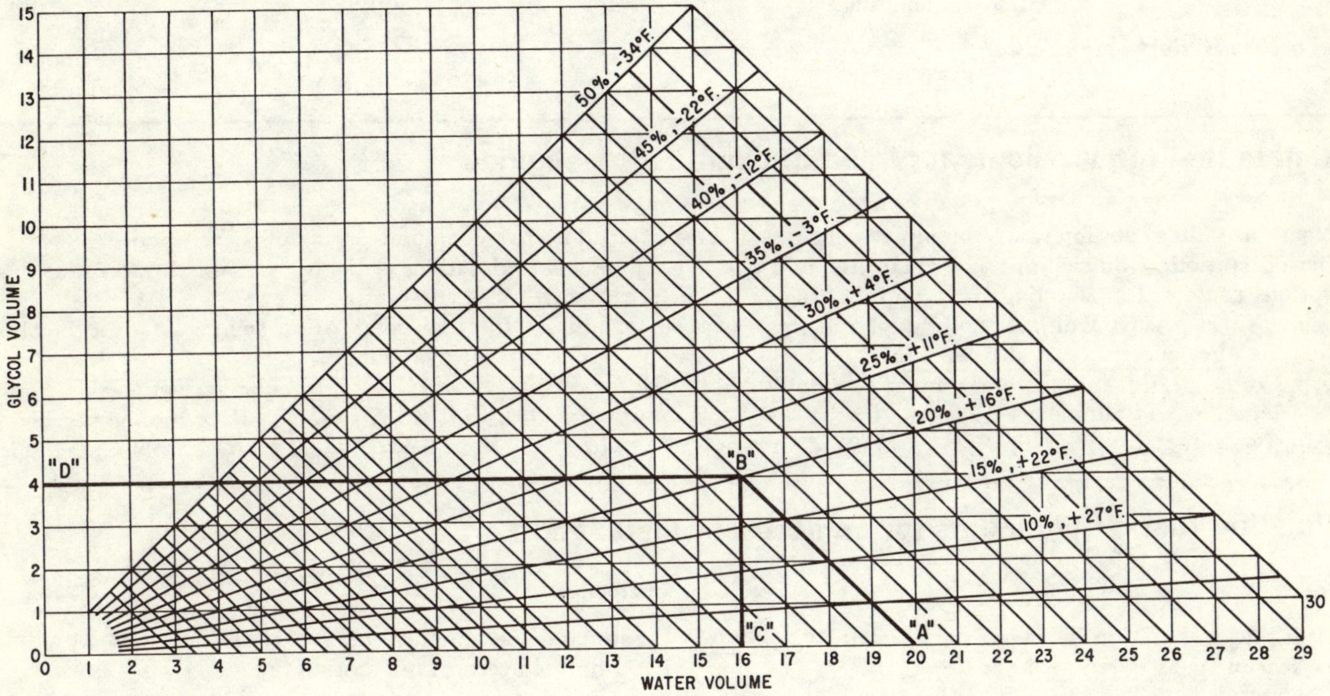

This plot of water volume versus glycol volume at various conditions of temperature and percent of glycol in the system makes winterizing field engines relatively easy.

Example:

Determine the amount of glycol to be used in a 200 gallon system for protection to 16° F.

To use the chart, first find the total system capacity, 200 gallons, at Point A on the water volume axis. Point A also can represent 2, 20, 2,000, etc. Proceed along the 45° line from Point A to Point B at the intersection with the 16° F line. Then, move from Point B to Point C and read 160 on the water volume axis. This is the amount of water the system should contain. Move from Point B to Point D on the glycol axis and read 40. This is the amount of glycol that must be added to the system.

How to determine glycol requirements to bring a system to a desired temperature protection level

Solve the equation below to find the number of gallons of aqueous solution that must be removed and replaced with glycol.

$$d = \frac{D}{B}$$

where: d = gallons of aqueous solution to be removed from the system and replaced with equal number of gallons of glycol.

D = total gallons of glycol required for protection at desired temperature less gallons of glycol in original system

B = 1.0 — fraction of glycol in system

Example: Find the number of gallons of existing aqueous solution that must be removed and replaced with glycol in a 1200 gallon system to give protection to —10° F. Glycol in the system determined by Hydrometer test is 28 percent (0.28 fraction). The published figure for —10° F protection is 519 gallons of glycol in the 1200 gallon system.

Then: Gallons of glycol in system = 1200 × 0.28 = 336 gallons

D = 519 — 336 = 183 gallons difference

and B = 1.0 — 0.28 = 0.72

Thus:

$$d = \frac{183}{0.72} = 254 \text{ gallons}$$

Therefore 254 gallons of existing aqueous solution needs to be removed and replaced with 254 gallons of glycol for the desired protection to —10° F.

Estimate the horsepower that can be transmitted by a shaft

1. Where there are no stresses due to bending, weight of the shaft, pulleys, gears or sprockets, use:

$$HP = \frac{D^3 N}{50}$$

Where D = diameter of shaft in inches, and
N = revolutions per minute.

2. For heavy duty service use:

$$HP = \frac{D^3 N}{125}$$

Example: What horsepower can be transmitted to an atmospheric cooling coil by a two-inch shaft turning at 1800 revolutions per minute?

$$HP = \frac{(2)^3 (1800)}{125}$$

$$= 115.2 \text{ horsepower}$$

Estimate fuel requirements for internal combustion engines

When installing an internal combustion engine at a gathering station, a quick approximation of fuel consumptions could aid in selecting the type fuel used.

Using Natural Gas: Multiply the brake hp at drive by 11.5 to get cubic feet of gas per hour.

Using Butane: Multiply the brake hp at drive by 0.107 to get gallons of butane per hour.

Using Gasoline: Multiply the brake hp at drive by 0.112 to get gallons of gasoline per hour.

These approximations will give reasonable accurate figures under full load conditions.

Example: Internal combustion engine rated at 50 bhp —3 types of fuel available.

Natural Gas:	50 × 11.5 = 575 cubic feet of gas per hour
Butane:	50 × 0.107 = 5.35 gal. of butane per hour
Gasoline:	50 × 0.112 = 5.60 gal. of gasoline per hour

Estimating fuel requirements for compressor installation

Rule:

Multiply installation horsepower by 240 to obtain cubic feet of fuel gas required per day.

Example:

Estimate the amount of fuel required to supply a 1,000-hp field compressor installation.

1,000 × 240 = 240 Mcfd

Weight in pounds of round steel shafting

Diameter of Shaft Inches	LENGTH IN FEET																	Weight Per Inch
	1	2	3	4	5	6	7	8	9	10	12	14	16	18	20	22	24	
15/16	2	5	7	9	12	14	16	19	21	23	28	33	37	42	47	51	56	.195
1 3/16	4	8	11	15	19	23	26	30	34	38	45	53	60	68	76	83	91	.314
1 7/16	6	11	17	22	28	33	39	44	50	55	66	77	88	99	110	121	133	.460
1 11/16	8	15	23	30	38	46	53	61	68	76	91	107	122	137	152	167	183	.634
1 15/16	10	20	30	40	50	60	70	80	90	100	120	140	161	181	201	221	241	.835
2 3/16	13	26	38	51	64	77	90	102	115	128	153	179	205	230	256	281	307	1.06
2 7/16	16	32	48	63	79	95	111	127	143	159	190	222	254	286	317	349	381	1.32
2 11/16	19	39	58	77	97	116	135	154	174	193	232	270	309	348	386	425	463	1.61
2 15/16	23	46	69	92	115	138	161	184	208	231	277	323	369	415	461	507	553	1 92
3 3/16	27	54	81	109	136	163	190	217	244	272	326	380	435	489	543	598	652	2.26
3 7/16	32	63	95	126	158	189	221	253	284	316	379	442	505	568	631	695	758	2.63
3 11/16	36	73	109	146	182	218	255	291	328	364	437	510	582	655	728	801	874	3.03
3 15/16	41	83	124	166	207	248	290	331	373	414	497	580	662	745	828	911	994	3.45
4 7/16	53	105	158	210	263	315	368	421	473	526	631	736	841	946	1052	1157	1262	4.38
4 15/16	65	130	195	260	326	391	456	521	586	651	781	911	1041	1172	1302	1432	1562	5.43
5	67	134	200	267	334	401	467	534	601	668	801	935	1068	1202	1335	1469	1602	5.56
5 7/16	79	158	237	316	395	474	553	632	711	790	947	1105	1263	1421	1579	1737	1894	6.58
5½	81	162	242	323	404	485	565	646	727	808	969	1131	1292	1454	1615	1777	1938	6.73
5 15/16	94	188	282	377	471	565	659	753	847	941	1130	1318	1506	1695	1883	2071	2259	7.85
6	96	192	288	385	481	577	673	769	865	961	1154	1346	1538	1731	1923	2115	2307	8.01
6½	113	226	338	451	564	677	790	903	1015	1128	1354	1579	1805	2031	2256	2482	2708	9.40
7	131	262	393	524	654	785	916	1047	1178	1309	1570	1832	2094	2355	2677	2878	3140	10.90
7½	150	300	451	601	751	901	1051	1202	1352	1502	1803	2103	2403	2704	3004	3305	3605	12.52
8	171	342	513	684	855	1024	1196	1367	1538	1709	2051	2303	2734	3076	3418	3760	4102	14.24

—Dodge Mfg. Corp.

Properties of shafting

Diameter of Shaft, Inches	Area of Shaft	Weight per Inch	Weight per Foot	SECTION MODULUS		MOMENT OF INERTIA		Radius of Gyration
				Bending	Torsion	Bending	Torsion	
15/16	.6903	.195	2.34	.0809	.1618	.0379	.0758	.2343
1 3/16	1.1075	.314	3.76	.1644	.3288	.0976	.1952	.2969
1 7/16	1.6230	.460	5.52	.2916	.5832	.2096	.4192	.3594
1 11/16	2.2365	.634	7.61	.4718	.9435	.3981	.7961	.4219
1 15/16	2.9483	.835	10.03	.7140	1.4281	.6971	1.3835	.4844
2 3/16	3.7583	1.06	12.78	1.0276	2.0553	1.1240	2.2480	.5469
2 7/16	4.6664	1.32	15.86	1.4218	2.8436	1.7328	3.4656	.6094
2 11/16	5.6727	1.61	19.29	1.9057	3.8113	2.5607	5.1215	.6719
2 15/16	6.7771	1.92	23.04	2.4885	4.9770	3.6549	7.3099	7344
3 3/16	7.9798	2.26	27.12	3.1794	6.3589	5.0672	10.1345	.7969
3 7/16	9.2806	2.63	31.56	3.9878	7.9755	6.8539	13.7079	.8594
3 11/16	10.680	3.03	36.31	4.9226	9.8452	9.0761	18.1521	.9219
3 15/16	12.177	3.45	41.40	5.9932	11.9865	11.7992	23.5984	.9844
4 7/16	15.466	4.38	52.58	8.5786	17.1571	19.0337	38.0674	1.1094
4 15/16	19.147	5.43	65.10	11.8174	23.6348	29.1742	58.3483	1.2344
5	19.635	5.56	66.76	12.2718	24.5437	30.6796	61.3592	1.2500
5 7/16	23.221	6.58	78.96	15.7833	31.5666	42.9108	85.8217	1.3594
5½	23.758	6.73	80.78	16.3338	32.6677	44.9180	89.8361	1.3750
5 15/16	27.688	7.85	94.14	20.5499	41.0999	61.0077	122.0153	1.4844
6	28.274	8.01	96.12	21.2058	42.4115	63.6172	127.2345	1.5000
6½	33.183	9.40	112.82	26.9612	53.9225	87.6240	175.2481	1.6250
7	38.485	10.90	130.84	33.6739	67.3479	117.8588	235.7176	1.7500
7½	44.179	12.52	150.21	41.4175	82.8350	155.3155	310.6311	1.8750
8	50.265	14.24	170.90	50.2655	100.5310	201.0619	402.1239	2.0000

—Dodge Mfg. Corp.

Steel shaft sizes calculated for strength

Base Formula: $d = \sqrt[3]{\frac{C.(H.P.)}{N}}$

Base Fiber Stress = 60,000 lbs. sq. in.

Diameter of Shaft, Inches		$\frac{H.P.}{N}$		
Fractional	Decimal	C=50 Very Little Bending Fac. Saf. 9.4	C=75 Average Conditions of Bending Fac. Saf. 14.1	C=100 Very Severe Conditions of Bending Fac. Saf. 18.8
1/16	0.0625	0.000004883	0.000003256	0.000002492
1/8	0.125	0.00003905	0.00002604	0.00001953
3/16	0.1875	0.0001318	0.00008788	0.00006592
1/4	0.25	0.0003124	0.0002083	0.0001563
5/16	0.3125	0.0006105	0.0004070	0.0003053
3/8	0.375	0.001054	0.0007030	0.0005272
7/16	0.4375	0.001675	0.001116	0.0008375
1/2	0.5	0.002500	0.001666	0.001250
9/16	0.5625	0.003558	0.002372	0.001779
5/8	0.625	0.004883	0.003256	0.002442
11/16	0.6875	0.006500	0.004333	0.003250
3/4	0.75	0.008439	0.005626	0.004220
13/16	0.8125	0.01074	0.007155	0.005367
7/8	0.875	0.01340	0.008932	0.006699
15/16	0.9375	0.01647	0.01098	0.008235
1	1.	0.02000	0.01333	0.01000
1 1/16	1.0625	0.02396	0.01598	0.01199
1 1/8	1.125	0.02846	0.01898	0.01423
1 3/16	1.1875	0.03353	0.02236	0.01677
1 1/4	1.25	0.03905	0.02604	0.01953
1 5/16	1.3125	0.04519	0.03019	0.02259
1 3/8	1.375	0.05199	0.03465	0.02599
1 7/16	1.4375	0.05940	0.03960	0.02970
1 1/2	1.5	0.06750	0.04500	0.03375
1 9/16	1.5625	0.07628	0.05086	0.03815
1 5/8	1.625	0.08578	0.05719	0.04289
1 11/16	1.6875	0.09524	0.06409	0.04807
1 3/4	1.75	0.1072	0.07144	0.05358
1 13/16	1.8125	0.1191	0.07939	0.05955
1 7/8	1.875	0.1318	0.08788	0.06592
1 15/16	1.9375	0.1456	0.09707	0.07281
2	2.	0.1600	0.1066	0.07998
2 1/16	2.0625	0.1755	0.1169	0.08774
2 1/8	2.125	0.1918	0.1278	0.09592
2 3/16	2.1875	0.2095	0.1397	0.1048
2 1/4	2.25	0.2278	0.1519	0.1140
2 5/16	2.3125	0.2474	0.1649	0.1237
2 3/8	2.375	0.2678	0.1785	0.1339
2 7/16	2.4375	0.2897	0.1932	0.1449
2 1/2	2.5	0.3124	0.2083	0.1563
2 9/16	2.5625	0.3363	0.2243	0.1682
2 5/8	2.625	0.3616	0.2411	0.1808
2 11/16	2.6875	0.3884	0.2589	0.1942
2 3/4	2.75	0.4159	0.2772	0.2079
2 13/16	2.8125	0.4449	0.2966	0.2225
2 7/8	2.875	0.4754	0.3170	0.2378
2 15/16	2.9375	0.5074	0.3383	0.2537
3	3.	0.5399	0.3683	0.2700
3 1/16	3.0625	0.5742	0.3836	0.2871
3 1/8	3.125	0.6105	0.4070	0.3053
3 3/16	3.1875	0.6479	0.4319	0.3240
3 1/4	3.25	0.6866	0.4577	0.3434
3 5/16	3.3125	0.7266	0.4844	0.3634
3 3/8	3.375	0.7685	0.5123	0.3842
3 7/16	3.4375	0.8127	0.5418	0.4061

Diameter of Shaft, Inches		$\frac{H.P.}{N}$		
Fractional	Decimal	C=50 Very Little Bending Fac. Saf. 9.4	C=75 Average Conditions of Bending Fac. Saf. 14.1	C=100 Very Severe Conditions of Bending Fac. Saf. 18.8
3 1/2	3.5	0.8567	0.5718	0.4288
3 9/16	3.5625	0.9038	0.6026	0.4520
3 5/8	3.625	0.9526	0.6351	0.4763
3 11/16	3.6875	1.003	0.6688	0.5017
3 3/4	3.75	1.054	0.7030	0.5272
3 13/16	3.8125	1.108	0.7382	0.5538
3 7/8	3.875	1.163	0.7759	0.5820
3 15/16	3.9375	1.222	0.8143	0.6108
4	4.	1.280	0.8535	0.6401
4 1/16	4.0625	1.340	0.8933	0.6701
4 1/8	4.125	1.404	0.9356	0.7018
4 3/16	4.1875	1.469	0.9792	0.7345
4 1/4	4.25	1.536	1.023	0.7678
4 5/16	4.3125	1.603	1.069	0.8019
4 3/8	4.375	1.675	1.116	0.8375
4 7/16	4.4375	1.748	1.165	0.8742
4 1/2	4.5	1.823	1.214	0.9111
4 9/16	4.5625	1.900	1.266	0.9497
4 5/8	4.625	1.978	1.319	0.9893
4 11/16	4.6875	2.059	1.373	1.030
4 3/4	4.75	2.143	1.429	1.072
4 13/16	4.8125	2.228	1.486	1.114
4 7/8	4.875	2.316	1.544	1.158
4 15/16	4.9375	2.407	1.604	1.203
5	5.	2.500	1.666	1.250

Haven-Swett "Treatise on Leather Belting", American Leather Belting Association.

Compressors

The following data are for use in the approximation of horsepower needed for compression of gas.

Definitions

The "N value" of gas is the ratio of the specific heat at constant pressure (C_p) to the specific heat at constant volume (C_v).

If the composition of gas is known, the value of N for a gas mixture may be determined from the molal heat capacities of the components. If only the specific gravity of gas is known an approximate N value may be obtained by using the chart in Figure 1.

Piston Displacement of a compressor cylinder is the volume swept by the piston with the proper deduction for the piston rod. The displacement is usually expressed in cu. feet per minute.

Clearance is the volume remaining in one end of a cylinder with the piston positioned at the end of the delivery stroke for this end. The clearance volume is expressed as a percentage of the volume swept by the piston in making its full delivery stroke for the end of the cylinder being considered.

Ratio of compression is the ratio of the absolute discharge pressure to the absolute inlet pressure.

Actual capacity is the quantity of gas compressed and delivered, expressed in cubic feet per minute, at the intake pressure and temperature.

Volumetric efficiency is the ratio of actual capacity, in cubic feet per minute, to the piston displacement, in cubic feet per minute, expressed in percent.

Adiabatic horsepower is the theoretical horsepower required to compress gas in a cycle in which there is no transfer of sensible heat to or from the gas during compression or expansion.

Isothermal horsepower is the theoretical horsepower required to compress gas in a cycle in which there is no change in gas temperature during compression or expansion.

Indicated horsepower is the actual horsepower required to compress gas, taking into account losses within the compressor cylinder, but not taking into account any loss in frame, gear or power transmission equipment.

Compression efficiency is the ratio of the theoretical horsepower to the actual indicated horsepower, required to compress a definite amount of gas. The efficiency, expressed in percent, should be defined in regard to the base at which the theoretical power was calculated, whether adiabatic or isothermal.

Mechanical efficiency is the ratio of the indicated horsepower of the compressor cylinder to the brake horsepower delivered to the shaft in the case of a power driven machine. It is expressed in percent.

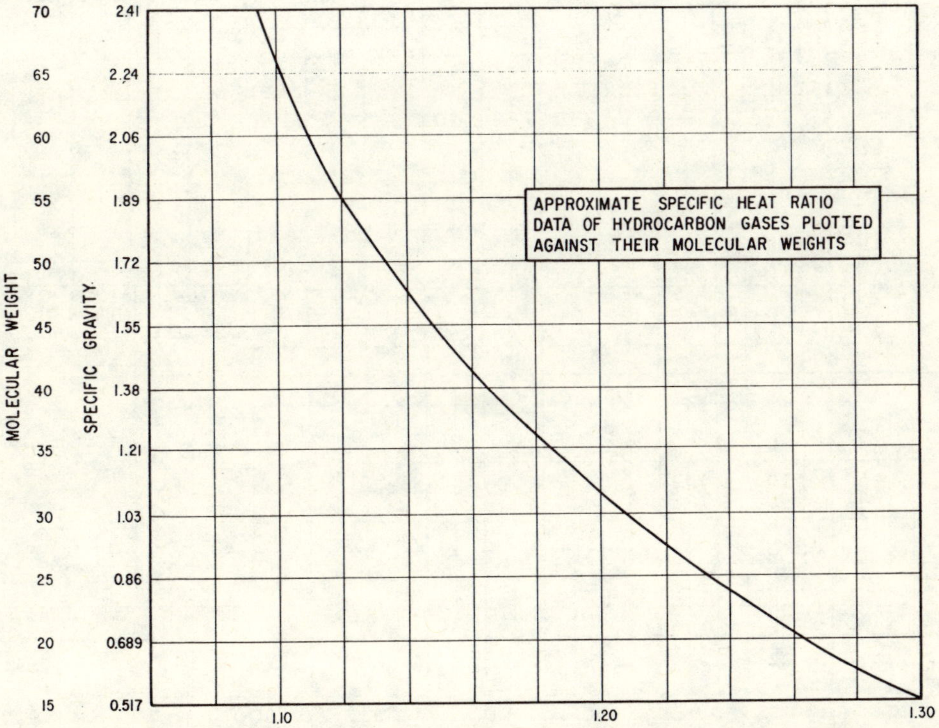

—NGSMA Engineering Data Book

Figure 1. Ratio of specific heat (n-value).

Overall efficiency is the product, expressed in percent, of the compression efficiency and the mechanical efficiency. It must be defined according to the base, adiabatic isothermal, which was used in establishing the compression efficiency.

Piston rod gas load is the varying, and usually reversing, load imposed on the piston rod and crosshead during the operation, by different gas pressures existing on the faces of the compressor piston.

The maximum piston rod gas load is determined for each compressor by the manufacturer, to limit the stresses in the frame members and the bearing loads in accordance with mechanical design. The maximum allowed piston rod gas load is affected by the ratio of compression and also by the cylinder design; i.e. whether it is single or double acting.

Compression horsepower determination

The method outlined below permits determination of approximate horsepower requirements for compression of gas.

1. From Figure 2, determine the atmospheric pressure in p.s.i.a. for the altitude above sea level, at which the compressor is to operate.

2. Determine intake pressure (P_s) and discharge pressure (P_d) by adding the atmospheric pressure to the corresponding gage pressure for the conditions of compression.

3. Determine total compression ratio $R = P_d/P_s$. If ratio R is more than 5 to 1, two or more compressor stages will be required. Allow for a pressure loss of approximately 5 p.s.i. between stages. Use the same ratio for each stage. The ratio per stage, so that each stage has the same ratio, can be approximated by finding the nth root of the total ratio, when n = number of stages. The exact ratio can be found by trial and error, accounting for the 5 psi. interstage pressure losses.

4. Determine the N value of gas (Chart, Figure 1).

5. The chart, Figure 3 gives horsepower requirements for compression of one million cu. ft. per day for the compression ratios and N values commonly encountered in oil producing operations.

6. If the suction temperature is not 60 deg. F, correct the curve horsepower figure in proportion to absolute temperature. This is done as follows:

$$\text{H.P.} \times \frac{460° + Ts}{460° + 60°F} = \text{H.P. (corrected for suction temperature)}$$

Where Ts is suction temperature in degrees Fahrenheit.

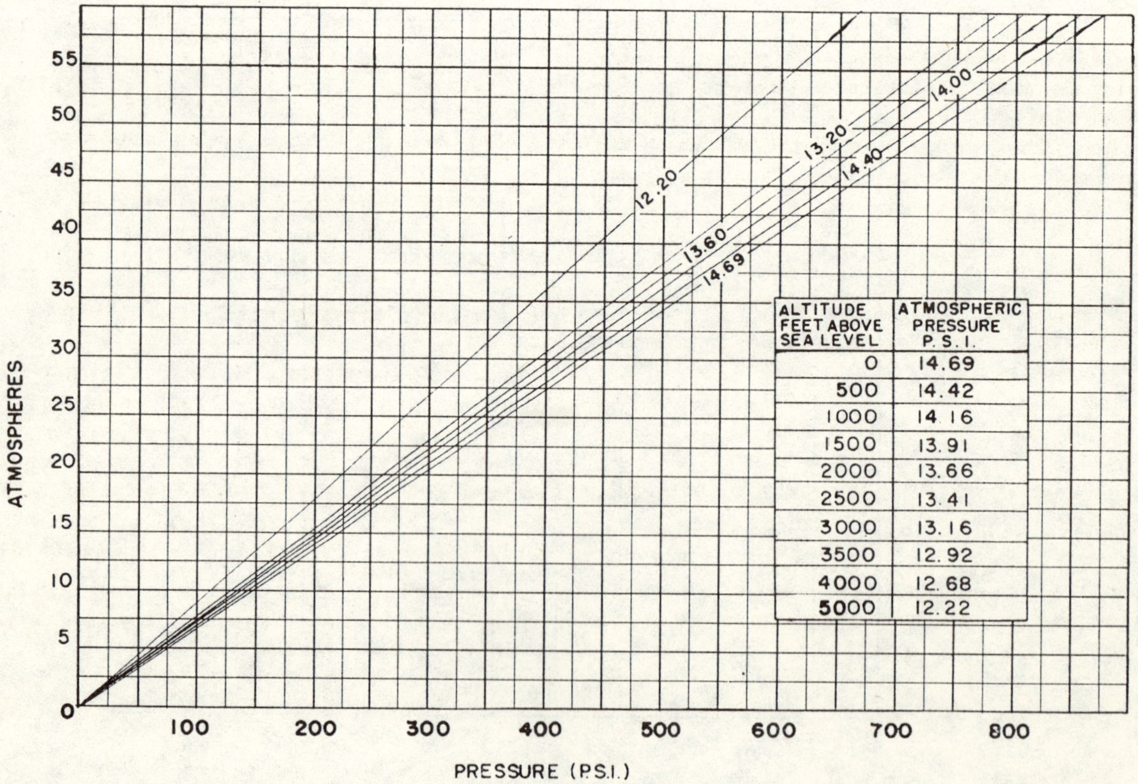

ALTITUDE FEET ABOVE SEA LEVEL	ATMOSPHERIC PRESSURE P.S.I.
0	14.69
500	14.42
1000	14.16
1500	13.91
2000	13.66
2500	13.41
3000	13.16
3500	12.92
4000	12.68
5000	12.22

Figure 2. Atmospheres at various atmospheric pressures. (From *Modern Gas Lift Practices and Principles,* Merla Tool Corp.)

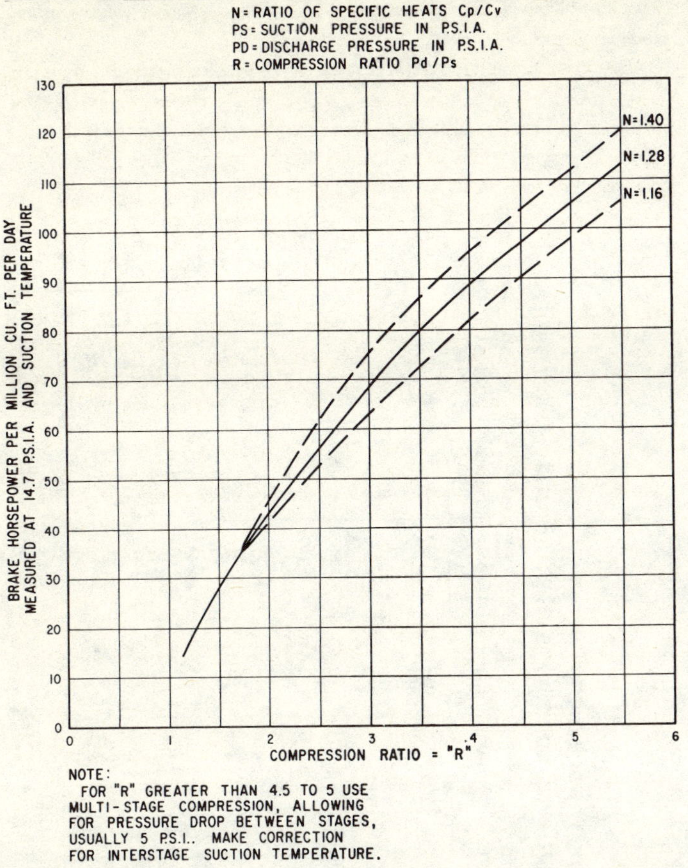

Figure 3. Brake horsepower required for compressing natural gas.

7. Add together the horsepower loads determined for each stage to secure the total compression horsepower load. For altitudes greater than 1500 ft. above sea level apply a multiplier derived from the following table to determine the nominal sea level horsepower rating of the internal combustion engine driver.

Altitude	Multiplier	Altitude	Multiplier
1500 ft.	1.000	4000 ft.	1.12
2000 ft.	1.03	4500 ft.	1.14
2500 ft.	1.05	5000 ft.	1.17
3000 ft.	1.07	5500 ft.	1.20
3500 ft.	1.10	6000 ft.	1.22

8. For a portable unit with a fan cooler and pump driven from the compressor unit, increase the horsepower figure by 7½ percent.

The resulting figure is sufficiently accurate for all purposes. The nearest commercially available size of compressor is then selected.

The method does not take into consideration the supercompressibility of gas and is applicable for pressures up to 1000 psi. In the region of high pressures, neglecting the deviation of behavior of gas from that of the perfect gas may lead to substantial errors in calculating the compression horsepower requirements. The enthalpy-entropy charts may be used conveniently in such cases. The procedures are given in references 1 and 2.

Example:

What is the nominal size of a portable compressor unit required for compressing 1,600,000 standard cubic feet of gas per 24 hours at a temperature of 85 deg. F from 40 p.s.i.g. pressure to 600 p.s.i.g. pressure? The altitude above sea level is 2500 ft. The N value of gas is 1.28. The suction temperature of stages, other than the first stage, is 130 deg. F.

Solution:

$P_s = 53.41$ P.S.I.A. $Pd = 613.41$ P.S.I.A.

$R_t = Pd/Ps = 11.5 = (3.4)^2 = (2.26)^3$

Try solution using 3.44 ratio and 2 stages.

1st stage: 53.41 P.S.I.A. × 3.44 = 183.5 P.S.I.A. discharge.

2nd stage: 178.5 P.S.I.A. × 3.44 = 614 P.S.I.A. discharge.

Horse power from curve, Figure 3-12 = 77 H.P. for 3.44 ratio

$$\frac{77 \text{ H.P.} \times 1{,}600{,}000}{1{,}000{,}000} = 123.1 \text{ (for } 60^\circ \text{ F. suction temp.)}$$

$$\text{1st stage: } 123.1 \text{ H.P.} \times \frac{460 + 85 \text{ deg.}}{460 + 60 \text{ deg.}} = 129.1 \text{ H.P.}$$

$$\text{2nd stage: } 123.1 \text{ H.P.} \times \frac{460 + 130 \text{ deg.}}{460 + 60 \text{ deg.}} = 139.7 \text{ H.P.}$$

1.05 (129.1 H.P. + 139.7 H.P.) = 282 H.P.

1.075 × 282 H.P. = 303 H.P.

Nearest nominal size compressor is 300 H.P.

Centrifugal compressors

The centrifugal compressors are inherently high volume machines. They have extensive application in gas transmission systems. Their use in producing operations is very limited.

References

1. *Engineering Data Book,* Natural Gasoline Supply Men's Association, 1957.

2. Dr. George Granger Brown: "A Series of Enthalpy-entropy Charts for Natural Gas," *Petroleum Development and Technology,* Petroleum Division AIME, 1945.

Surface handling of gas

In the preceding parts of this and other chapters of this handbook references were made to questions relating to gas. Before problems pertaining to surface handling of gas are presented in more detail, certain definitions and basic principles will be given, which are useful to understanding and solution of gas problems encountered in producing operations.

Definitions and principles

Definitions

The *critical temperature* is the temperature, above which a single component gas cannot be liquefied by pressure alone. In fundamental equations of gas relations temperature is expressed in degrees absolute (Rankine) which is degrees Fahrenheit plus 460.

The *critical pressure* is the pressure causing liquefaction of the gas at or about its critical temperature.

The *reduced temperature* or *pressure* is the actual temperature or pressure to which the gas is subjected, divided by the critical temperature or pressure respectively.

The term *"pseudo"* denotes a gas mixture as distinguished from a single component gas. For instance, the pseudo-critical temperature of a certain gas is the molecular average critical temperature of the gas components. The critical temperatures and pressures of the components may be obtained from any of several available tables of physical properties of natural gas hydrocarbons, such as Tables 1 and 2. The manner of determining the pseudo-critical temperatures and pressure is illustrated in Table 3.

The *average molecular weight* of a gas may be calculated as indicated in Table 4.

The *specific gravity* of gas, referred to air, is computed by dividing the molecular weight of gas by the molecular weight of air (28.9).

One *pound-mol* is the pounds of a gas equal in number to the number that expresses the molecular weight of gas.

The *vapor pressure* of a liquid is the pressure of the vapor of that liquid at any given temperature at which the vapor and liquid phase of the substance can exist in equilibrium.

The *latent heat* of a hydrocarbon is the amount of heat required to vaporize it completely, at its boiling point, without raising the temperature of the liquid or the vapor.

The *heat of combustion* of a hydrocarbon is the total amount of heat released by the complete combustion of a unit weight of the hydrocarbon. The unit generally employed is the British thermal unit which is the amount of heat required to raise 1 lb. of water 1° F. at a standard temperature.

The *compressibility factor* is a dimensionless factor correcting for the difference in behavior between and ideal and an actual gas. If the composition of gas is known then its pseudo-critical temperature and pressure may be determined in the manner illustrated in Table 3. Pseudo-reduced temperature and pressure are then calculated for the conditions under which the gas exists and the compressibility factor is found from the chart in Figure 1. If only the specific gravity of gas is available then the chart in Figure 2 may be used, which shows the approximate relationship between specific gravity of gas and pseudo-reduced temperature and pressure.

The approximate compressibility factor may be then determined by the method given above.

Table 1
Physical constants of hydrocarbons
Values from CNGA and NGAA (modified for standard conditions of 60° F and 14.65 psia.)

	Methane	Ethane	Propane	Iso-Butane	Normal-Butane	Iso-Pentane	Normal Pentane	Normal Hexane	Normal Heptane	Heptanes+ C_7^+
Molecular Formula	CH_4	C_2H_6	C_3H_8	$i-C_4H_{10}$	$n-C_4H_{10}$	$i-C_5H_{12}$	$n-C_5H_{12}$	$n-C_6H_{14}$	$n-C_7H_{16}$	....
Molecular Weight	16.042	30.068	44.094	58.120	58.120	72.146	72.146	86.172	100.198	M
Critical Temp. °F. Abs.	344	549	666	732	766	829	846	914	972	....
Critical Pressure, psia	673	712	617	544	551	483	485	434	397	....
Critical Density, lb./gal.	1.351	1.695	1.888	1.945	1.89[illegible]	1.955	1.935	1.958	1.957	....
Critical Volume (cu. ft./lb. mol.)	1.586	2.371	3.123	3.990	4.130	4.930	4.98	5.88	6.84	....
Specific Gravity of Gas (Air = 1.00)	0.554	1.038	1.522	2.006	2.006	2.491	2.491	2.975	3.459	M/28.966
Cu. Ft. of Gas per pound	23.73	12.66	8.63	6.55	6.55	5.28	5.28	4.42	3.80	380.69/M
Cu. Ft. of Gas/Gal. of Liquid	59.18	39.43	36.56	30.75	31.89	27.50	27.76	24.46	21.77	3170S/M
Lb. per MCF of Gas	42.14	78.98	115.83	152.67	152.67	189.51	189.51	226.36	263.20	2.6268M
Gal. of Liquid per MCF of Gas	16.87	25.36	27.38	32.57	31.41	36.41	36.07	40.94	45.94	.3155M/S
Bbl. of Liquid/MMCF of Gas	401.6	603.9	651.9	775.4	747.5	867.0	858.8	974.8	1094	7.511M/S
Specific Gravity of Liquid (60/60)	0.3	0.374	0.508	0.563	0.584	0.625	0.631	0.664	0.688	S
Lb./Gal. of Liquid at 60° F.	2.5	3.11	4.23	4.69	4.86	5.20	5.25	5.53	5.73	8.327S
Lb./Bbl. of Liquid at 60° F.	105	130.6	177.7	197.0	204.1	218.4	220.5	232.3	240.7	349.7S
Lb./per Cu. Ft. of Liquid at 60° F.	18.70	23.26	31.64	35.08	36.35	38.90	39.27	41.36	42.86	62.29S
Cu. Ft. Liquid/Lb. at 60° F.	.0535	.0430	.0316	.0285	.0275	.0257	.0254	.0242	.0233	.01605/S
Gal./Lb. Mol. at 60° F.	6.4	9.64	10.41	12.38	11.94	13.84	13.71	15.57	17.47	.120M/S
MCF/Barrel	2.49	1.62	1.53	1.28	1.34	1.15	1.16	1.02	0.914	....

Constants Used in Calculations

1 cu. ft. = 28.316 liters
1 cu. ft. = 7.480 gallons
1 gal. = 3785.53 milliliters
760 mm. Hg = 14.696 psi

0° F. = 459.58° Rankin
Density of Water at 60° F. = 8.327 lb./gal.
Molecular weight of air = 28.966
1 mol. = 380.69 cu. ft. of perfect gas at 14.65 psia and 60° F.

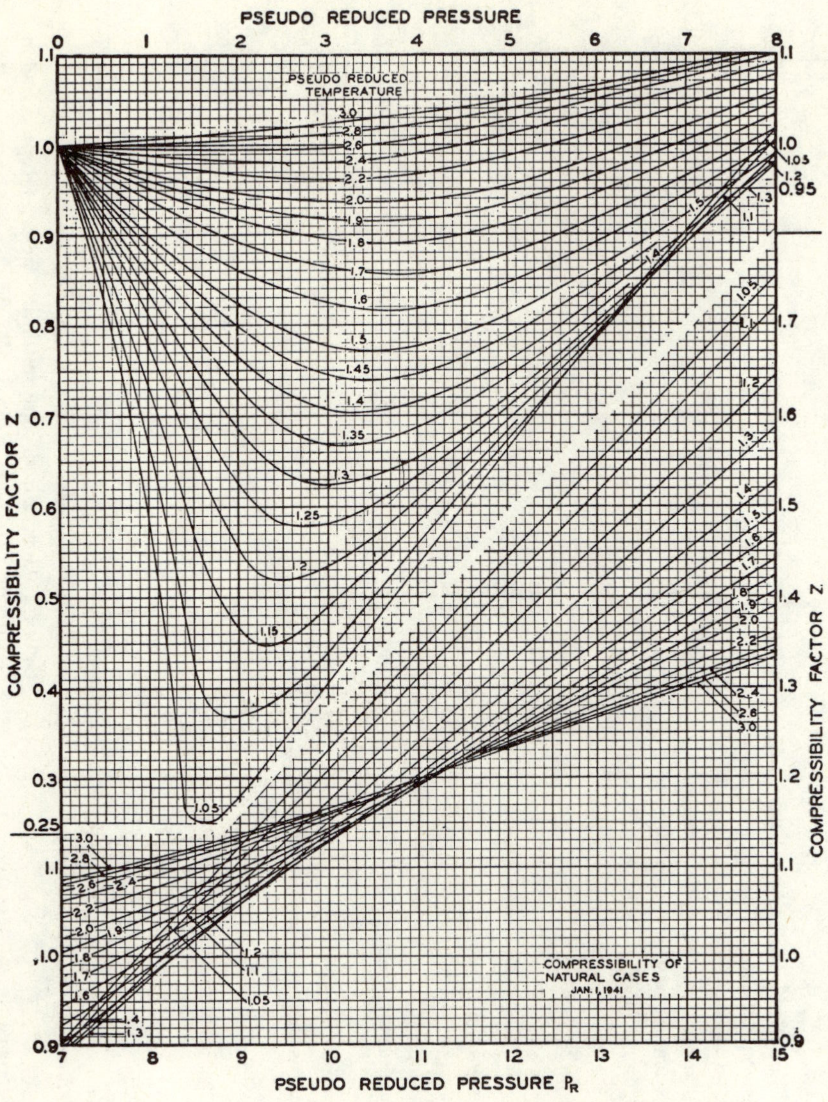

Figure 1. Compressibility factor as a function of pseudo-reduced temperature and pressure. (Reference 1).

Table 2
Physical properties of natural gas hydrocarbons (from reference 1)

COMPOUND	Formula	Mol. Weight	Normal Boiling Point	Critical Temperature		Critical Pressure
			°F.	°F.	°Abs.	psi., Abs.
Methane	CH_4	16.04	—258.5	—116	344	673
Ethane	C_2H_6	30.07	—127.5	89	549	712
Propane	C_3H_8	44.09	— 43.9	205	666	617
n-Butane	C_4H_{10}	58.12	31.1	306	766	551
iso-Butane	C_4H_{10}	58.12	10.2	272	732	544
n-Pentane	C_5H_{12}	72.15	96.9	386	846	485
iso-Pentane	C_5H_{12}	72.15	82.3	370	829	483
n-Hexane	C_6H_{14}	86.17	155.8	454	914	435
n-Heptane	C_7H_{16}	100.2	209.1	512	972	397
n-Octane	C_8H_{18}	114.22	258.2	564	1024	362

Table 3
Computations of pseudo-critical temperature and pressure of natural gas, "A" (from reference 1)

1	2	3	4	5	6
COMPONENT	Mole Fraction	Individual Absolute Critical Temperature T_c	2×3 Absolute Pseudo-Critical Temperature $\underline{T_c}$	Individual Absolute Critical Pressure P_c	2×5 Absolute Pseudo-Critical Pressure $\underline{P_c}$
CH_4	0.8319	344	286	673	560
C_2H_6	0.0848	549	46.5	712	60.4
C_3H_8	0.0437	666	29.1	617	27.0
$i\text{-}C_4H_{10}$	0.0076	732	5.56	544	4.13
$n\text{-}C_4H_{10}$	0.0168	766	12.86	551	9.26
$i\text{-}C_5H_{12}$	0.0057	829	4.72	483	2.75
$n\text{-}C_5H_{12}$	0.0032	846	2.71	485	1.55
C_6H_{14}	0.0063	914	5.75	435	2.74
	......	...	393.2° Abs.	...	667.83 psi Abs.

Table 4
Computation of molecular weight of a natural gas, "A" (from reference 1)

COMPONENT		Percent by Gaseous Volume, or Molecular Percent		Molecular Weight of Components		Sum of the Products is the Average Molecular Weight
Methane	CH_4	83.19%	×	16.04	=	13.31
Ethane	C^2H_6	8.48	×	30.07	=	2.54
Propane	C_3H_8	4.37	×	44.09	=	1.92
iso-Butane	$i\text{-}C_4H_{10}$	0.76	×	58.12	=	0.44
Normal Butane	$n\text{-}C_4H_{10}$	1.68	×	58.12	=	0.97
iso-Pentane	$i\text{-}C_5H_{12}$	0.57	×	72.15	=	0.41
Normal Pentane	$n\text{-}C_5H_{12}$	0.32	×	72.15	=	0.23
Hexane	C_6H_{14}	0.63	×	86.17	=	0.54
Average Molecular Weight						20.36

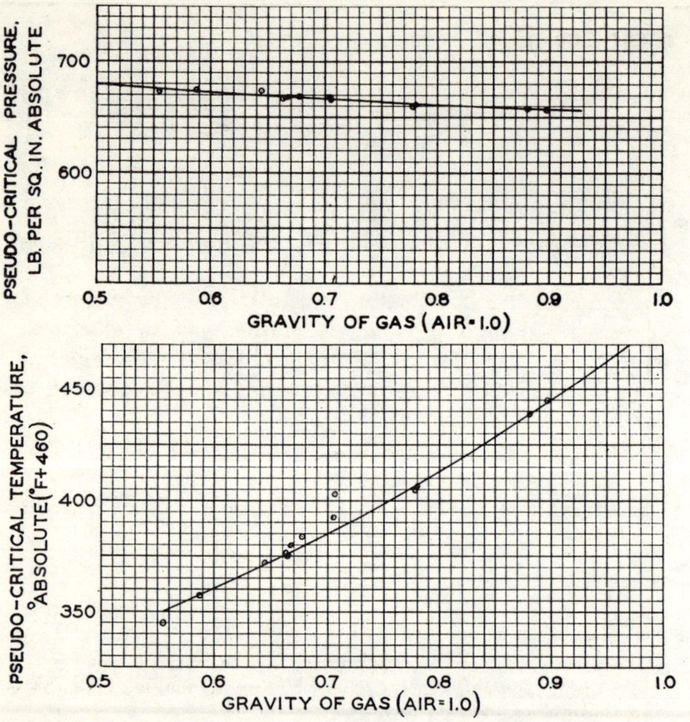

Figure 2. Approximate pseudo-critical temperature and pressure vs. gas gravity. (Reference 1).

Example

Estimate compressibility factor for gas with specific gravity 0.65 at 1000 psia pressure and 150 deg. F. temperature. From Figure 2 for gas with sp. gravity 0.65 the pseudo-critical temperature is 373 deg. abs. and the pseudo-reduced pressure is 670 psia. Then

$$\text{pseudo-reduced temperature is } \frac{150 + 460}{373} = 1.63$$

and

$$\text{pseudo-reduced pressure is } \frac{1000}{670} = 1.49$$

from Chart Figure 1 compressibility factor
Z = 0.89

Reference

1. *Engineering Data Book,* Natural Gasoline Supply Men's Association, 1957.

Generalized compressibility factor

The nomogram (Figure 1) is based on a generalized compressibility chart.[1] It is based on data for 26 gases, excluding helium, hydrogen, water and ammonia. The accuracy is about one percent for gases other than those mentioned.

To use the nomogram, the values of the reduced temperature (T/T_c) and reduced pressure (P/P_c) must be calculated first.

Where:

T = temperature in consistent units
T_c = critical temperature in consistent units
P = pressure in consistent units
P_c = critical pressure in consistent units

Example

Pr = 0.078, Tr = 0.84, what is the compressibility factor, z?

Connect Pr with Tr and read z = 0.948.

Source

Davis, D. S., *Petroleum Refiner,* 37, No. 11, (1961).

Reference

1. Nelson, L. C., and Obert, E. F., *Chem. Engr.,* 203 (1954).

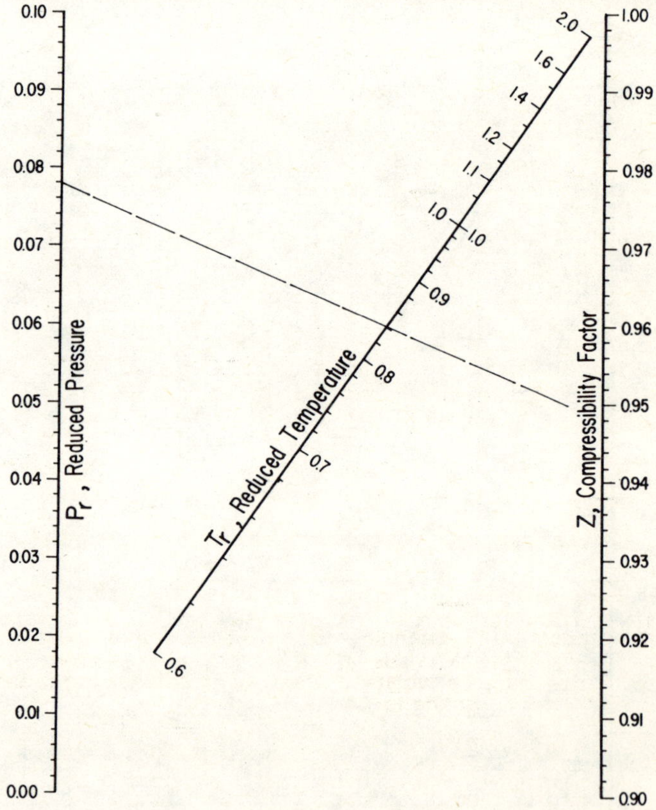

Figure 1. Generalized compressibility factor. (Reproduced by permission *Petroleum Refiner,* Vol. 37, No. 11, copyright 1961, Gulf Publishing Co., Houston.)

Nomograph aids in diagnosing compressor cylinder ills

B. W. Robertson, Tennessee Gas Pipeline Co., Houston

Compressor cylinders, like people, develop a fever when they are sick. Unlike humans, however, the normal temperature (suction) is not constant. The trick of using temperatures to determine malfunctions lies in determining the correct starting position.

To determine the starting position, it is necessary to know the true "k" value of the product being pumped. The "k" value may be obtained from measured pressures and temperatures according to the formula: $T2/T1 = (P2/P1)\frac{k\text{-}I}{k}$ where the subfix 2 refers to discharge conditions and the 1 to suction, and all values of P and T are absolute.[1]

The preferred time to obtain this "k" would be with new equipment, or from data obtained when the equipment was new. The next best time would be with newly overhauled equipment. Lacking either of this data, any values of "k" must be considered to contain some malfunctions, and any unnecessarily high numbers should be discarded.

Nomograph 1 has been successfully used for a number of years, although it is slowly giving way to electronic equipment. Nomograph 2 may be finished by completing the suction and discharge temperature scales according to the following formulas:

Suction: $t_i = 8.557\ \rho \text{ Log } (t_s + 460) - 23.135\ \rho + 2.51$ inches above base line.

Discharge: $t = 23.067\ \rho \text{ Log } (t_d + 460) - 62.65\ \rho$ inches above base line.

Where $\rho = \frac{k}{k - 1},$

and Logs are to the base 10.

Reference

1. Virgil Moring Faires, *Applied Thermodynamics,* 11th Printing, 1946, Page 43.

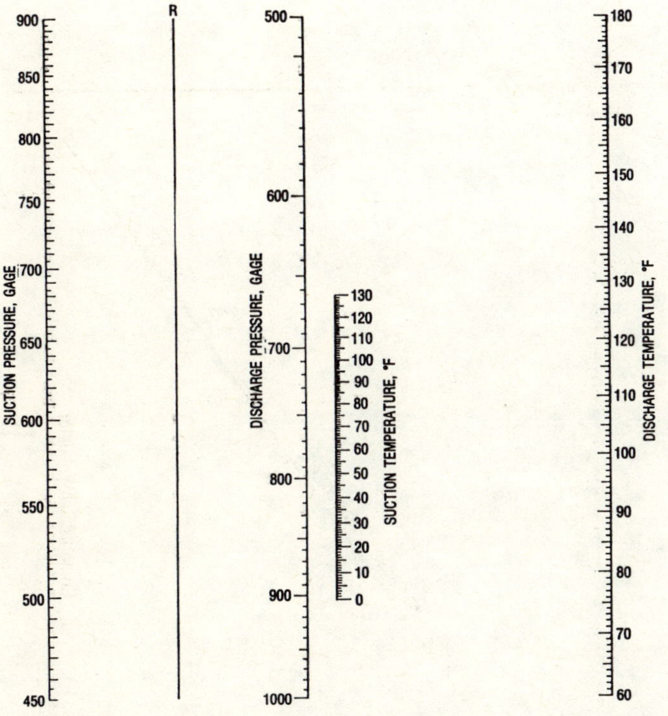

Nomograph 1—To determine starting position, it is necessary to know the true "k" value of the product being pumped. The "k" value may be obtained from measured pressure and temperatures according to the formula given.

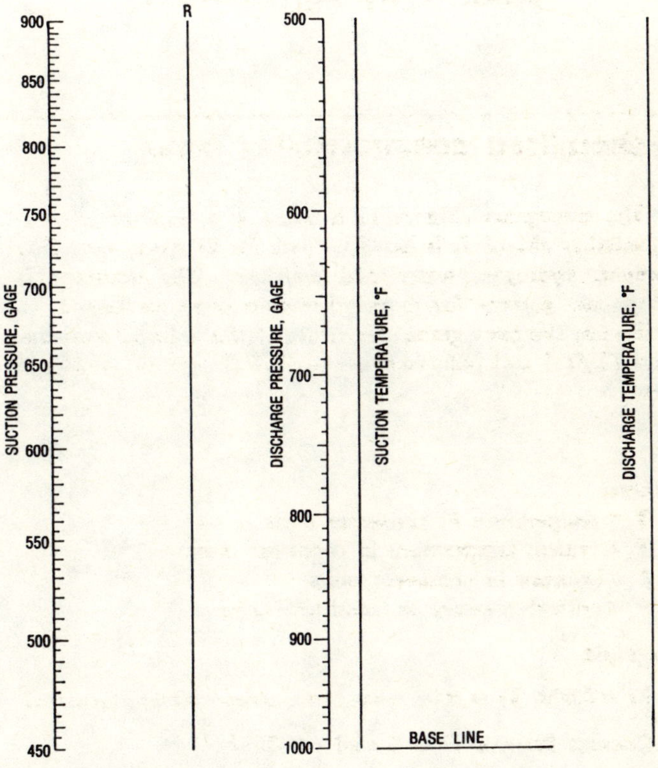

Nomograph 2—This nomograph may be finished by completing the suction and discharge temperature scales according to the formulas given.

9. Gas—Design

Allowable working pressure for carbon steel pipe

Based on formula No. 7 of the ASA Code for Pressure Piping, Section 3B, this table was computed for 6⅝-inch O.D. to 36-inch O.D. carbon steel seamless pipe, API Specifications 5L Grade A, 5L Grade B, and API 5LX Grade X-42.

The formula is:

$$t = \frac{PD}{2(S + YP)} + u + c$$

where:

t = wall thickness, inches
P = allowable working pressure, psi
D = outside diameter, inches
S = allowable stress, psi
Y = a coefficient
u = under-thickness tolerance, inches
c = allowance for threading, mechanical strength, and corrosion, inches

In calculation of the tables the values of Y, U and c were:

Y = 0.4
u = 0.125 t (u = 12½% of t)
c = 0.050-inches

Substituting in formula and solving for P:

$$P = \frac{1.75\ S\ (t - 0.057)}{D + 0.04 - 0.7\ t}$$

The S-values are

Spec. API 5L Grade A — 25,500 psi
Spec. API 5L Grade B — 29,750 psi
Spec. API 5LX Grade X-42 — 35,700 psi.

Example—Find pipe specification for a 12-inch line operating at 1,078 psi: In the 12¾-inch column these conditions will be met using Grade X-42 pipe with 0.281-inch wall thickness, Grade B pipe with 0.330-inch wall thickness, or Grade A pipe with 0.375-inch wall thickness.

O.D. / Wall Thickness	6⅝	8⅝	10¾	12¾	14	16	18	20	22	24	26	28	30	32	34	36	O.D. / Wall Thickness
0.188	895 1044 1253	685 799 959	548 640 768	....	....	....	....	....	....	....	....	....	....	....	....	....	0.188
0.203	.....	1070	857	721	.....	.....	.....	.....	.....	.....	.....	.....	.....	.....	.....	.....	0.203
0.219	1110 1295 1554	849 991 1189	680 793 952	572 667 801	 729	 637	....	....	....	....	....	....	....	....	....	....	0.219
0.250	1327 1548 1858	1014 1184 1420	811 947 1136	683 797 956	621 725 870	543 633 760	482 562 675	434 506 607	 551	 505	 466	 433	 404	 378	 356	 336	0.250
0.281	....	....	....	794 926 1111	722 842 1010	631 736 883	560 654 784	504 588 705	 641	 587	 542	 503	 469	 439	 414	 390	0.281
0.312	1765 2059 2471	1347 1572 1886	....	905 1056 1267	823 961 1153	719 839 1007	638 745 894	574 670 804	521 608 730	478 557 669	441 514 617	 573	 534	 501	 471	 445	0.312
0.344	1994 2326 2791	1520 1774 2128	1214 1416 1700	1021 1190 1429	928 1083 1299	810 946 1135	720 839 1007	647 755 906	587 685 823	538 628 753	 695	 645	 602	 564	 530	 501	0.344
0.375	2216 2568 3103	1689 1970 2364	....	1133 1322 1586	1030 1202 1442	900 1049 1259	798 931 1118	718 837 1005	652 760 912	597 696 836	551 642 771	 715	 667	 625	 588	 555	0.375
0.406	.....	.....	.....	1744	.....	.....	1228	1104	1002	918	847	786	733	687	646	610	0.406
0.438	....	2034 2373 2848	1622 1892 2271	1362 1589 1907	1238 1444 1733	1081 1261 1513	959 1119 1342	862 1005 1206	782 913 1095	716 836 1003	661 771 925	 858	 801	 750	 706	 666	0.438
0.500	....	 3328	 2651	 2225	1444 1685 2022	1260 1470 1764	1118 1304 1565	1004 1171 1406	911 1063 1276	834 974 1168	770 898 1077	 1000	 932	 873	 822	 775	0.500
Wall Thickness / O.D.	6⅝	8⅝	10¾	12¾	14	16	18	20	22	24	26	28	30	32	34	36	Wall Thickness / O.D.

Find the stress in pipe wall due to internal pressure

Multiply ½ by diameter by pressure and divide by wall thickness to get stress in psi.

Examples:

Find the wall stress due to internal pressure in a 24-inch pipeline having a ¼-inch wall thickness if the pressure gage reads 800 psi: $\frac{1}{2} \times 24 \times 800 \div \frac{1}{4} =$ 38,400 psi.

What wall thickness should be specified for a 30-inch discharge line where pressure runs 1000 psi and wall stress is to be limited to 30,000 psi? $\frac{1}{2} \times 30 \times 1000 \div$ thickness = 30,000.

Wall thickness = ½ inch.

This method of calculating transverse tensile stress due to internal pressure on the pipe is the Barlow formula. Actually, there are two accepted variations, and some designers use inside diameter of the pipe in calculating this stress. This gives a slightly lower figure for the transverse tensile stress and consequently gives less safety factor. The method used here is based on the assumption that maximum stress occurs at outside diameter of the pipe. The results give a conservative value for safety factor calculations.

10. Gas—Engineering

Nomograph for determining pipe-equivalent factors

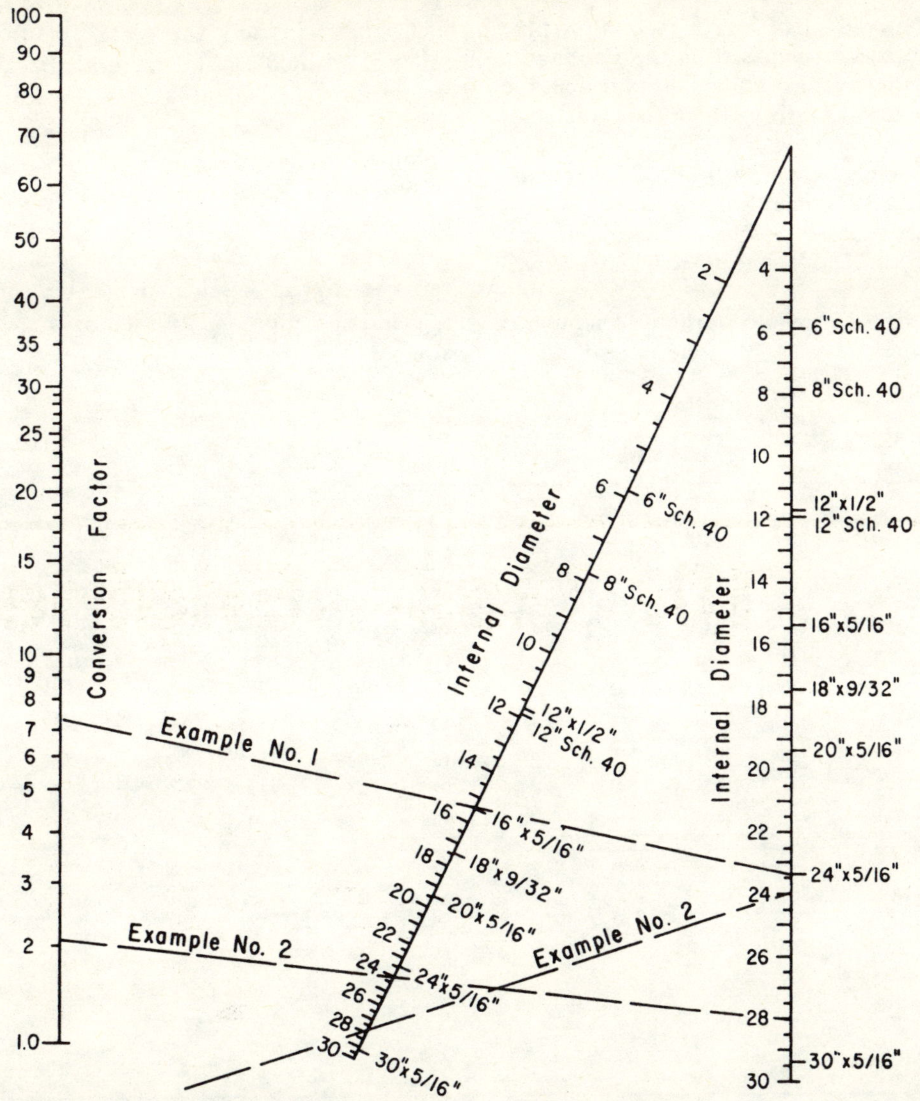

For turbulent flow, this handy nomograph will save a great deal of time in flow calculations where different sizes of pipe are connected. The advantage over a table of values is in having all sizes of pipe expressed in relation to any other size.

Example 1. 16″ × 5/16″ in terms of 24″ × 5/16″

$$\left(\frac{23.375}{15.375}\right)^{4.735} = 1.5203^{4.735} = 7.26.$$

Example 2. Reciprocal values can be found when chart limits are exceeded. 28″ I.D in terms of 24″ I.D. falls beyond the lower limits of the chart, however 24″ I.D. in terms of 28″ I.D. gives a conversion factor of 2.1.

How much gas is contained in a given line section

Multiply the square of the inside diameter, in inches, by the gauge pressure, in pounds per square inch; multiply this by 0.372; the answer is the approximate number of cubic feet of gas (standard conditions) in 1000 feet of line.

Example: How much gas in 1000 feet of 16″ schedule 30 pipe at 350 pounds pressure?
$15.25 \times 15.25 \times 350 \times 0.372 = 30{,}280$ cubic feet of gas.

Example: Approximately how much gas in eight miles of .250 wall 24-inch pipe if the pressure is 400 psi?
Cubic feet = $(23.5)(23.5)(400)(8)(5.28)(0.372) =$ 3,470,000 cubic feet.

When a section of line is blown down from one pressure to another, the total gas lost may be computed by the difference in the contents at the two pressures, using the above rule.

How to estimate equivalent length factors for gas lines

This table shows the equivalent length factors for pipe lines of different diameters. It is based on the Panhandle Formula and the factors were obtained by varying the lengths of the pipe and keeping the other variables constant.

Example: Find the equivalent length of a 24-inch pipe line as compared with a 12¾-inch line.

Enter the chart at the top—24-inch—(23.00 ID) and proceed downward to the 12.250 parameter. The equivalent length is 21.2.

This means that under the same conditions of temperatures, pressures, specific gravities, etc. one mile of 12¾-inch pipe will flow the same amount of gas that 21.2 miles of 24-inch will flow. Another way of putting it, the pressure drop in one mile of 24-inch pipe will be the same as that in .048 miles of 12¾-inch if the pressures, volumes etc. are the same.

Where a line is composed of several sizes of pipe diameters or where the lines are looped, this chart should prove useful. In solving problems of this sort, the line is reduced to an equivalent length of some arbitrarily selected size pipe and treated as a single line of uniform size.

Internal Diameter	Nominal Diameter	36″	34″	30″	26″	24″	24″	22″	20″	18″	16″	12¾″	10¾″	8⅝″
		35.250	33.250	29.375	25.375	23.375	23.000	21.500	19.500	17.500	15.500	12.250	10.250	8.125
35.250″	36″	1	0.758	0.416	0.204	0.137	0.126	0.092	0.056	0.034	0.019	0.006	0.002	0.0008
33.250	34	1.32	1	0.55	0.270	0.181	0.167	0.121	0.075	0.044	0.025	0.008	0.003	0.001
29.375	30	2.41	1.82	1	0.492	0.330	0.305	0.22	0.136	0.081	0.045	0.014	0.006	0.002
25.375	26	4.90	3.71	2.03	1	0.671	0.622	0.45	0.278	0.164	0.092	0.029	0.012	0.004
23.375	24	7.25	5.54	3.04	1.49	1	0.922	0.67	0.415	0.246	0.137	0.044	0.018	0.006
23.000	24	7.90	6.00	3.29	1.62	1.08	1	0.72	0.450	0.267	0.148	0.048	0.020	0.0064
21.500	22	10.93	8.31	4.55	2.24	1.50	1.37	1	0.624	0.369	0.205	0.065	0.027	0.009
19.500	20	17.60	13.33	7.31	3.59	2.41	2.22	1.60	1	0.590	0.328	0.105	0.044	0.014
17.500	18	29.8	22.5	12.35	6.08	4.075	3.78	2.72	1.69	1	0.555	0.177	0.074	0.024
15.500	16	52.8	40.5	22.2	10.9	7.345	6.75	4.88	3.04	1.80	1	0.319	0.134	0.043
12.250	12¾	168	127	69.8	34.3	23.01	21.2	15.35	9.55	5.65	3.13	1	0.420	0.136
10.250	10¾	400	303	167	81.7	54.67	50.6	36.5	22.7	13.4	7.47	2.38	1	0.324
8.125	8⅝	1235	935	513	252	168.8	156.2	112.5	70.5	41.5	23.0	7.35	3.08	1
6.125	6⅝	4860	3680	2020	992	666	615	443	276	163	90.7	29.0	12.2	3.94

Estimating comparative capacities of gas pipe lines

To estimate capacity of one diameter line in terms of another read downward in the appropriate column to desired diameter line. Example: A 24-inch (23.375-inch ID) pipe line has a capacity 160 percent that of a 20-inch (19.500-inch ID) line.

This chart is based on diameter factors using the Panhandle equation. Equivalent inlet and outlet pressures, lengths of line section, and uniform flow are assumed for both lines.

(Figures give capacity of one diameter pipe as percent of another.)

Internal Diameter	Nominal Diameter	36″	34″	30″	26″	24″	24″	20″	18″	16″	12¾″
		35.250″	33.250″	29.3125″	25.375″	23.375″	23.000″	19.500″	17.500″	15.500″	12.250″
35.250″	36″	100	85.7	62.7	43.6	35.2	33.7	22.0	16.5	12.0	6.5
33.250″	34″	116.8	100	73.1	50.8	41.0	39.3	25.7	19.2	14.0	7.6
29.3125″	30″	159.7	137	100	69.6	56.1	53.8	35.7	26.3	19.2	10.4
25.375″	26″	229	197	149	100	80.7	77.5	50.2	37.8	27.6	14.8
23.375″	24″	285	244	178	124	100	96.9	62.3	46.9	34.2	18.5
23.000″	24″	298	255	186	129	104	100	65.1	49.0	35.7	19.2
19.500″	20″	458	392	287	199	160	154	100	75.3	55.0	29.6
17.500″	18″	608	519	379	264	213	204	132	100	73.2	39.3
15.500″	16″	835	715	522	363	293	281	182	137	100	53.8
12.250″	12¾″	1545	1322	968	673	543	521	238	254	185	100

Determination of leakage from gas line using pressure drop method

To determine the leakage in Mmcf per year at a base pressure of 14.4 psia and a temperature of 60° F, take as a basis a one mile section of line under one hour test. Use the formula:

$$Ly = 9 \times D^2 \left[\frac{P_1}{460 + t_1} - \frac{P_2}{460 + t_2} \right]$$

where:

D is the inside diameter of the pipe in inches.

P_1 and t_1 are the pressure and temperature at the beginning of the test in psia and °F respectively.

and

P_2 and t_2 are the pressure and temperature at end of the test.

The error in using this formula can be about 4%.

Example: A section of a 10-inch pipe line one mile in length is tested using the pressure drop method. I.D. is equal to 10.136 inches. Pressure and temperature at the start of the test were 190.5 psig and 52° F. After one hour the readings were 190.0 psig and 52° F. Calculate the leakage in Mmcf/year.

The atmospheric pressure for the area is 14.4 psia. Thus

$P_1 = 190.5 + 14.4 = 204.9$ psia

$P_2 = 190.0 + 14.4 = 204.4$ psia

$$Ly = 9\ (10.136)^2 \left[\frac{204.9}{460 + 52} - \frac{204.4}{460 + 52} \right]$$

$$= \frac{(9)\ (102.7)\ (0.5)}{512}$$

$= 0.903$ Mmcf per mile per year.

A quick way to determine size of gas gathering lines

Here is a short cut way to estimate gas flow in gathering lines. For small gathering lines, the answer will be within 10 percent of that obtained by more difficult and more accurate formulas.

$$Q = \frac{(500)\ (d^3)\ \sqrt{P_1^2 - P_2^2}}{\sqrt{L}}$$

Q = cubic feet of gas per 24 hours

d = pipe I.D. in inches

P_1 = psi (absolute) at starting point

P_2 = psi (absolute) at ending point

L = length of line in miles

Example: How much gas would flow through four miles of six-inch I.D. pipe if the pressure at the starting point is 485 psi and if the pressure at the downstream terminus is 285 psi?

Answer:

$$Q = \frac{(500)\ (6^3)\ \sqrt{(485 + 15)^2 - (285 + 15)^2}}{\sqrt{4}}$$

$$Q = \frac{(500)\ (216)\ \sqrt{500^2 - 300^2}}{2}$$

$$Q = \frac{(500)\ (216)\ (400)}{2}$$

Q = 21,600,000 cu. ft. per 24 hrs.

Using a more accurate formula (a version of Weymouth's)

$$Q = \frac{(871)\ (d^{8/3})\ \sqrt{P_1^2 - P_2^2}}{\sqrt{L}}$$

$$Q = \frac{(871)\ (120)\ (400)}{2}$$

Q = 20,900,000 cu. ft. per 24 hrs.

The error in this case is 700,000 cu. ft. per 24 hours or about 3 percent.

Energy conversion data for estimating

The accompanying table gives the high heating values of practically all commonly used fuels along with their relation one to the other.

Example: If light fuel oil will cost a potential industrial customer 18 cents per gallon and natural gas will cost him 75 cents per Mcf, which fuel will be more economical?

Solution: Assume the operator would use 10,000 gallons of light fuel oil per year at 18 cents per gallon. The cost would be $1,800. From the table, 10,000 gallons of fuel oil is equivalent to 10,000 x 134 = 1,340,000 cubic feet of natural gas. At 75 cents per Mcf this would cost 1,340 x .75 = $1,005.

The burning efficiency of both fuels is the same. The savings by using natural gas would come to $795 per year.

FUEL	Heating Value	PERCENT BURNING EFFICIENCY			Unit	NUMBER OF CUBIC FEET OF NATURAL GAS EQUIVALENT TO 1 UNIT OF OTHER FUEL		
		Domestic	Commercial	Industrial		Domestic	Commercial	Industrial
Natural Gas	1000 B.T.U. per Cubic Foot	75	75	78	Cubic Foot	1	1	1
Propane	91,600 B.T.U. per U.S. gallon	75	75	78	Gallon	91.5	91.5	91.5
Butane	102,000 B.T.U. per U.S. gallon	......	75	78	Gallon	....	102	102
Light Fuel Oil	134,000 B.T.U. per U.S. gallon	65–75	75	78	Gallon	117–134	134	134
Heavy Fuel Oil	146,000 B.T.U. per U.S. gallon	......	75	78	Gallon	....	146	146
Hydro-electricity	3,412 B.T.U. per Kilowatt Hour	100	100	100	Kilowatt Hour	4.549	4.549	4.549
*Coal	8700—11,000 B.T.U. per pound	55-60	65	73	Pound	6.37—8.07	7.55—9.57	8.45—10.7

* Coal qualities vary considerably. The figures given are average for Western Coals; slack, run-of-the-mine; potentially available at lowest cost.

How to estimate time required to get a shut-in test on gas transmission lines and approximate a maximum acceptable pressure loss for new lines

These two rules may prove helpful for air or gas testing of gas transmission lines. They are not applicable for hydrostatic tests. Values of the constants used by different transmission companies may vary with economic considerations, line conditions and throughputs.

To determine the minimum time required to achieve a good shut-in test after the line has been charged and becomes stabilized, use the following formula:

$$H_m = \frac{3 \times D^2 \times L}{P_1}$$

where,

H_m = Minimum time necessary to achieve an accurate test in hours

D = Internal diameter of pipe in inches

L = Length of section under test in miles

P_1 = Initial test pressure in pounds per square inch gage

Example: How long should a 14-mile section of 26-inch (25.375-inch ID) pipe line be shut in under a 1050 psi test pressure to get a good test?

$$H_m = \frac{3 \times (25.375)^2 \times 14}{1050} = 25.75 \text{ hours or 25 hours and 45 minutes}$$

When a new line is being shut in for a test of this duration or longer, the following formula may be used to evaluate whether or not the line is "tight":

$$P_{ad} = \frac{H_t \times P_1}{D \times 949}$$

where,

P_{ad} = Acceptable pressure drop in pounds per square inch gage

H_t = Shut-in test time in hours

D = Internal diameter of pipe in inches

P_1 = Initial test pressure in pounds per square inch gage

Example: What would be the maximum acceptable pressure drop for a new gas transmission line under air or gas test when a 44-mile section of 30-inch (29.250-inch ID) pipe line has been shut in for 100 hours at an initial pressure of 1100 psig?

$$P_{ad} = \frac{100 \times 1100}{29.250 \times 949} = 4 \text{ pounds per square inch}$$

If the observed pressure drop after stabilization is less than 4 psig, the section of line would be considered "tight."

Note: Corrections must be made for the effect of temperature variations upon pressure during the test.

How to determine the relationship of capacity increase to investment increase

when increasing pipe diameter, while keeping the factors P(in), P(out) and L constant.

Solution: The top number in each square represents the percent increase in capacity when increasing the pipe size from that size shown in the column at the left to the size shown by the column at the bottom of the graph. The lower numbers represent the percent increase in investment of the bottom size over the original pipe size. All numbers are based on pipe capable of 1,000 psi working pressure.

(1966 Cost Data)

34																	17/12
32																17/13	37/27
30															19/13	39/28	62/43
28														20/11	43/26	67/41	
26													23/11	47/23	75/39		
24												24/11	52/23	82/37			
22											26/14	56/26	91/40				
20										29/17	62/33	101/47					
18									34/14	72/33	117/51						
16								38/23	85/39	138/63							
14							44/14	98/39	165/58								
12						29/16	85/32	156/62									
10					61/22	108/42	198/61										
8				82/26	193/53	278/78											
6			105/22	274/54	502/87												
4	0/0	190/38	497/69	988/112													
PIPE SIZE	4	6	8	10	12	14	16	18	20	22	24	26	28	30	32	34	36

XXX (top) — Top Number represents % increase in capacity when increasing pipe size from size on left to size at bottom

XXX (lower) — Lower Number represents % increase in investment for pipe size at bottom over pipe size at left.

Estimate pipe size requirements for increasing thruput volumes of natural gas

Reflects the capacity to pipe size relationship when L (length), P_1 (inlet pressure) and P_2 (outlet pressure) are constant.

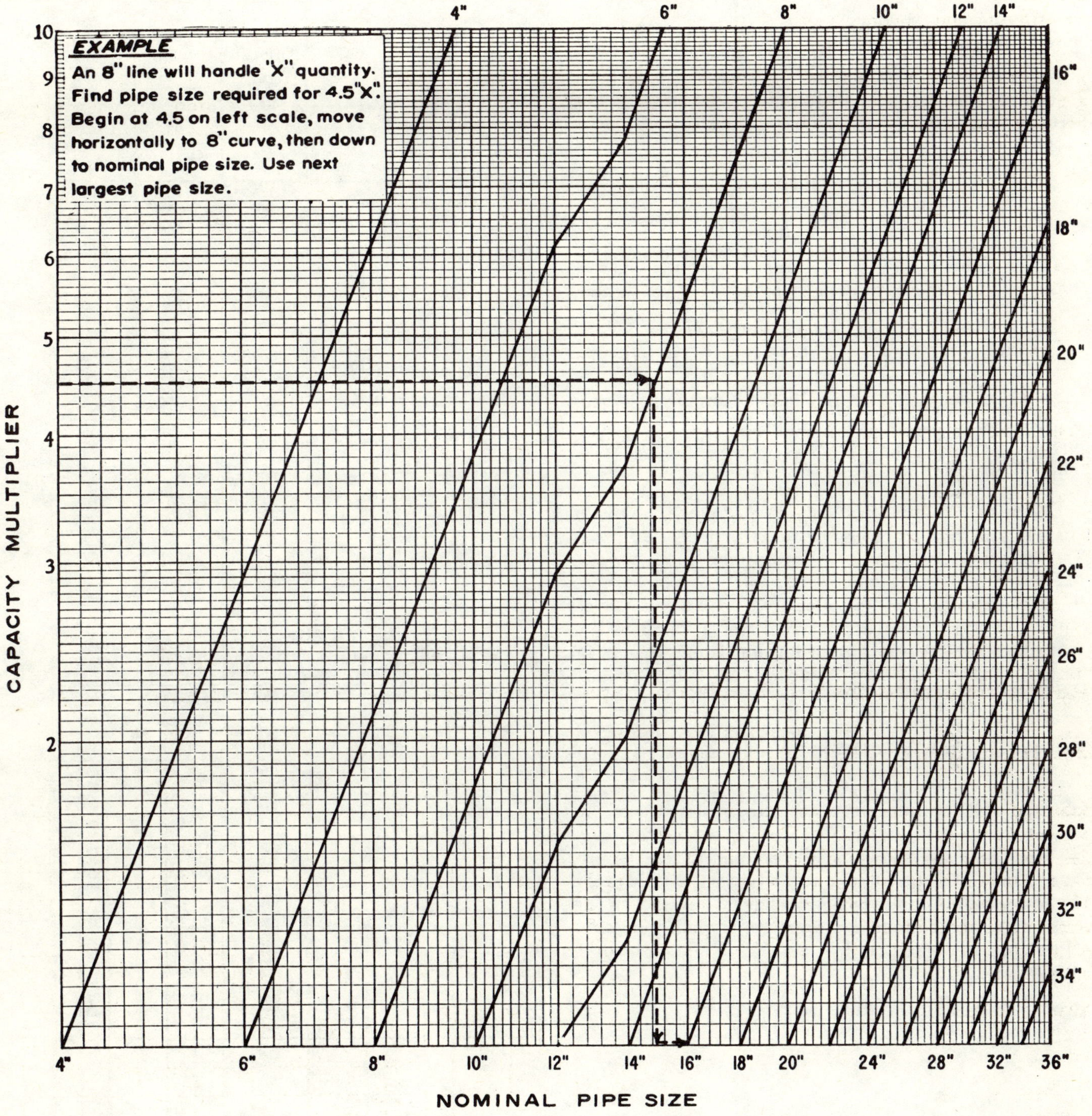

Calculate line loss using cross-sectional areas table when testing mains with air or gas

$$\text{Initial volume} = \frac{P_i}{P_B}(AL)$$

$$\text{Final volume} = \frac{P_f}{P_B}(AL)$$

When: P_i = initial pressure, or pressure at start of test, psia

P_f = final pressure, or pressure at end of test, psia

P_B = base pressure, psia

A = cross-sectional area of inside of pipe in square feet

L = length of line in feet

The loss then would be the difference of the volumes, found by the above formula, which could be written:

$$\text{Loss during test} = AL\,\frac{(P_i - P_f)}{P_B}$$

To facilitate calculations, Table 1 contains the cross-sectional areas in square feet for some of the more popular wall thickness pipe from 1-inch through 24-inch nominal size.

Example: Assume 3,562-feet of 8.125-inch inside diameter pipe is tested with air for a period of 26-hours. The initial pressure was 116 psig, the final pressure 112 psig, and the barometer was 14.4 psia in both instances. It is desired to calculate the volume of loss during the test using a pressure base of 14.65 psia.

If the area "A" is obtained from Table 1, and information as given above is substituted in the formula, we would have:

$$\text{Loss during test} = 0.360 \times 3562 \times \frac{[(116+14.4)-(112+14.4)]}{14.65}$$

$$= 350.1 \text{ Cubic Feet}$$

Nominal Pipe Size D_n	Outside Diameter	Wall Thickness	Inside Diameter	"A" Inside Area in Sq. Feet
1	1.315	.133	1.049	.006
		.179	.957	.005
1¼	1.660	.140	1.380	.010
		.191	1.278	.009
1½	1.900	.145	1.610	.014
		.200	1.500	.012
2	2.375	.154	2.067	.023
		.218	1.939	.021
2½	2.875	.203	2.469	.033
		.276	2.323	.029
3	3.500	.125	3.250	.058
		.156	3.188	.055
		.188	3.124	.053
		.216	3.068	.051
		.250	3.000	.049
3½	4.000	.125	3.750	.077
		.156	3.688	.074
		.188	3.624	.072
		.226	3.548	.069
		.250	3.500	.067
4	4.500	.125	4.250	.099
		.156	4.188	.096
		.188	4.124	.093
		.219	4.062	.090
		.237	4.026	.088
		.250	4.000	.087
5	5.563	.156	5.251	.150
		.188	5.187	.147
		.219	5.125	.143
		.258	5.047	.139
6	6.625	.188	6.249	.213
		.219	6.187	.209
		.250	6.125	.205
		.280	6.065	.201
8	8.625	.188	8.249	.371
		.219	8.187	.366
		.250	8.125	.360
		.277	8.071	.355
10	10.750	.188	10.374	.587
		.219	10.312	.580
		.250	10.250	.573
		.279	10.192	.567
		.307	10.136	.560
12	12.750	.219	12.312	.827
		.250	12.250	.818
		.281	12.188	.810
		.312	12.126	.802
		.330	12.090	.797
14	14.000	.250	13.500	.994
		.281	13.438	.985
		.312	13.376	.976
		.344	13.312	.966
		.375	13.250	.958
16	16.000	.250	15.500	1.310
		.281	15.438	1.300
		.312	15.376	1.289
		.344	15.312	1.279
		.375	15.250	1.268
18	18.000	.250	17.500	1.670
		.281	17.438	1.658
		.312	17.376	1.647
		.344	17.312	1.635
		.375	17.250	1.623
20	20.000	.250	19.500	2.074
		.281	19.438	2.061
		.312	19.376	2.048
		.344	19.312	2.034
		.375	19.250	2.021
22	22.000	.312	21.376	2.492
		.344	21.312	2.477
		.375	21.250	2.463
24	24.000	.312	23.376	2.980
		.344	23.312	2.964
		.375	23.250	2.948

TABLE 1

Nomograph for calculating density and specific volume of gases and vapors

The density of a gas or vapor can be calculated from the equation for an ideal gas:

$$\rho = \frac{144P}{RT} = \frac{MP}{10.72T} = \frac{2.70\ PSg}{T}$$

where:

ρ = density of gas, lbs./cu./ft.
P = absolute pressure, psi (psig + 14.7)
R = individual gas constant (10.72/M)
M = molecular weight of gas
T = temperature, °R
Sg = specific gravity of individual gas relative to air (= to ratio of gas molecular weight to that of air)

The above equation and corresponding nomograph (Figure 1) can be used to calculate the density (or specific volume) of gases at low pressures and high temperatures, where the ideal gas law holds. Note that the pressure scale is calibrated in psi gauge so that the correction of 14.7 psi is not needed when using the nomograph.

Example

What is the density of dry methane if its temperature is 100° F and its pressure 15 psig.

Connect	With	Mark or Read
M = 16	t = 100° F	Index
Index Mark	P = 15 psig	ρ = 0.08 lbs./cu. ft.

Source

"Fluid Flow Through Pipe, Fittings and Valves," *Technical Paper No. 410,* A-11, Crane Company, Chicago, Illinois (1957).

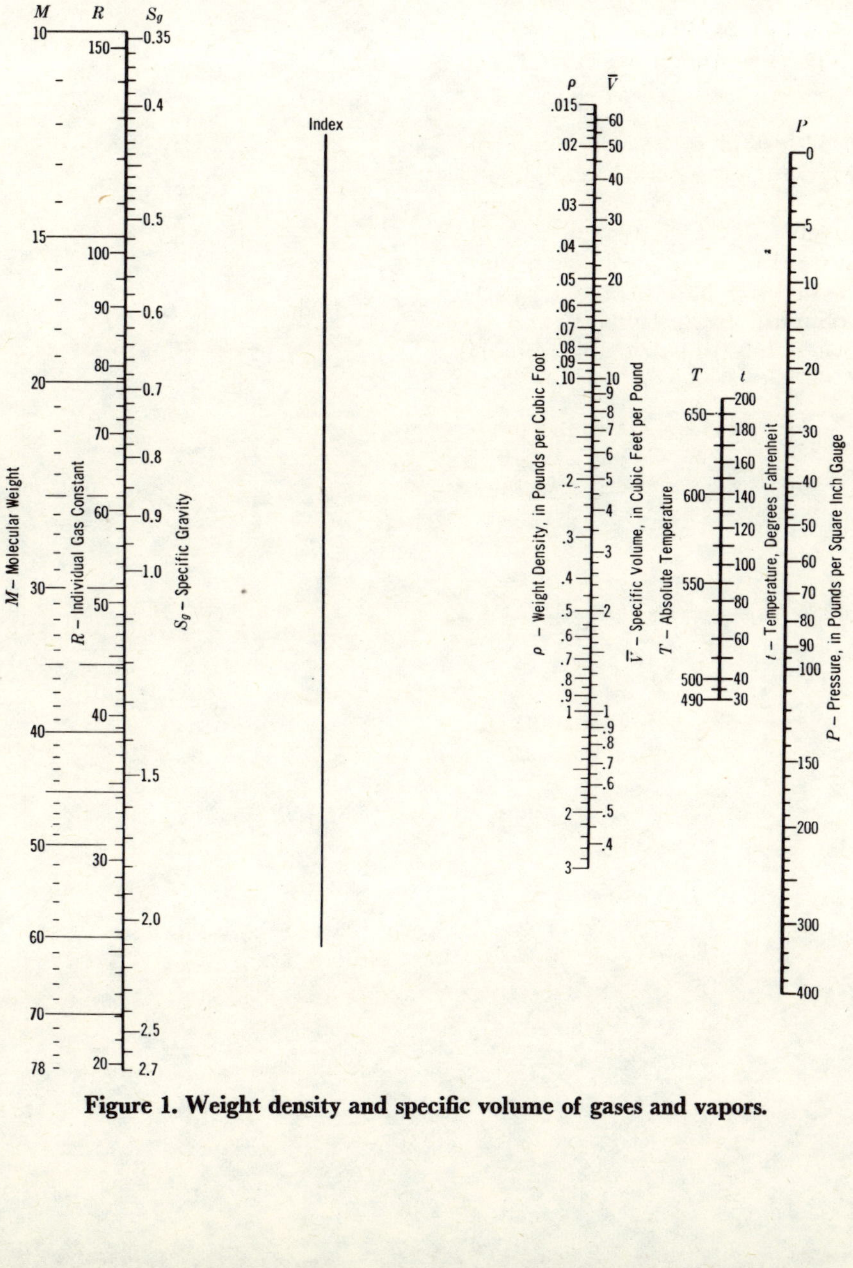

Figure 1. Weight density and specific volume of gases and vapors.

11. Gas—Flow

Flow of fuel gases in pipe lines

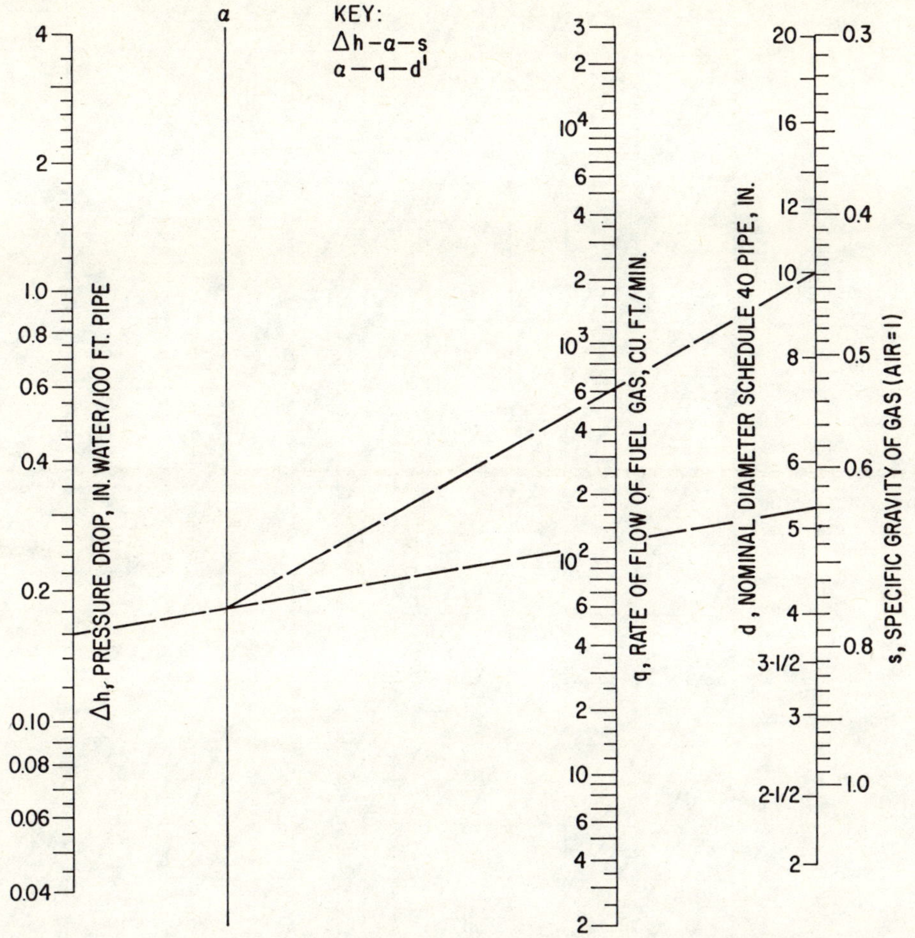

D. S. Davis,
Special Projects, Inc.,
Bailey Island, Maine

THE RATE OF FLOW of fuel gases at low pressures, ordinary temperatures, and under turbulent conditions in pipe lines is reflected by an equation[1] based on sound theoretical considerations:

$$q = \frac{3.185\,\Delta h^{0.543}\,d^{2.63}}{s^{0.468}}$$

where:

q = rate of flow of gas, cubic feet per minute

Δh = pressure drop, inches of water per 100 feet of pipe

d = actual inner diameter of Schedule 40 steel pipe, inches

s = specific gravity of gas at the prevailing temperature and pressure relative to air at 68°F and 30 inches of mercury.

The equation can be solved readily and accurately by means of the accompanying monograph constructed in accordance with methods described previously.[2]

The use of the chart is illustrated as follows. At what rate will fuel gas with a specific gravity of 0.640 flow in a 10-inch Schedule 40 pipe line if the pressure is 0.160 inch of water per 100 feet of pipe? Following the key, connect 0.160 on the Δh scale and 0.640 on the s scale with a straight line and mark the intersection with the α axis. Connect this point with 10 on the d' scale for nominal diameters and read the rate of flow as 624 cubic feet per minute on the q scale.

Some authorities give the exponent of s as 0.408 when for the data given the rate of flow as read from the chart would be decreased by 2.7 percent (17 cubic feet per minute) to attain 607 cubic feet per minute.

LITERATURE CITED

[1] Perry, J. H., ed., "Chemical Engineers' Handbook," 3rd ed., p. 1589, McGraw-Hill Book Co., Inc., New York (1950).

[2] Davis, D. S., "Nomography and Empirical Equations," 2nd ed., Chapter 6, Reinhold Publishing Corp., New York, 1962.

How to calculate the velocity of gas within a pipe line section

The nomograph presented for obtaining the velocity of a gas within a pipe line is based on the equation:

$$V = \frac{0.75\ Q}{d^2\ P}$$

where:

V = velocity expressed in feet/second

Q = gas volume, Mcf/hour at standard conditions

d = inside diameter of the pipe, inches

P = pressure, psia

Example: Given a pipe line having an inside diameter of 4.026 inches handling gas at a flow rate of 1,000 Mcf/hour at a pressure of 100 psig. Find the velocity of the gas.

Solution: Using the nomograph, draw a line vertically from the point on the lower left section marked "inside diameter of the pipe" of 4.026 inches to intersect with the diagonal "volume" line at 1,000 Mcf. From the point of intersection, draw a line horizontally to intersect the "pressure" line marked 100 psig. Drop from this point vertically down to intersect the base axis marked "velocity of gas."

Read the answer at the point of intersection as 400 feet/second.

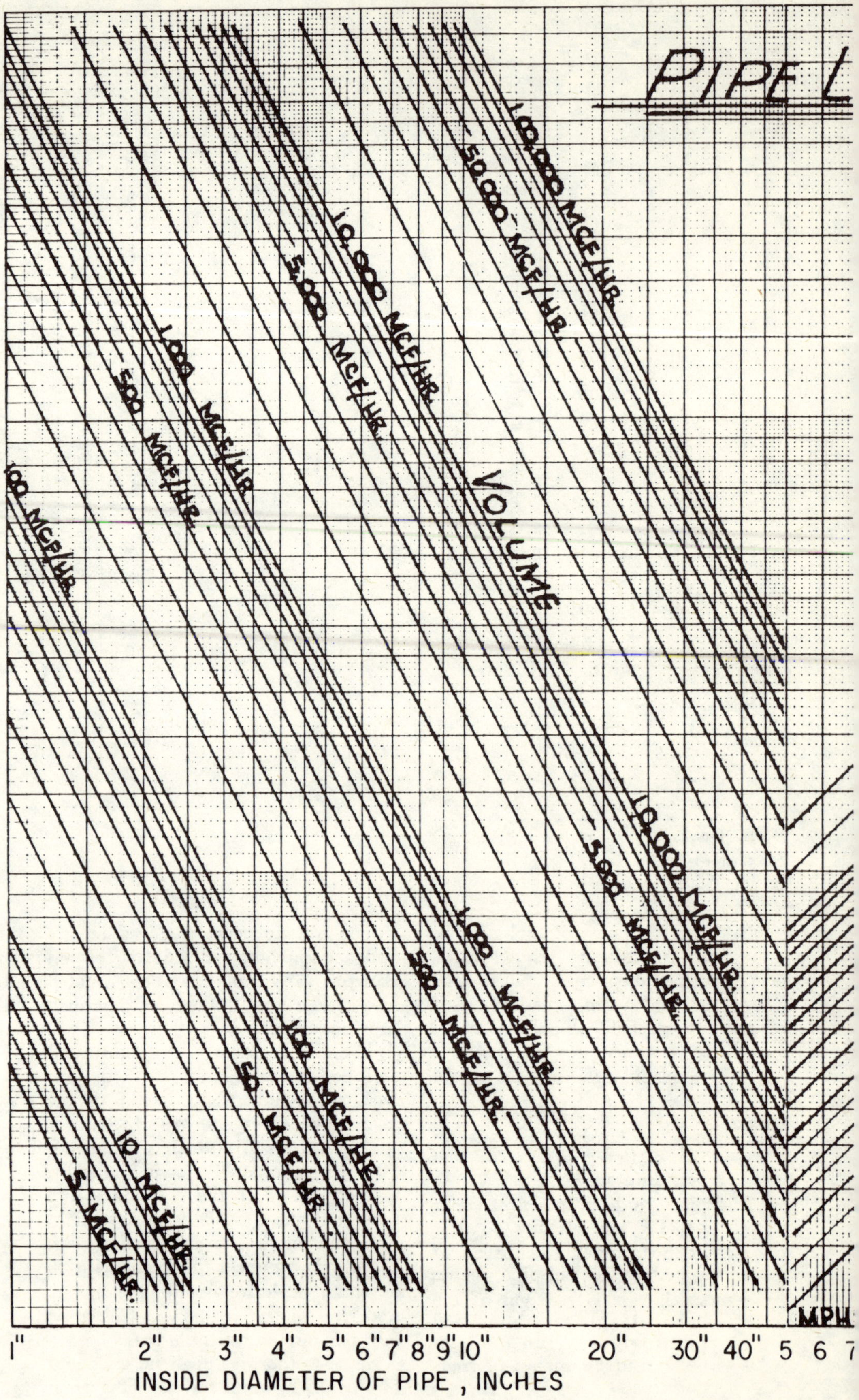

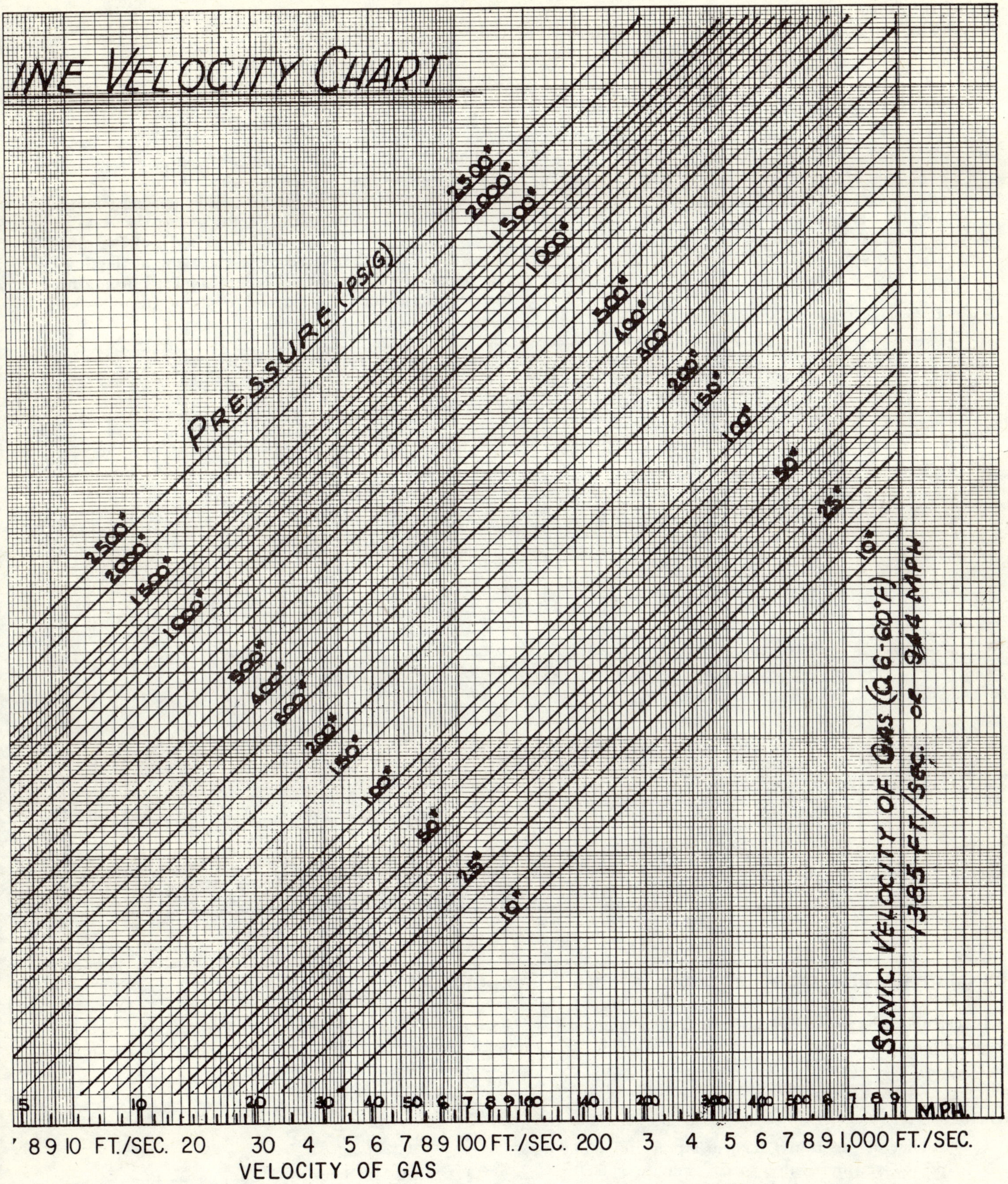
INE VELOCITY CHART
PRESSURE (PSIG)
2,500#
2,000#
1,500#
1,000#
500#
400#
300#
200#
150#
100#
50#
25#
10#
SONIC VELOCITY OF GAS (0.6-60°F)
1385 FT./SEC. OF 944 MPH
5 10 20 30 40 50 6 7 8 9 100 140 200 300 400 500 6 7 8 9 M.P.H.
' 8 9 10 FT./SEC. 20 30 4 5 6 7 8 9 100 FT./SEC. 200 3 4 5 6 7 8 9 1,000 FT./SEC.
VELOCITY OF GAS

Determining throat pressure in a blow-down system

In designing a gas blow-down system, it is necessary to find the throat pressure in a blow-off stack of a given size with a given flow at critical velocity. The accompanying nomograph is handy to use for preliminary calculations. It was derived from the equation:

$$P_t = \frac{W}{d^2} \times \frac{\sqrt{\dfrac{RT_1}{K\,(K-1)}}}{11{,}400}$$

where:

P_t = pressure in the throat—psia
W = weight of flowing gas—lbs/hr.
d = actual I.D. of stack—inches
R = gas constant—ft lbs/°F
$K = \frac{C_p}{C_v}$
C_p = specific heat at constant pressure—Btu/lb/°F
C_v = specific heat at constant volume—Btu/lb/°F
T_1 = absolute temperature of gas flowing—°R = (°F + 460)

Using the following values for natural gas R = 88.888 and K = 1.31 the above equation reduces to:

$$P_t = \frac{W}{d^2} \times .0012978\ \sqrt{T_1}$$

Example: To use the accompanying nomograph, assume that the rate of flow is known to be 2.00 × 10⁶ lbs/hr., the I.D. of the stack is 12 inches and the flowing temperature is 480° R. Start from the abscissa at the Rate of Flow, W, and proceed upward to the I. D. of the pipe. From this intersection proceed to the right to the flowing temperature of 480° R. Turn upward from this intersection to the P_t scale and read the answer—395 psia.

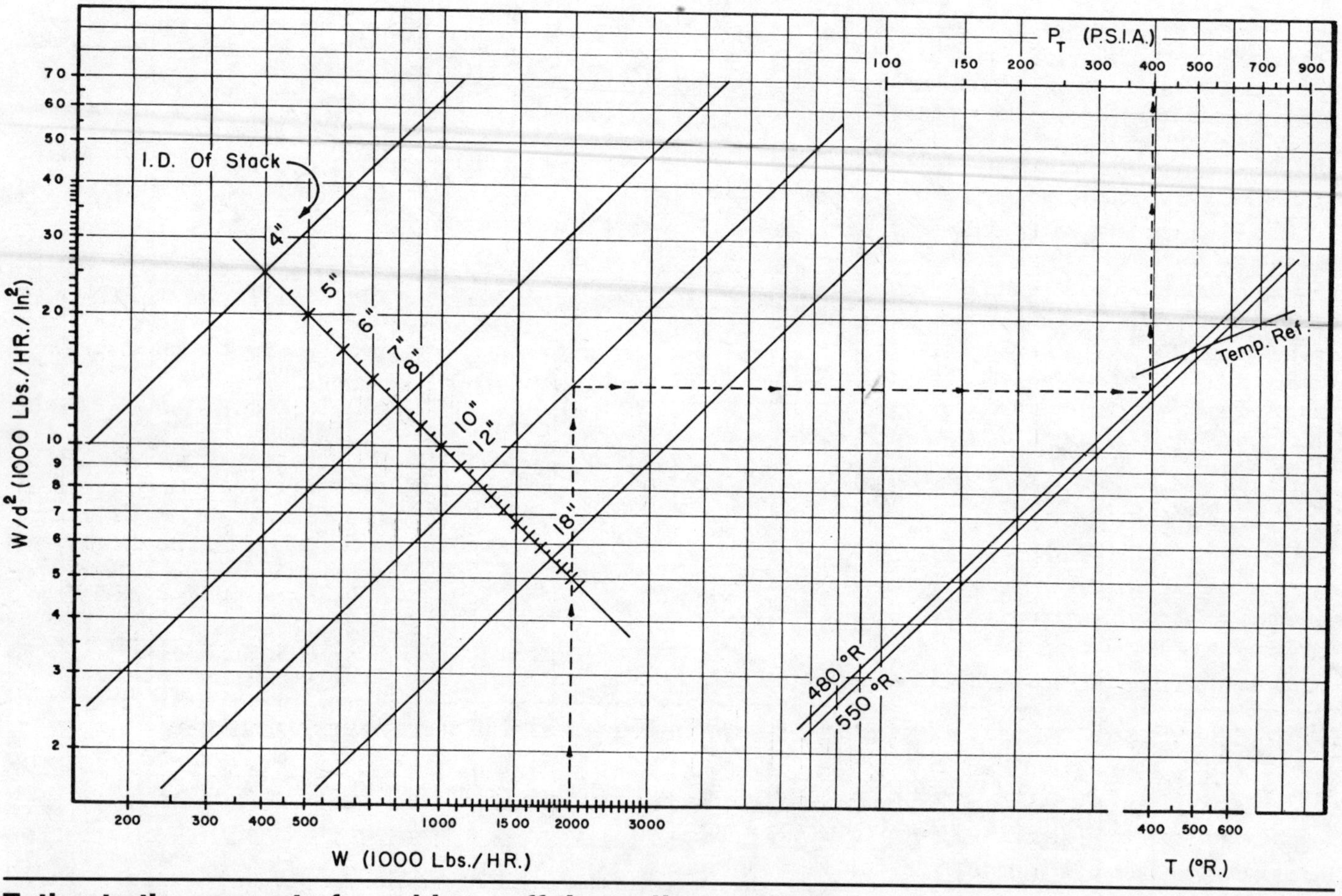

Estimate the amount of gas blown off thru a line puncture

To calculate the volume of gas lost from a puncture or blowoff, use the formula:

$$Q = D^2P_1$$

Where

Q = volume of gas in Mcf/hour at a pressure of 14.9 psi, 60° F. and with a specific gravity of 0.60;
D = diameter of the nipple or orifice in inches; and
P = absolute pressure in pounds per square inch at some nearby point upstream from the opening.

Example: How much gas will be lost during a five minute blowoff through a two-inch nipple if the upstream pressure is 1000 psi absolute?

$$Q = D^2P_1$$

$$Q = (2)^2 \times 1000$$

$$Q = 4000 \text{ Mcf/hour}$$

$$Q_{5\,min} = 4000 \times \frac{5}{60} = 333 \text{ Mcf.}$$

A practical way to calculate gas flow for pipe lines

Here is a short cut way to calculate gas flow in pipe lines. It is based on the Weymouth formula. At 60° F. and specific gravity of 0.60, the answer will be accurate. For every 10 degrees variation in temperature, the answer will be 1 percent in error. For every 0.01 variation in specific gravity, the answer will be three-fourths percent in error.

Formula:

$$Q = \frac{(871)\,(d^{8/3})\,\sqrt{P_1^2 - P_2^2}}{\sqrt{L}}$$

$Q =$ cubic feet of gas per 24 hours. 871 is a constant.

$d =$ pipe I.D. in inches.

$P_1 =$ psi (absolute) at starting point.

$P_2 =$ psi (absolute) at ending point.

$L =$ length of line in miles.

Example: How much gas would flow through one mile of six-inch I.D. pipe if the pressure at the starting point is 485 psi and if the pressure at the downstream terminus is 285 psi?

Answer:

$$Q = \frac{(871)\,(6^{8/3})\,\sqrt{(485+15)^2 - (285+15)^2}}{\sqrt{1}}$$

$$= \frac{(871)\,(120)\,\sqrt{500^2 - 300^2}}{1}$$

$$= (871)\,(120)\,(400)$$

$$= 41{,}800{,}000 \text{ cu. ft. per 24 hours.}$$

Example: What would be the flow through four miles of 10¾-inch quarter-inch wall pipe using the same pressures?

$$Q = \frac{(871)\,(10.25^{8/3})\,\sqrt{500^2 - 300^2}}{\sqrt{4}}$$

$$Q = \frac{(871)\,(495.7)\,(400)}{2}$$

$$Q = 86{,}350{,}000 \text{ cu. ft. per day.}$$

How to calculate the weight of gas in a pipe line

Rule: Find the volume of the pipe in cubic feet and multiply by the weight of the gas per cubic foot. To find the latter, multiply the absolute pressure of the gas times 3 and divide by 1000:

The basis for the latter is that gas with a specific gravity of .60 at 70° F. weighs 3.06 lb./cu. ft. at 1000 psia. And everything else remaining equal, weight of the gas is proportional to absolute pressure. Thus to find the weight of gas at, say, 500 psia, multiply $3 \times \frac{500}{1000}$ = 1.5 pounds. (1.53 is more accurate.)

Example: Find the weight of gas in a 1250-foot aerial river crossing where the average pressure reads 625.3 on the gauges; and temperature of the gas is 70° F. and specific gravity of gas is .60.

Solution: Change psig to psia: 625.3 + 14.7 = 640 psia

Weight of gas per cu. ft.

$$3 \times \frac{640}{1000} = 1.92 \text{ lb./cu. ft.}$$

Volume of 100 feet of 23¼-inch ID pipe = 3688 cu. ft.

Weight of gas = 7081 pounds.

This method is fairly accurate; here is the same problem calculated with the formula:

$$W = \frac{(V)\,(144)\,(P_{abs})}{RT}$$

Where W = weight of gas in pounds; V = volume of pipe; P_{abs} = absolute pressure of gas; R = universal constant 1544 ÷ molecular weight of gas; and T = temperature of gas in degrees Rankine. (To find the molecular weight of gas multiply specific gravity × molecular weight of air, or in this case, .6 × 28.95 = 17.37)

$$W = \frac{(3688)\,(144)\,(640)}{(1544 \div 17.37)\,(70 + 460)}$$

$$W = 7213 \text{ pounds}$$

Estimate average pressure in gas pipe line using up and downstream pressures

To find the average gas pressure in a line, first divide downstream pressure by upstream pressure and look up the value of the factor X in the table shown. Then multiply X times the upstream pressure to get the average pressure.

Example: Find the average pressure in a pipe line if the upstream pressure is 775 psia and the downstream pressure is 420 psia.

$$\frac{P_d}{P_u} = \frac{420}{775} = 0.54$$

Interpolating from the table,

$$X = 0.78 + 4/5\,(0.02) = 0.796$$

$$P_{average} = 0.796 \times 775 = 617 \text{ psia.}$$

This method is accurate to within 2 or 3 psi. Here is the same problem calculated with the formula:

$$P_{average} = \frac{2}{3}\left(P_u + P_d - \frac{P_u P_d}{P_u + P_d}\right)$$

$$= \frac{2}{3}\left(1195 - \frac{325{,}500}{1195}\right)$$

$$P_{average} = 615 \text{ psia.}$$

P_d/P_u	X
.10	.67
.15	.68
.20	.69
.25	.70
.30	.71
.35	.725
.40	.74
.45	.76
.50	.78
.55	.80
.60	.82
.65	.84
.70	.86
.75	.88
.80	.90
.85	.925
.90	.95
.95	.975
1.00	1.00

Chart for determining viscosity of natural gas

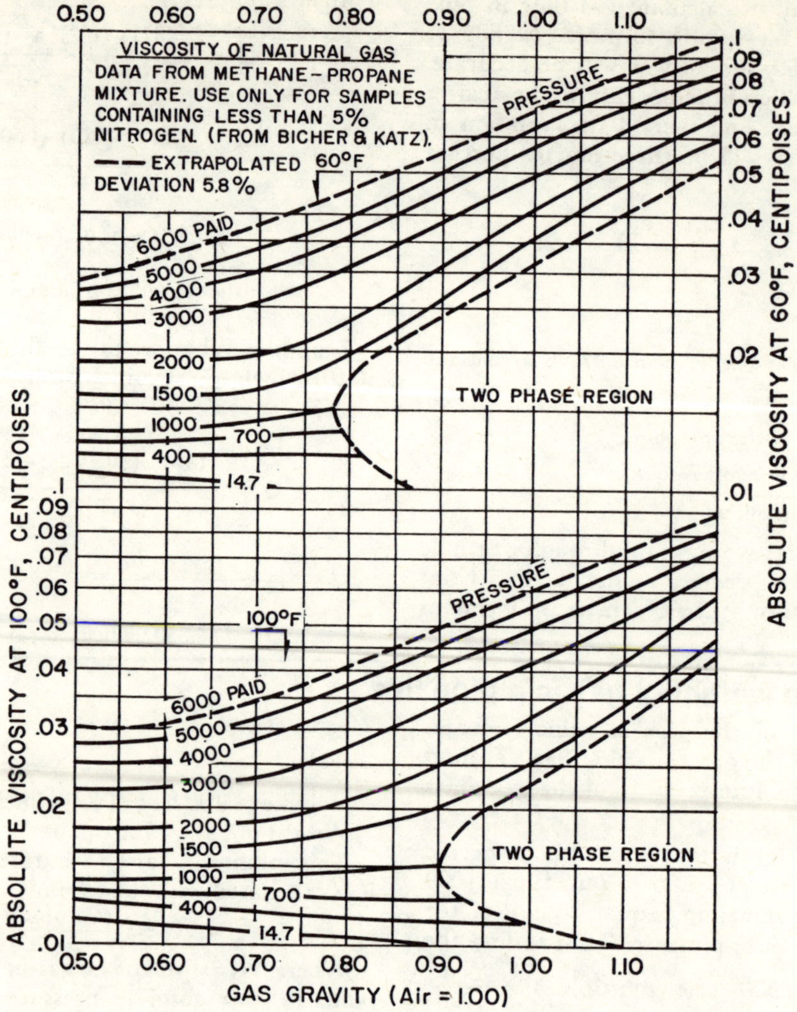

From: Carlton Beal, "The Viscosity of Air, Water, Natural Gas, Crude Oil and Its Associated Gases at Oil Field Temperatures and Pressures," *Petroleum Development and Technology,* AIME, 1946.

Flow of gas

For gas flow problems as encountered in oil field production operations, the Fanning equation for pressure drop may be used. A modified form of this equation employing units commonly used in oil field practice is:[1, 2, 3, 4, 5, 6]

$$P = \frac{FL\,(MCFD)^2\,STZ}{20{,}000\, d^5\, P_{av}}$$

and for determining of the friction factor "F":

$$Re = \frac{20.14\,(MCFD)\,S}{d\,\mu}$$

Where:

P = pressure drop, psia
F = friction factor, dimensionless
L = length of pipe, feet
MCFD = gas flow at standard condition
S = specific gravity of gas
T = absolute temperature, deg. F. plus 460
$P_{av.}$ = average flowing pressure, psia
d = internal pipe diameter, inches
Re = Reynolds number, dimensionless
μ = viscosity, centipoises

In actual practice, empirical flow formulas are used by many to solve the gas flow problems of field production operations. The Weymouth formula is the one most frequently used since results obtained by its use agree quite closely with actual valves. Recent modification of the formula, by including the compressibility Factor, Z, made the formula applicable for calculation of high pressure flow problems.

The modified formula is as follows:

$$Q_s = 433.45 \frac{T_s}{P_s} \times d^{2.667} \times \left(\frac{P_1^2 - P_2^2}{LSTZ} \right)^{1/2}$$

Q_s = rate of flow of gas in cu. feet per 24 hours measured at standard conditions
d = internal diameter of pipe in inches
P_1 = initial pressure, psi absolute
P_2 = terminal pressure, psi absolute
L = length of line in miles
S = specific gravity of flowing gas (air = 1.0)
T = absolute temperature of flowing gas
standard conditions for measurements:
T_s = standard absolute temperature
P_s = standard pressure, psi absolute
Z = compressibility factor

Values of expression $d^{2.667}$ for different inside diameter of pipe are given in Table 1.

Another formula quite commonly used is the one determined by F. H. Oliphant:

$$Q = 42a\sqrt{\frac{P_1^2 - P_2^2}{L}}$$

where

Q = discharge in cubic feet per hour
42 = a constant
P_1 = initial pressure in lbs. per square inches, absolute
P_2 = final pressure in pounds per square inches, absolute
L = length of line in miles
a = diameter coefficient

This formula assumes a specific gravity of gas of 0.6. For any other specific gravity multiply the final result by

$$\sqrt{\frac{0.6}{\text{sp. gr. gas}}}$$

Table 1
Values of $d^{2.667}$ for use in Weymouth's gas-flow formula

(Increments of tenths of an inch with certain standard pipe sizes inserted)

d	$d^{2.667}$	d	$d^{2.667}$	d	$d^{2.667}$	d	$d^{2.667}$
1.0	1.0	6.2	129.9	11.5	673.8	16.9	1881
1.1	1.3	6.3	135.4	11.6	689.5	17.0	1911
1.2	1.6	6.4	141.2	11.7	705.5	17.1	1941
1.3	2.0	6.5	147.1	11.8	721.7	17.2	1971
1.4	2.4	6.6	153.3	11.9	738.1	17.250	1988
1.5	2.9	6.7	159.5	12.000	754.8	17.3	2002
1.6	3.5	6.8	166.0	12.1	771.7	17.4	2033
1.7	4.1	6.9	172.6	12.2	788.8	17.5	2064
1.8	4.8	7.0	179.3	12.3	806.1	17.6	2096
1.9	5.5	7.1	186.2	12.4	823.7	17.7	2128
2.0	6.3	7.2	193.3	12.5	841.6	17.8	2160
2.067	6.9	7.3	200.5	12.6	859.7	17.9	2193
2.1	7.2	7.4	207.9	12.7	878.0	18.0	2225
2.2	8.2	7.5	215.5	12.8	896.5	18.1	2259
2.3	9.2	7.6	223.3	12.9	915.3	18.2	2292
2.4	10.3	7.7	231.2	13.0	934.4	18.3	2326
2.5	11.5	7.8	239.3	13.1	953.7	18.4	2360
2.6	12.8	7.9	247.6	13.2	973.2	18.5	2394
2.7	14.1	7.981	254.5	13.250	983.9	18.6	2429
2.8	15.6	8.0	256.0	13.3	993.0	18.7	2464
2.9	17.1	8.1	264.6	13.4	1013	18.8	2499
3.0	18.7	8.2	273.4	13.5	1033	18.9	2535
3.068	19.9	8.3	282.4	13.6	1054	19.0	2570
3.1	20.4	8.4	291.6	13.7	1075	19.1	2607
3.2	22.2	8.5	300.9	13.8	1096	19.182	2639
3.3	24.1	8.6	310.5	13.9	1117	19.2	2643
3.4	26.1	8.7	320.2	14.0	1138	19.250	2664
3.5	28.2	8.8	330.1	14.1	1160	19.3	2680
3.6	30.4	8.9	340.2	14.2	1182	19.4	2717
3.7	32.7	8.941	344.6	14.3	1205	19.5	2755
3.8	35.2	9.0	350.5	14.4	1227	19.6	2793
3.9	37.7	9.1	361.0	14.5	1250	19.7	2831
4.0	40.3	9.2	371.6	14.6	1273	19.8	2869
4.026	41.0	9.3	382.5	14.7	1297	19.9	2908
4.1	43.1	9.4	393.6	14.8	1321	20.0	2947
4.2	45.9	9.5	404.8	14.9	1344	20.1	2987
4.3	48.9	9.6	416.3	15.0	1368	20.2	3026
4.4	52.0	9.7	428.0	15.1	1393	20.3	3067
4.5	55.2	9.8	439.8	15.2	1418	20.4	3107
4.6	58.5	9.9	451.9	15.250	1431	20.5	3148
4.7	62.0	10.0	464.2	15.3	1443	20.6	3196
4.8	65.6	10.020	467.0	15.4	1468	20.7	3230
4.9	69.3	10.1	476.6	15.5	1494	20.8	3272
5.0	73.1	10.2	489.3	15.6	1519	20.9	3314
5.1	77.1	10.3	502.2	15.7	1545	21.0	3357
5.2	81.2	10.4	515.3	15.8	1572	21.1	3400
5.3	85.4	10.5	528.7	15.9	1599	21.1	3443
5.4	89.8	10.6	542.2	16.0	1625	21.250	3468
5.5	94.3	10.7	555.9	16.1	1653	21.3	3486
5.6	98.9	10.8	569.9	16.2	1680	21.4	3530
5.7	103.7	10.9	584.1	16.3	1708	21.5	3574
5.8	108.6	11.000	598.5	16.4	1736	21.6	3619
5.9	113.7	11.1	613.1	16.5	1765	21.7	3663
6.0	118.9	11.2	627.9	16.6	1793	21.8	3709
6.065	122.4	11.3	643.0	16.7	1822	21.9	3754
6.7	124.2	11.4	658.3	16.8	1851	22.0	3800

—The Oil & Gas Journal

Table 2
Temperature—specific gravity multipliers

Specific Gravity	FLOWING TEMPERATURE—DEGREES F. 0	10	20	30	40	50	60	70	80	90	100	110	129
0.60...........	1.15	1.14	1.12	1.11	1.10	1.09	1.08	1.07	1.06	1.05	1.04	1.03	1.02
0.70...........	1.06	1.05	1.04	1.03	1.02	1.01	1.00	.991	.981	.972	.964	.955	.948
0.80...........	.995	.984	.974	.964	.954	.945	.935	.927	.918	.910	.901	.894	.886
0.90...........	.938	.928	.918	.909	.899	.891	.882	.874	.965	.858	.850	.842	.835
1.00...........	.890	.880	.871	.862	.853	.845	.837	.829	.821	.814	.806	.799	.792

The value of diameter coefficient "a" for different sizes of pipes are given below:

¼ inch	...... .0317	2½ inches	......10.37	8 inches	 198
½ inch	...... .1810	3 inches	......16.50	10 inches	 350
¾ inch	...... .5012	4 inches	......34.10	12 inches	 556
1 inch	......1.0000	5 inches	......60.00	16 inches	1160
1½ inches	....2.9300	5⅝ inches	......81.00	18 inches	1570
2 inches	....5.9200	6 inches	......95.00		

For pipes greater than 12 inches in diameter the measure is taken from the outside and for pipes of ordinary thickness the corresponding inside diameters and multipliers are as follows:

Outside	Inside	Multiplier
15 inch	14¼ inch	863
16 inch	15¼ inch	1025
18 inch	17¼ inch	1410
20 inch	19¼ inch	1860

For determining gas flow rates for specific pressures drops for pipe sizes used in production operations, Tables 3, 4, and 5 may be used. The tables are calculated by the Weymouth formula for listed inside diameters of standard weight threaded line pipe for the sizes shown.

Tables 3, 4, and 5 give the hourly rates of flow of 0.70 specific gravity gas, flowing at 60 deg. F. and measured at a standard temperature of 60 deg. F. and at a standard pressure base of 14.4 psi. + 4 oz. gauge pressure of 14.65 psi. For specific gravity of gas other than 0.70 and for flowing temperature other than 60 deg. F. correction can be made by use of the factors shown in Table 2. For instance, to determine the flow of gas of specific gravity 0.06 flowing at the temperature of 20 deg. F. the value obtained from the gas flow Tables 3, 4, and 5 would be multiplied by 1.12 the factor obtained from Table 2.

To find from Tables 3, 4, and 5, the volume of gas delivered through any length "L", the volume found from the table should be multiplied by $1/\sqrt{L}$. "L" must be expressed as a multiple of the length heading the table in use. For instance if the delivery is to be determined from a 4500 ft. line, the volume found for a 1000 ft. line in Table 3 should be multiplied by $1\sqrt{4.5}$. At the bottom of each of the two tables in question multipliers are given to be used as correction factors.

Use of Tables 3, 4, and 5 may be illustrated by the following example:

Example: What is the delivery of a 2″ gas line, 3600 ft. long with an upstream pressure of 200 psi and a downstream pressure of 50 psi? The specific gravity of gas is 0.07; the flowing temperature, 70 deg. F.
From Table 3 the volume for 1,000 ft. of line, for 0.70 sp. gravity gas and 60 deg. F. flowing temperature for 200 and 50 psi up- and down-stream pressure respectively is 109.5 MCF per Hour. Correction factor for the pipe length, from the same table is .53. Correction factor, from Table 2 for specific gravity and flowing temperature is .991. Therefore:

$$\text{Delivery of Gas} = 109.5 \times .53 \times .991 = 57{,}512 \text{ cu. ft./hour.}$$

Table 3
Volume of gas delivered—thousands of cubic feet per hour
1000 foot & 2000 foot lines

0.7 SP. GR. 60° F.

Gage Pressure Up Stream P.S.I.	Down Stream P.S.I.	1000 Foot Line: 2	2½	3	4	6	2000 Foot Line: 2	2½	3	4	6
		Nominal Pipe Size* — Thousands of Cu. Ft. per Hour					Nominal Pipe Size* — Thousands of Cu. Ft. per Hour				
400	10	221.5	355.8	635.0	1311.	3910.	156.6	251.6	449.0	926.8	2765.
	50	219.2	352.1	628.4	1297.	3869.	155.0	248.9	444.3	917.2	2736.
	100	213.2	342.5	611.4	1262.	3764.	150.8	242.2	432.3	892.4	2662.
	350	105.6	169.7	302.9	625.2	1865.	74.70	120.0	214.2	442.1	1319.
350	10	194.7	312.7	558.1	1152.	3436.	137.6	221.1	394.6	814.6	2430.
	50	192.0	308.5	550.6	1136.	3390.	135.8	218.1	389.3	803.6	2397.
	100	185.2	297.5	531.0	1096.	3270.	131.0	210.4	375.5	775.1	2312.
	300	98.63	158.4	282.8	583.7	1741.	69.74	112.0	200.0	412.8	1231.
300	10	167.8	269.6	481.2	993.1	2962.	118.7	190.6	340.2	702.3	2095.
	50	164.8	264.6	472.4	975.0	2908.	116.5	187.1	334.0	689.4	2056.
	100	156.8	251.9	449.5	927.9	2768.	110.9	178.1	317.9	656.1	1957.
	250	91.08	146.3	261.1	539.0	1608.	64.40	103.5	184.7	381.1	1137.
250	10	141.0	226.4	404.1	834.2	2488.	99.67	160.1	285.8	589.8	1759.
	50	137.3	220.5	393.6	812.5	2424.	97.08	155.9	278.3	574.5	1714.
	100	127.6	205.0	365.9	755.3	2253.	90.24	145.0	258.7	534.0	1583.
	200	82.84	133.1	237.5	490.3	1462.	58.58	94.09	167.9	346.7	1034.
200	10	114.0	183.2	327.0	674.9	2013.	80.64	129.5	231.2	477.2	1424.
	50	109.5	175.9	313.9	647.9	1933.	77.42	124.4	222.0	458.2	1367.
	100	97.08	155.9	278.3	574.5	1714.	68.65	110.3	196.8	406.3	1212.
	150	73.68	118.4	211.3	436.0	1301.	52.10	83.69	149.4	308.3	919.7
150	10	87.04	139.8	249.6	515.1	1537.	61.55	98.87	176.5	364.2	1086.
	25	85.45	137.3	245.0	505.7	1508.	60.42	97.06	173.2	357.6	1067.
	50	80.98	130.1	232.2	479.3	1430.	57.26	91.98	154.2	338.9	1011.
	100	63.21	101.5	181.2	374.1	1116.	44.70	71.80	128.2	264.5	789.0

Gage Pressure Up Stream P.S.I.	Down Stream P.S.I.	1000 Foot Line: 2	2½	3	4	6	2000 Foot Line: 2	2½	3	4	6
		Nominal Pipe Size* — Thousands of Cu. Ft. Per Hour					Nominal Pipe Size* — Thousands of Cu. Ft. per Hour				
100	10	59.84	96.12	171.6	354.1	1056.	42.31	67.97	121.3	250.4	746.9
	25	57.50	92.36	164.9	340.3	1015.	40.66	65.31	116.6	240.6	717.7
	50	50.62	81.32	145.1	299.6	893.6	35.79	57.50	102.6	211.8	631.9
	75	38.22	61.39	109.6	226.2	674.6	27.02	43.41	77.5	159.9	477.0
80	5	49.46	79.45	141.8	292.7	873.1	34.97	56.18	100.3	207.0	617.4
	10	48.82	78.43	140.0	288.9	861.8	34.52	55.46	98.98	204.3	609.4
	30	44.50	71.64	127.9	264.0	787.3	31.54	50.66	90.42	186.6	556.7
	55	34.26	55.03	98.23	202.8	604.8	24.23	38.91	69.46	143.4	427.6
60	5	38.45	61.77	110.3	227.6	678.8	27.19	43.68	77.96	160.9	480.0
	10	37.63	60.45	107.9	222.7	664.3	26.61	42.74	76.29	157.5	469.7
	20	35.32	56.73	101.3	209.0	623.5	24.97	40.12	71.61	147.8	440.9
	40	27.17	43.65	77.91	160.8	479.7	19.21	30.86	55.09	113.7	339.2
40	5	27.21	43.71	78.01	161.0	480.3	19.24	30.91	55.16	113.9	339.6
	10	26.03	41.81	74.63	154.1	549.5	18.41	29.57	52.77	108.9	324.9
	15	24.50	39.36	70.26	145.0	432.6	17.33	27.83	49.68	102.5	305.9
	25	20.08	32.26	57.58	118.8	354.5	14.20	22.81	40.71	84.04	250.7
30	5	21.38	34.35	61.30	126.5	377.4	15.12	24.29	43.35	89.48	266.9
	10	19.86	31.90	56.94	117.5	350.6	14.04	22.56	40.26	83.11	247.9
	15	17.81	28.61	51.07	105.4	314.5	12.60	20.23	36.11	74.54	222.4
	20	15.03	24.14	43.09	88.94	265.3	10.63	17.07	30.47	62.89	187.6
20	5	15.21	24.43	43.61	90.00	268.5	10.75	17.28	30.83	63.65	189.8
	10	12.98	20.85	37.22	76.83	229.2	9.180	14.75	26.32	54.33	162.1
10	5	7.923	12.73	22.72	46.89	138.9	5.603	9.000	16.06	33.16	98.90

MULTIPLIERS TO CONVERT 1000 FOOT LINE VALUES TO LENGTHS LISTED:

Length of Line—Feet	200	400	600	800	1000	1200	1400	1600	1800	2000	2200	2400	2600	2800	3000	3200	3400	3600	3800	4000	4200	4400	4600	4800	5000	5200	Feet—Length of Line
Multipliers	2.24	1.58	1.29	1.12	1.00	.91	.85	.79	.75	.71	.67	.65	.62	.60	.58	.56	.54	.53	.51	.50	.49	.48	.47	.46	.45	.44	Multipliers

For remarks regarding use of the Table refer to page 78.

Courtesy Armco Steel Corporation.

Table 4
1 mile & 2½ mile lines

0.7 SP. GR. 60° F.

Gage Pressure Up Stream P.S.I.	Down Stream P.S.I.	1 Mile Line: 2	2½	3	4	6	2½ Mile Line: 2	2½	3	4	6
		Nominal Pipe Size* — Thousands of Cu. Ft. per Hour					Nominal Pipe Size* — Thousands of Cu. Ft. per Hour				
400	10	96.39	154.8	276.4	570.4	1701.	60.96	97.92	174.8	360.8	1076.
	50	95.38	153.2	273.5	564.5	1684.	60.32	96.90	173.0	357.0	1065.
	100	92.80	149.1	266.1	549.2	1638.	58.69	94.28	168.3	347.3	1036.
	350	45.98	73.85	131.8	272.1	811.6	29.08	46.71	83.37	172.1	513.3
350	10	84.71	136.1	242.9	501.3	1495.	53.58	86.06	153.6	317.1	945.8
	50	83.57	134.2	239.6	494.6	1475.	52.85	84.90	151.5	312.8	933.0
	100	80.61	129.5	231.1	477.1	1423.	50.98	81.90	146.2	301.7	900.0
	300	42.92	68.95	123.1	254.0	757.7	27.15	43.61	77.84	160.7	479.2
300	10	73.03	117.3	209.4	432.2	1289.	46.19	74.20	132.4	273.4	815.4
	50	71.70	115.2	205.6	424.3	1266.	45.35	72.84	130.0	268.4	800.5
	100	68.23	109.6	195.6	403.8	1204.	43.15	69.32	123.7	255.4	761.8
	250	38.64	63.67	113.6	234.6	699.7	25.07	40.27	71.88	148.4	442.5
250	10	61.34	98.53	175.9	363.0	1083.	38.80	62.32	111.2	229.6	684.8
	50	59.75	95.98	171.3	353.6	1055.	37.79	60.70	108.3	223.6	667.1
	100	55.54	89.21	159.2	328.7	980.4	35.13	56.42	100.7	207.9	620.1
	200	36.05	57.91	103.4	213.4	636.4	22.80	36.63	65.37	134.9	402.5
200	10	49.63	79.72	142.3	293.7	876.1	31.39	50.42	90.00	185.8	554.1
	50	47.65	76.54	136.6	282.0	841.1	30.13	49.41	86.40	178.3	532.0
	100	42.25	67.87	121.1	250.0	745.8	26.72	42.92	76.61	158.1	471.7
	150	32.06	51.51	91.94	189.8	566.0	20.28	32.58	58.14	120.0	358.0
150	10	37.88	60.85	108.6	224.2	668.7	23.96	38.48	68.69	141.8	422.9
	25	37.19	59.74	106.6	220.1	656.5	23.52	37.78	67.44	139.2	415.2
	50	35.24	56.61	101.0	208.6	622.1	22.29	35.80	63.91	131.9	393.5
	100	27.51	44.19	78.87	162.8	485.6	17.40	27.95	49.88	103.0	307.1

Gage Pressure Up Stream P.S.I.	Down Stream P.S.I.	1 Mile Line: 2	2½	3	4	6	2½ Mile Line: 2	2½	3	4	6
		Nominal Pipe Size* — Thousands of Cu. Ft. per Hour					Nominal Pipe Size* — Thousands of Cu. Ft. per Hour				
100	10	26.04	41.83	74.67	154.1	459.7	16.47	26.46	47.22	97.47	290.7
	25	25.02	40.20	71.75	148.1	441.7	15.83	25.42	45.38	93.66	279.4
	50	22.03	35.39	63.16	130.4	388.9	13.93	22.38	39.95	82.46	246.0
	75	16.63	26.72	47.68	98.4	293.6	10.52	16.90	30.16	62.25	185.7
80	5	21.53	34.58	61.72	127.4	380.0	13.61	21.87	39.03	80.57	240.3
	10	21.25	34.13	60.92	125.7	375.1	13.44	21.59	38.53	79.53	237.2
	30	19.41	31.18	55.65	114.9	342.6	12.28	19.72	35.20	72.65	216.7
	55	14.91	23.95	42.75	88.24	263.2	9.430	15.15	27.04	55.81	166.5
60	5	16.74	26.88	47.98	99.04	285.4	10.58	17.00	30.35	62.64	186.8
	10	16.38	26.31	46.95	96.91	289.1	10.36	16.64	29.70	61.29	182.8
	20	15.37	24.69	44.07	90.96	271.3	9.721	15.62	27.87	57.53	171.6
	40	11.83	19.00	33.91	69.98	208.8	7.479	12.01	21.44	44.26	132.0
40	5	11.84	19.02	33.95	70.08	209.0	7.489	12.03	21.47	44.32	132.2
	10	11.33	18.20	32.48	67.04	200.0	7.165	11.51	20.54	42.40	126.5
	15	10.66	17.13	30.58	63.11	188.3	6.745	10.83	19.34	39.92	119.1
	25	8.740	14.04	25.06	51.72	154.3	5.527	8.879	15.85	32.71	97.57
30	5	9.315	14.95	26.68	55.07	164.3	5.885	9.453	16.87	34.83	103.9
	10	8.643	13.88	24.78	51.15	152.6	5.466	8.781	15.67	32.35	96.49
	15	7.752	12.45	22.23	45.88	136.8	4.903	7.875	14.06	29.02	86.55
	20	6.540	10.51	18.75	38.71	115.5	4.137	6.645	11.86	24.48	73.02
20	5	6.619	10.63	18.98	39.17	116.8	4.186	6.724	12.00	24.77	73.90
	10	5.650	9.076	16.20	33.44	99.74	3.573	8.740	10.25	21.15	63.08
10	5	3.448	5.539	9.886	20.41	60.87	2.181	3.503	6.253	12.91	38.50

MULTIPLIERS TO CONVERT 1 MILE LINE VALUES TO LENGTHS LISTED:

Length of Line—Miles	1.2	1.4	1.6	1.8	2.0	2.2	2.4	2.6	2.8	3.0	3.2	3.4	3.6	3.8	4.0	4.2	4.4	4.6	4.8	Miles—Length of Line
Multipliers	.91	.85	.79	.75	.71	.67	.65	.62	.60	.58	.56	.54	.53	.51	.50	.49	.48	.47	.46	Multipliers

For remarks regarding use of the Table refer to page 78.

Courtesy of Armco Steel Corporation.

Table 5
5 mile & 10 mile lines

0.7 SP. GR. 60° F.

GAGE PRESSURE Up Stream P.S.I.	GAGE PRESSURE Down Stream P.S.I.	5 MILE LINE 2	5 MILE LINE 2½	5 MILE LINE 3	5 MILE LINE 4	5 MILE LINE 6	10 MILE LINE 2	10 MILE LINE 2½	10 MILE LINE 3	10 MILE LINE 4	10 MILE LINE 6
		Thousands of Cu. Ft. Per Hour					Thousands of Cu. Ft. Per Hour				
400	10	43.10	69.24	123.6	255.1	760.9	30.48	48.96	87.39	180.4	538.0
	50	42.66	68.52	122.3	252.4	753.0	30.16	48.45	86.48	178.5	532.4
	100	41.50	66.67	119.0	245.6	732.6	29.35	47.14	84.14	173.7	518.0
	350	20.56	33.03	58.95	121.7	363,0	14.54	23.35	41.69	86.04	256.7
350	10	37.88	60.86	108.6	224.2	668.8	26.79	43.03	76.81	158.5	472.9
	50	37.37	60.03	107.2	221.2	659.7	26.43	42.45	75.77	156.4	466.5
	100	36.05	57.91	103.4	213.3	636.4	25.49	40.95	73.09	150.9	450.0
	300	19.20	30.84	55.04	113.6	338.9	13.57	21.80	38.92	80.33	239.6
300	10	32.66	52.46	93.64	193.3	576.6	23.09	37.10	66.22	136.7	407.7
	50	32.07	51.51	91.94	189.8	566.0	22.67	36.42	65.01	134.2	400.2
	100	30.51	49.02	87.49	180.6	538.7	21.58	34.66	61.86	127.7	380.9
	250	17.73	28.47	50.82	104.9	312.9	12.53	20.13	35.94	74.18	221.3
250	10	27.43	44.07	78.65	162.3	484.3	19.40	31.16	55.62	114.8	342.4
	50	26.72	42.92	76.61	158.1	471.7	18.89	30.35	54.17	111.8	333.5
	100	24.84	39.90	71.21	147.0	438.5	17.56	28.21	50.36	103.9	310.0
	200	16.12	25.90	46.23	95.41	284.6	11.40	18.31	32.69	67.47	201.2
200	10	22.20	35.65	63.64	131.4	391.8	15.69	25.21	45.00	92.88	277.0
	50	21.31	34.23	61.10	126.1	376.2	15.07	24.20	43.20	89.17	266.0
	100	18.89	30.35	54.17	111.8	333.5	13.36	21.46	38.31	79.07	235.8
	150	14.34	23.03	41.11	84.86	253.1	10.14	16.29	29.07	60.01	179.0
150	10	16.94	27.21	48.57	100.3	299.0	11.98	19.24	34.35	70.89	211.5
	25	16.63	26.72	47.68	98.42	293.6	11.76	18.89	33.72	69.60	207.6
	50	15.76	25.32	45.19	93.27	278.2	11.14	17.90	31.95	65.96	196.7
	100	12.30	19.76	35.27	72.81	217.2	8.699	13.97	24.94	51.48	153.6

0.7 SP. GR. 60° F.

GAGE PRESSURE Up Stream P.S.I.	GAGE PRESSURE Down Stream P.S.I.	5 MILE LINE 2	5 MILE LINE 2½	5 MILE LINE 3	5 MILE LINE 4	5 MILE LINE 6	10 MILE LINE 2	10 MILE LINE 2½	10 MILE LINE 3	10 MILE LINE 4	10 MILE LINE 6
		Thousands of Cu. Ft. Per Hour					Thousands of Cu. Ft. Per Hour				
100	10	11.65	18.71	33.39	68.92	205.6	8.235	13.23	23.61	48.74	145.4
	25	11.19	17.98	32.09	66.23	197.5	7.913	12.71	22.69	46.83	139.7
	50	9.852	15.83	28.25	55.31	173.9	6.966	11.19	19.97	41.23	123.0
	75	7.438	11.95	21.32	44.02	131.3	5.259	8.448	15.08	31.12	92.84
80	5	9.626	15.46	27.60	56.97	169.9	6.807	10.93	19.52	40.28	120.2
	10	9.502	15.26	27.24	56.23	167.7	6.719	10.79	19.26	39.76	118.6
	30	8.680	13.94	24.89	51.37	153.2	6.138	9.860	17.60	36.32	108.4
	55	6.668	10.71	19.12	39.46	117.7	4.715	7.574	13.52	27.90	83.23
60	5	7.484	12.02	21.46	44.29	132.1	5.292	8.501	15.17	31.32	93.42
	10	7.324	11.76	21.00	43.34	129.3	5.179	8.319	14.85	30.65	91.42
	20	6.873	11.04	19.71	40.68	121.3	4.861	7.808	13.94	28.77	85.80
	40	5.288	8.495	15.16	31.30	93.36	3.740	6.007	10.72	22.13	66.01
40	5	5.296	8.507	15.18	31.34	93.48	3.745	6.015	10.74	22.16	66.10
	10	5.066	8.138	14.53	29.98	89.43	3.582	5.755	10.27	21.20	63.24
	15	4.769	7.661	13.67	28.22	84.19	3.372	5.417	9.669	19.96	59.53
	25	3.908	6.278	11.21	23.13	69.00	2.764	4.439	7.924	16.36	48.79
30	5	4.161	6.685	11.93	24.63	73.46	2.943	4.727	8.437	17.41	51.94
	10	3.865	6.209	11.08	22.87	68.23	2.733	4.390	7.836	16.17	48.25
	15	3.467	5.569	9.940	20.52	61.20	2.451	3.938	7.029	14.51	43.27
	20	2.925	4.698	8.386	17.31	51.63	2.068	3.322	5.930	12.24	36.51
20	5	2.960	4.755	8.487	17.52	52.25	2.093	3.362	6.001	12.39	36.95
	10	2.527	4.059	7.244	14.95	44.60	1.787	2.870	5.123	10.57	31.54
10	5	1.542	2.477	4.421	9.126	27.22	1.090	1.752	3.126	6.453	19.25

MULTIPLIERS TO CONVERT 10 MILE LINE VALUES TO LENGTHS LISTED.

Length of Line—Miles..........	5.2	5.4	5.6	5.8	6.0	6.2	6.4	6.6	6.8	7.0	7.2	7.4	7.6	7.8	8.0	8.2	8.4	8.6	8.8	9.0	9.2	9.4	9.6	9.8	Miles—Length of Line
Multipliers.......	1.39	1.36	1.34	1.31	1.29	1.27	1.25	1.23	1.21	1.20	1.18	1.16	1.15	1.13	1.12	1.10	1.09	1.08	1.07	1.05	1.04	1.03	1.02	1.01	Multipliers

For remarks regarding use of the Table refer to page 78.

Courtesy of Armco Steel Corporation.

Multi-phase flow

As stated at the beginning of this chapter, the problems of the simultaneous oil flow and gas or of oil, gas, and water through one pipe have become increasingly important in oil field production operations. The problems are complex because the properties of two or more fluids must be considered and because of the different patterns of fluid flows, depending upon the flow conditions. These patterns, in any given line, may change as the flow conditions change, and they may be coexistent at different points of the line.[3,5,7]

Different investigators recognize different flow patterns and use different nomenclature. Those generally recognized are:

1. Bubble flow—bubbles of gas flow along the upper part of the pipe with about the same velocity as the liquid.
2. Plug flow—the bubbles of gas coalesce into large bubbles and fill out the large part of the cross-sectional area of the pipe.
3. Laminar flow—the gas-liquid interface is relatively smooth with gas flowing in the upper part of the pipe.
4. Wavy flow—same as above except that waves are formed on the surface of the liquid.
5. Slug flow—the tops of some waves reach the top of the pipe. These slugs move with high velocity.
6. Annular flow—the liquid flows along the walls of the pipe and the gas moves through the center with high velocity.
7. Spray flow—the liquid is dispersed in the gas.

The above description of flow patterns has been given to emphasize the first reason why the pressure drop must be higher for the multi-phase than the single-phase flow. In the latter the pressure drop is primarily the result of friction. In the multi-phase flow, in addition to friction, the energy losses are due to gas accelerating the fluid, due to compression of gas and formation of waves and slugs. These losses are different for different patterns, with patterns changing as a result of changes in flow conditions. The second reason is the fact that with two phases present in the pipe, less cross-sectional area is available for each of the phases. As previously discussed, the pressure drop is inversely proportional to the fifth power of pipe diameter.

The literature on multi-phase flow is quite extensive. Reference 7 gives 72 references on the subject.

Two-Phase Horizontal Flow

Over twenty correlations have been developed by different investigators for predicting the pressure drop in the two-phase flow. Under the project, sponsored jointly by the American Petroleum Institute and American Gas Association at the University of Houston, five of these correlations were tested by comparing them with 2,620 experimental measurements culled from more than 15,000 available.

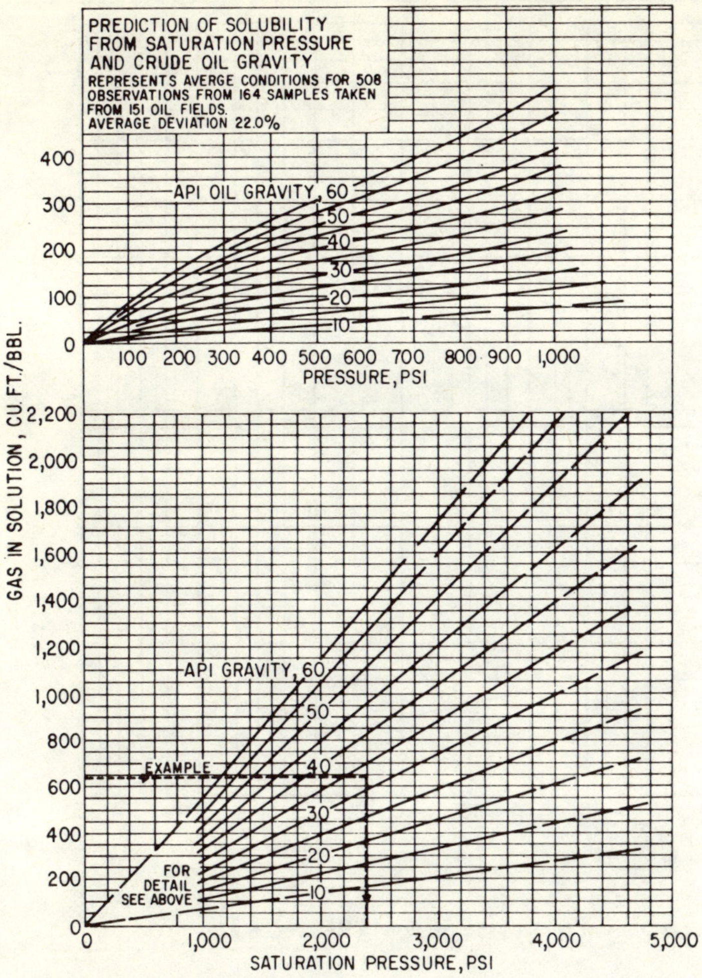

Figure 1. Prediction of solubility from saturation pressure and gravity of crude oil. (Reference 12.)

The statistical part of the report on the project concludes that of the five correlations tested, the Lockhart-Martinelli correlation shows the best agreement with the experimental data, particularly for pipe sizes commonly used in oil field production operations.[8] The method may be summarized as follows:[3,4]

1. Determine the single-phase pressure drops for the liquid and the gas phase as if each one of these were flowing alone through the pipe. This is done by use of previously given equations.

2. Determine the dimensionless parameter $X = \sqrt{\frac{\sqrt{\Delta P_L}}{\Delta P_G}}$ where ΔP_L and ΔP_G are single-phase pressure drops for liquid and gas respectively.

3. The method recognizes four regimes of the two-phase flow as follows:

Flow Regimes

Gas	Liquid
Turbulent	Laminar
Turbulent	Turbulent
Laminar	Laminar
Laminar	Turbulent

The type of flow of each of the phases is determined by its Reynolds number (Figure 2).

4. Determine Factor Φ from a chart as a function of X for the appropriate flow regime.

5. The two-phase pressure drop is then:

$$\Delta P_{TP} = \Phi^2 P_L \text{ of } G$$

The single-phase pressure drop of either gas or liquid phase can be used. The results are the same.

For the two-phase flow problems in oil field production operations, certain simplifications can be used if only approximate estimates are desired:

1. Information which would permit determining the volume of oil and gas under flow conditions from equilibrium flash calculation is usually unavailable. The following procedure may be used:

Gas volume is determined by use of the chart in Figure 1 which shows gas in solution for oil of different gravities and different saturation pressures (in this case the flowline pressures). If the gas-oil ratio is known for one pressure the ratio for another pressure can be determined from the chart as follows:

Assume gas oil ratio of 600 cu. ft./bbl. at 900 psi for a 30° API oil. Determine the gas oil ratio at 1300 psi. From the chart, at 900 psi, the gas in solution is 205 cu.ft./bbl. and at 1300 psi is 300 cu.ft./bbl., an increase of 95 cu.ft./bbl. Therefore, at 1300 psi the gas oil ratio will be 600 - 95 = 505 cu.ft./bbl.

Increase in volume of oil because of the increase of the dissolved gas can be obtained from charts for calculation of formation volume of bubble point liquids. However, for approximate estimates in problems involving field flowlines, this step frequently can be omitted without seriously affecting the validity of results.

2. In a majority of cases here under consideration, the flow regime is turbulent-turbulent. Therefore, the chart in Figure 2 shows the factor for only this regime.

The procedure is illustrated by the following examples:

Example 1. Determine two-phase pressure drops for different flow ratios and different flow pressures under the following conditions:

Oil	31.° API
Viscosity	5 c.p. at 100° F.
Separator pressure	900 psig
Gas oil ratio	670 cu.ft./bbl.
Flow line	2 inches

Wellhead pressure of 1300 psig was assumed and calculations were made for flow rates of 100 and 50 bbls. per day. The single phase pressure drops were calculated for the gas and liquid phases, each of them flowing alone in the line, according to the Fanning equations:

Calculation of pressure drop for liquid phase:

$$H_f = \frac{f\ L\ (\text{bbls/day})^2\ (\text{lbs/gal})}{181{,}916\ D^5}$$

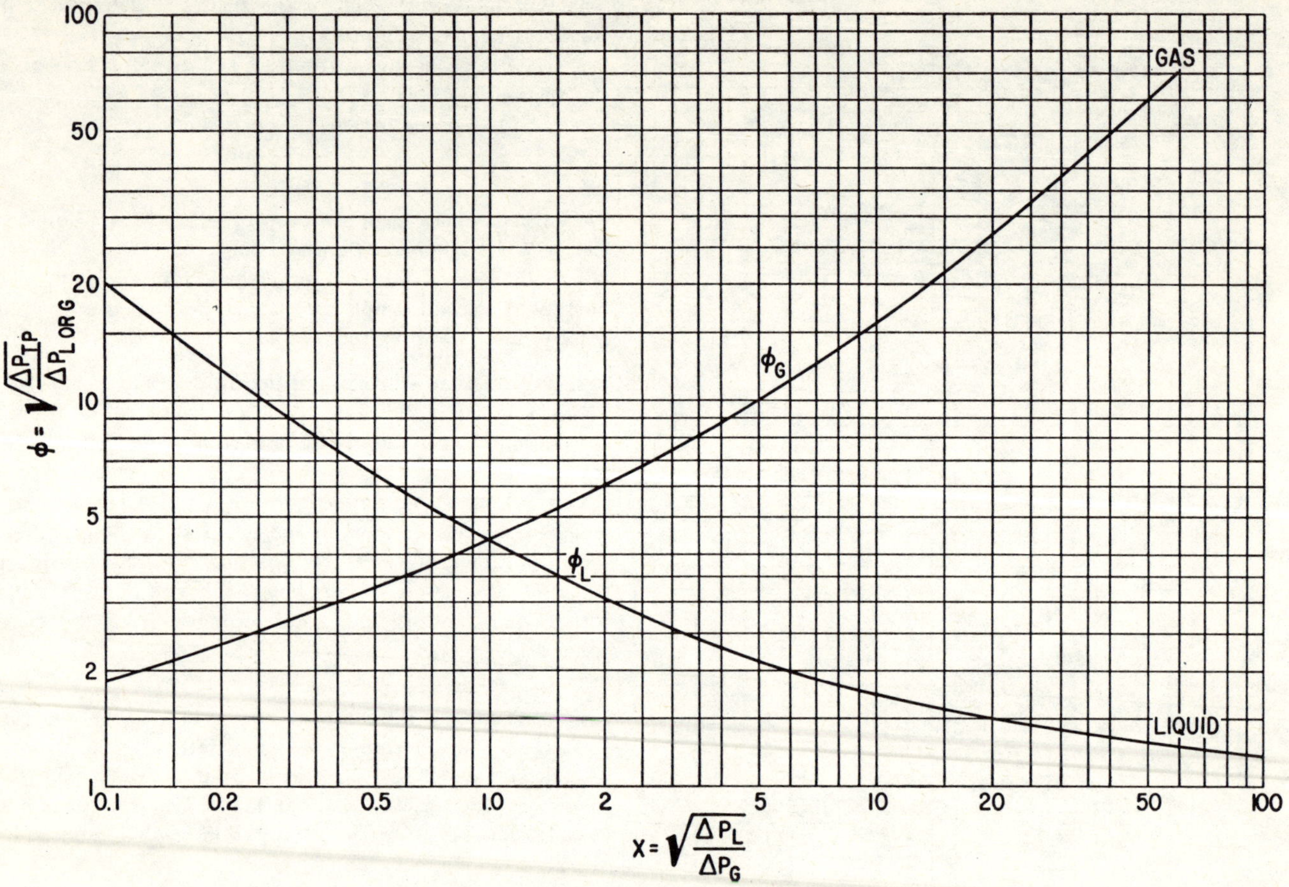

Figure 2. Parameter Φ for Lockhart-Martinelli correlation for two-phase flow. (Reference 9.)

where

f = friction factor

L = length of pipe, feet

D = inside diameter of pipe, inches

Calculation of pressure drop for gas phase: see pages 76-77.

For the two flow rates and the two pressures, these single-phase pressure drops were calculated to be as follows:

Flow Rate bbls./day	**100**		**50**	
Pressure, psia	1,300	900	1,300	900
ΔP_G psi/100 ft.	7.9×10^{-4}	2.0×10^{-3}	2.4×10^{-4}	7.7×10^{-4}
ΔP_L psi/100 ft.	$2. \times 10^{-2}$	2.7×10^{-2}	1.3×10^{-2}	1.3×10^{-2}
Then				
$X = \sqrt{\frac{\Delta P_L}{\Delta P_G}}$	5.9	3.8	7.4	4.2
From Figure 2 ϕ	2.0	2.4	1.9	2.3
and the two-phase pressure drop				
$\Delta P_{TP} = \phi_L \times \Delta P_L$	0.12	0.16	0.048	0.069

With two pressure drops determined for two flow line pressures for each of the flow rates, lines can be drawn showing pressure drop for different flow line pressures for these two rates (Figure 3). From these two lines relationships can be established for pressure drops for other rates and flow line pressures. The procedure is as follows:

Start with, for instance, 800 psia line. For the rate of 50 b.p.d. at 800 psia flow line pressure, the pressure drop from the chart is .074 psia/100 ft. From the 50 b.p.d. rate on the scale in the upper part of the chart, go downward to the line representing .074 psia. pressure drop. This gives one point of the 800 psia line.

For the rate of 100 b.p.d. the chart indicates that at 800 psia flowline pressure, the pressure drop is .175 psia/100 ft. From the 100 b.p.d. rate in the upper part of the chart, go downward to the .175 psia/100 ft. line. This gives the second point of the 800 psia line. The line can now be drawn through these two points. Lines for other pressures can be constructed in a similar way.

The chart is used as follows: For conditions represented by this particular chart what is the two-phase pressure drop for the rate of 60 b.p.d. and average flow line pressure of 1,000 psia?

Start at the rate of 60 b.p.d. on the upper scale of the chart. Go downward to the intersection with the 1,000 psia pressure, then to the left, and read the two-phase pressure drop as .075 psia/100 ft.

Example 2. If pipe diameter is to be determined for the desired pressure drop and the known length of line, flow rate,

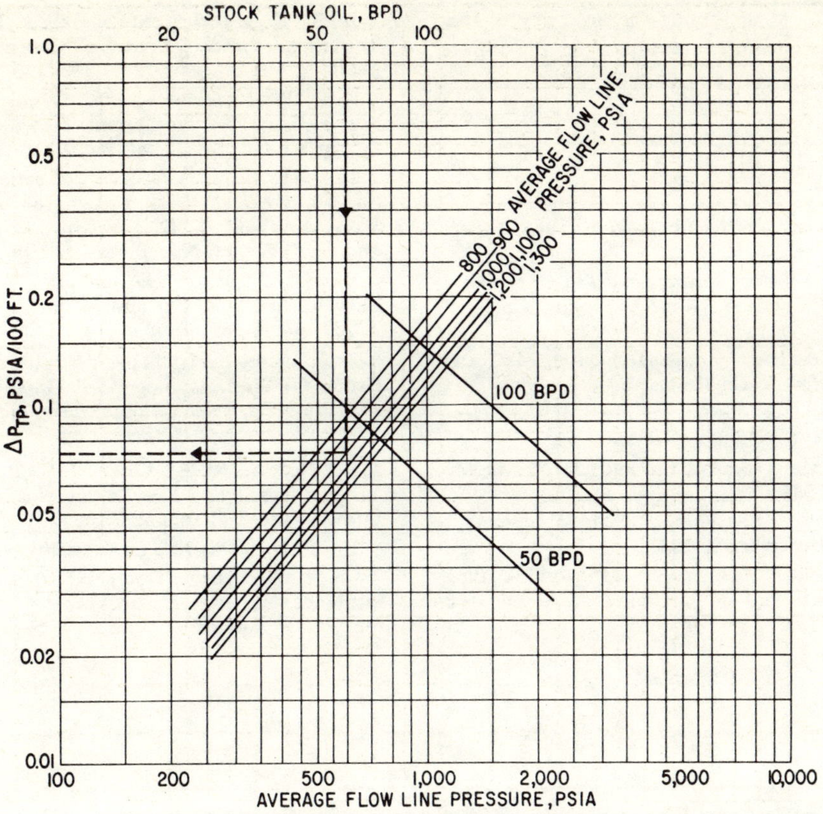

Figure 3. Determining two-phase flow pressure drops.

and the properties of fluids, then the problem cannot be solved directly. The procedure is as follows:

An estimate is made of the diameter of the two-phase line. The best approach is to determine, for the conditions, the diameter of the line for each of the phases flowing separately. In a majority of cases, the sum of these two diameters will be a good estimate of the diameter for the two-phase line. Calculations are then made for the two-phase pressure drop using the assumed diameter. If this drop is reasonably close to the desired one, the estimate is correct. If not, a new estimate must be made and the calculations repeated.

Inclined two-phase flow

In hilly terrain additional pressure drop can be expected in case of the two-phase flow. Little published information is available on the subject. Ovid Baker suggested the following empirical formula:[10]

$$\Delta P_{TPH} = \Delta P_{TP} + \frac{1.61\ NH\ \rho_L}{144\ V_g^{0.7}}$$

where

ΔP_{TPH} = pressure drop, corrected for hilly terrain in psi.
ΔP_{TP} = calculated two-phase pressure drop for horizontal pipe, psi.
N = number of hills.
H = average height of hills, ft.
ρ_L = density of liquid phase, lbs/ft.^{3}
V_g = velocity of gas, ft./sec. (based on 100 percent pipe area).

Horizontal three-phase flow

Very little is known about the effect on the pressure drop of addition to the gas-oil system of a third phase, an immissible liquid, water. Formation of an emulsion results increased viscosities. Formulas are available for approximate determination of viscosity of emulsion, if viscosity of oil and percent of water content of oil is known. The point is, however, that it is not known what portion of the water is flowing in emulsified form.

From one published reference[11] and from some unpublished data, the following conclusions appear to be well founded:

1. Within the range of water content of less than 10 percent or more than 90 percent, the flow mechanism appears to approach that of the two-phase flow.
2. Within the range of water content of from 70 to 90 percent, the three-phase pressure drop is considerably higher than for the two-phase flow.

For solution of three-phase flow problems of oil field flow lines, the following approach has been used. The oil and water are considered as one phase and the gas as the other phase. Previously given calculation of a two-phase drop is used. The viscosity of the oil-water mixture is determined as follows:

$$\mu_L = \frac{V_o \times \mu_o + V_w \times \mu_w}{V_o + V_w}$$

where μ_L, μ_o and μ_w are viscosities of mixture of oil and of water respectively, and V_o and V_w represent the corresponding volumes of oil and water.

No test data are available to determine the accuracy of this approach. Until more information is developed on the subject the method may be considered for approximate estimates.

References

1. *Cameron Hydraulic Data,* Thirteenth Edition, 1962. Ingersoll-Rand Company, New York, 1962.

2. *Flow of Fluids Through Valves, Fittings and Pipe,* Technical Paper No. 410, Engineering Division, Crane Company, Chicago, 1957.

3. Victor L. Streeter: *Handbook of Fluid Dynamics,* McGraw-Hill Book Company, Inc., New York, 1961.

4. Prof. W. E. Durand's Discussions in *Handbook of the Petroleum Industry,* David T. Day; John Wiley & Sons, Inc.

5. Ovid Baker: "Simultaneous Flow of Oil and Gas," *The Oil and Gas Journal,* July 26, 1954.

6. Dr. John M. Campbell: "Elements of Field Processing," *The Oil and Gas Journal,* December 9, 1957.

7. Dr. John M. Campbell: "Problems of Multi-phase Pipe Line Flow," in "Flow Calculations in Pipelining," *The Oil and Gas Journal,* November 4, 1959.

8. A. E. Duckler, Moye Wicks, III, and R. G. Cleveland, "Frictional Pressure Drop in Two-Phase Flow, *A.I.Ch.E. Journal,* January, 1964.

9. R. W. Lockhart and R. C. Martinelli: "Proposed Correlation of Data for Isothermal Two-Phase, Two-Component Flow in Pipes," *Chemical Engineering Progress* 45: 39-48, (1949).

10. Ovid Baker's discussion of article "How Uphill and Downhill Flow Affect Pressure Drop in Two-Phase Pipe Lines," by W. E. Brigham, E. D. Holstein and R. L. Huntington; *The Oil and Gas Journal,* November 11, 1957.

11. D. P. Sobocinski and R. D. Huntington: "Concurrent Flow of Air, Gas, Oil and Water in a Horizontal Line," *Trans-ASME 80:* 1,252, 1958.

12. Carlton Beal: "The Viscosity of Air, Water, Natural Gas, Crude Oil and Its Associated Gases at Oil Field Temperatures and Pressures," *Petroleum Development and Technology,* AIME, 1946.

13. F. G. Switzer: *Pump Fax*—Goulds Pumps, Inc.

14. *Pump Talk,* The Engineering Department, Dean Brothers Pumps, Inc., Indianapolis, Ind.

15. F. G. Switzer: *The Centrifugal Pump,* Goulds Pumps, Inc.

16. A. A. Zalis: "Don't Be Confused by Rotary Pump Performance Curves," *Hydrocarbon Processing and Petroleum Refiner,* Gulf Publishing Company, September, 1961.

Nomograph for calculating Reynolds number for compressible flow friction factor for clean steel and wrought iron pipe

Background

The nomograph (Figure 1) permits calculation of the Reynolds number for compressible flow and the corresponding friction factor. It is based on the equation:

$$Re = 6.31 \frac{W}{d\mu} = 0.482 \frac{q_s \times Sg}{d \cdot \mu}$$

where:

W = flow rate, lbs./hr.
Sg = specific gravity of gas relative to air (from Figure 3.1)
q_s = volumetric flow rate, cu. ft./sec. (at 14.7 psia and 60° F)
μ = fluid viscosity, centipoise

If the flow rate is given in lbs./hr., the nomograph can be used directly without resorting to Figure 2 to obtain q_s. Figure 2 converts volume flow to weight flow rates if the specific gravity of the fluid is known.

Example

Natural gas at 250 psig and 60° F with a specific gravity of 0.75, flows through an 8-inch schedule 40 clean steel pipe at a rate of 1,200,000 cu. ft./hr. at standard conditions.

Find the Reynolds number, and the friction factor.

At Sg = 0.75 obtain q = 69,000 from Figure 1, μ = 0.011.

Connect	With	Mark or Read
W = 69,000	μ = 0.011	Index
Index	d = 8 inches	Re = 5,000,000
Re = 5,000,000	horizontally to 8 inches	f = 0.014

Source

"Flow of Fluids Through Valves, Fittings, and Pipes," *Technical Paper No. 410,* 3-19, Crane Company, Chicago, Illinois (1957).

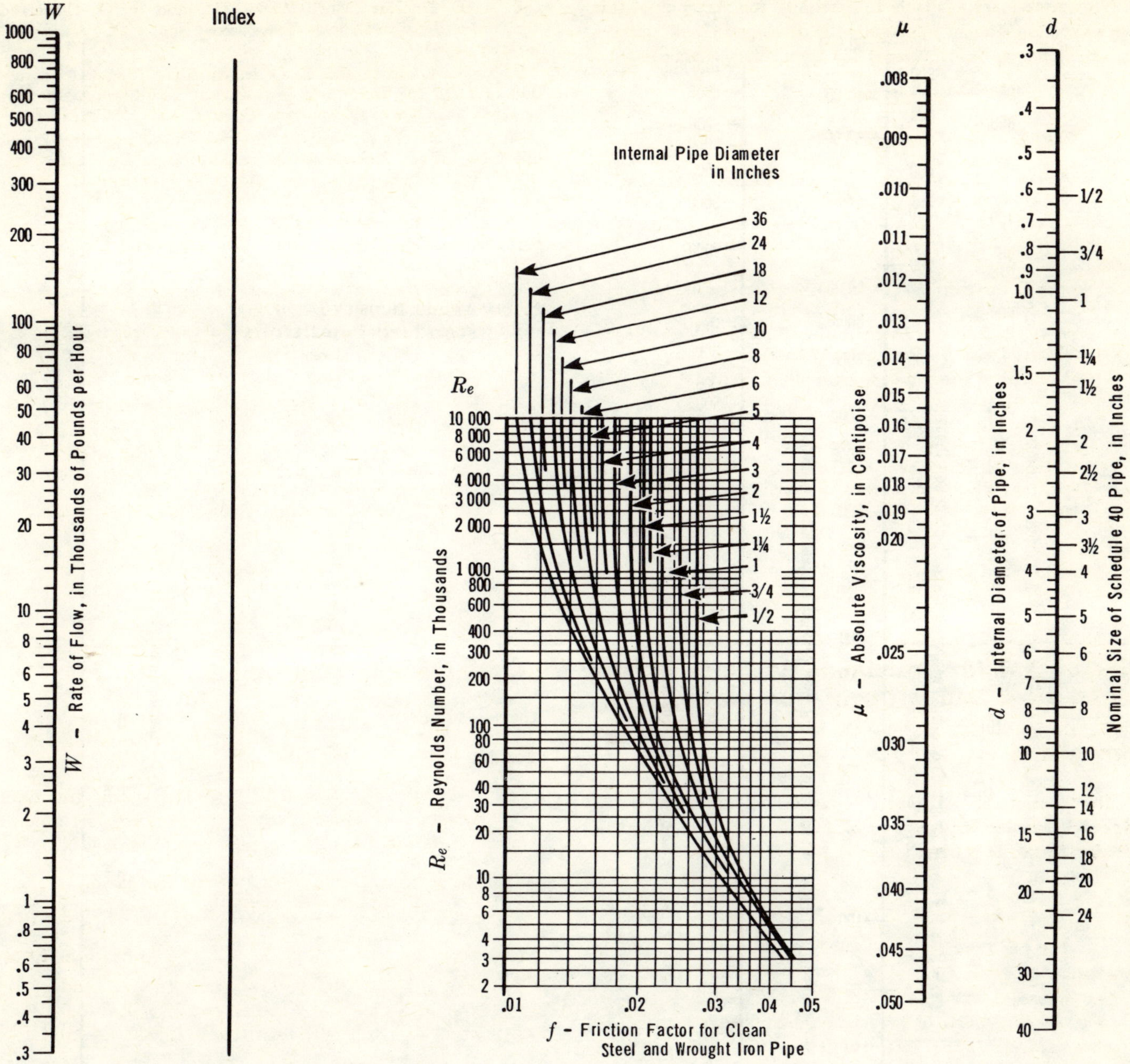

Figure 1. Reynolds number for compressible flow friction factor for clean steel and wrought iron pipe.

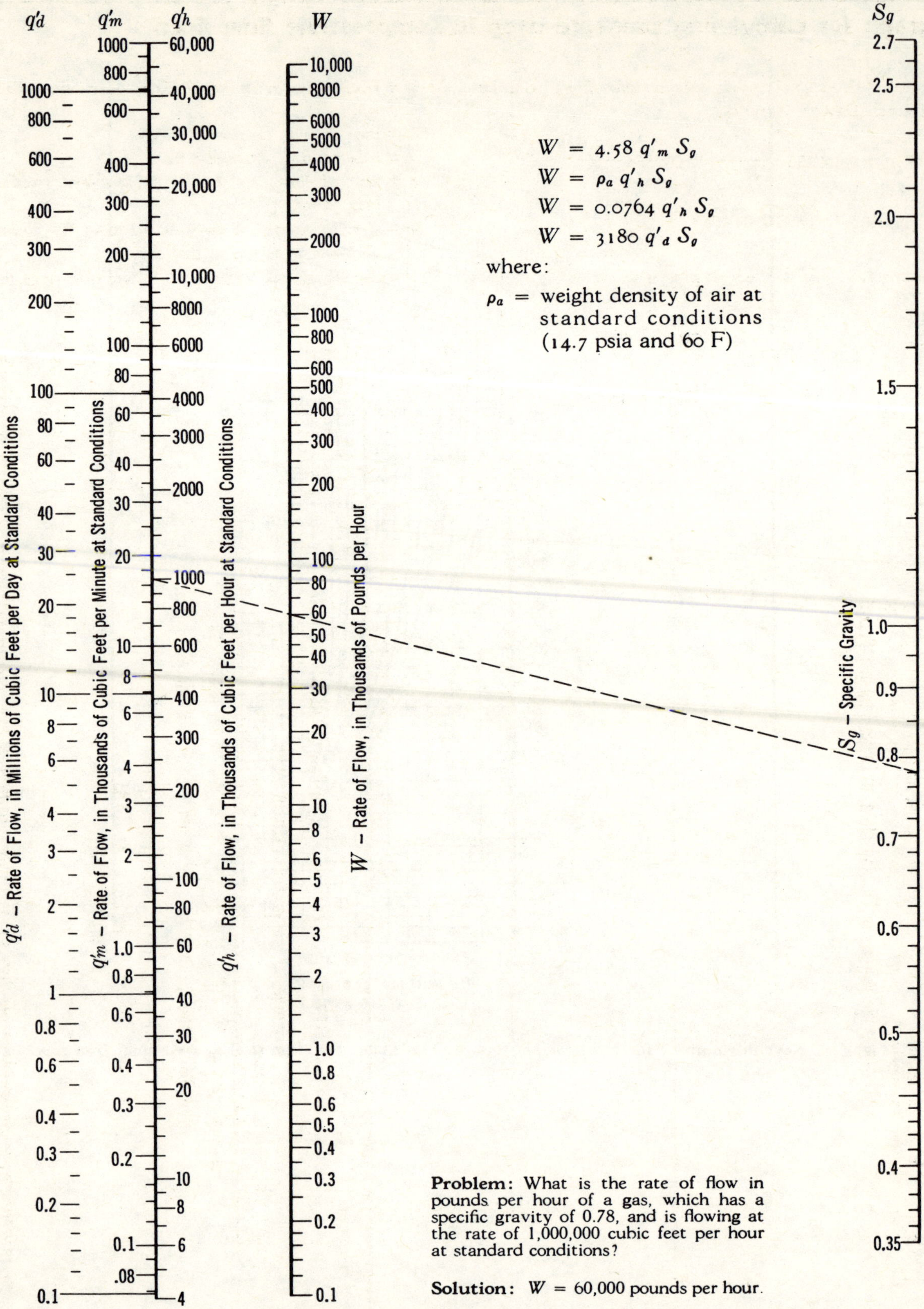

$$W = 4.58\, q'_m S_g$$
$$W = \rho_a q'_h S_g$$
$$W = 0.0764\, q'_h S_g$$
$$W = 3180\, q'_d S_g$$

where:

ρ_a = weight density of air at standard conditions (14.7 psia and 60 F)

Problem: What is the rate of flow in pounds per hour of a gas, which has a specific gravity of 0.78, and is flowing at the rate of 1,000,000 cubic feet per hour at standard conditions?

Solution: W = 60,000 pounds per hour.

Figure 2. Equivalent volume and weight flow rates of compressible fluids.

Nomograph for calculating pressure drop in compressible flow lines

The pressure drop of flowing compressible fluids can be calculated from Darcy's formula.

$$P_{100} = 0.000336 \frac{f\,w^2\,\overline{v}}{d^5} = 0.000336 \frac{f\,w^2}{d^5}$$

$$P_{100} = 0.000\,000\,01959 \frac{f\,(q^1)^2\,Sg^2}{d^5\,\rho}$$

where:

P_{100} = pressure drop per 100 feet of pipe, psig

f = friction factor to be obtained from the nomograph on page 134.

W = rate of flow, lbs./hr.

d = internal pipe diameter, in.

q^1 = rate of flow, cu. ft./hr. at standard conditions (14.7 psig, 60° F)

Sg = specific gravity of gas relative to air (= ratio of molecular weight of gas to that for air)

ρ = density of gas lbs./cu. ft. (Figure 1, page 69)

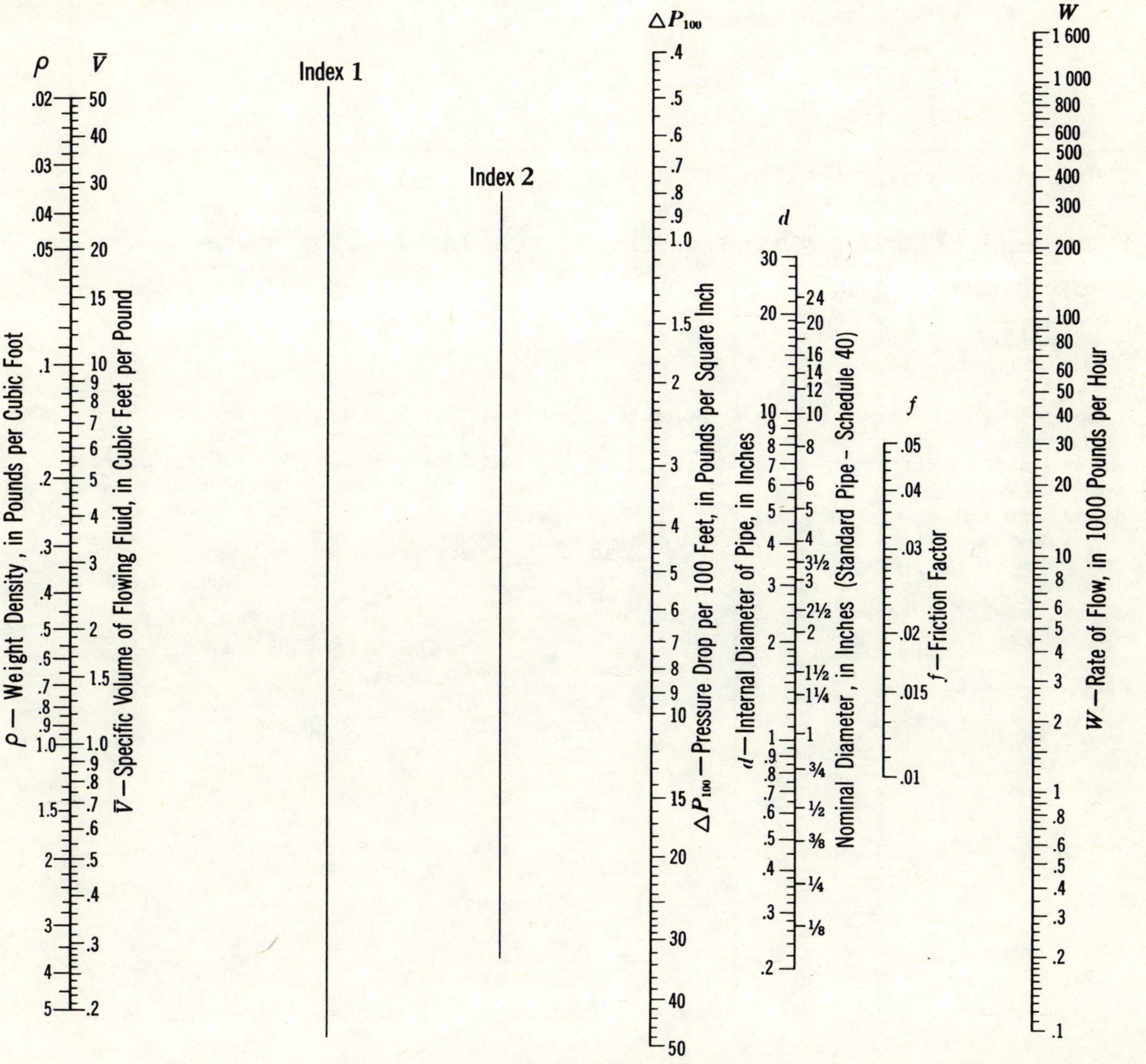

Figure 1. Pressure drop in compressible flow lines.

The pressure drop can be obtained from the nomogram (Figure 1) for any of the formulas shown above.

Example

Natural gas at 250 psig and 60° F flows through an 8-inch schedule 40 pipe at a rate of 1,200,000 cu. ft./hr. (14.7 psia and 60° F), its specific gravity is 0.75. Find the flow rate in lbs./hr. and the pressure drop per 100 feet of pipe.

1,200,000 standard cu. ft./hr. = 69,000 lbs./hr. (from Figure 2, page 86)
μ = 0.011 available data
f = 0.014 from Figure 2, page 134
ρ = 1.03 available data

Connect	With	Mark or Read
W = 69,000	d = 8-inch schedule 40	Index 2
Index 2	f = 0.014	Index 1
Index 1	ρ = 1.03	P_{100} = 0.68 lbs./sq. in.

Source

"Flow of Fluids Through Valves, Fittings, and Pipes," *Technical Publication No. 410,* 3-20, Crane Company, Chicago, Illinois, (1957).

Sizing control valves for thruput

These useful charts give data essential to calculation of pressure drops through regulating stations

W. G. Birkhead, Gas Measurement Engineer
United Gas Pipe Line Co., Shreveport, La.

PRESSURE LOSSES through piping fittings, once given little consideration in the sizing of control valves, have become important as more efficient control valves have been developed. With the development of high capacity valves and the desire for a more exacting flow formula and efficiency, designers and users look to consideration of minor losses through elbows, tees, swages or reduced ported valves. The capacity of control valves and regulators was once calculated using the basic hydraulic head equation with an assumed efficiency. This is no longer the case.

The head loss in any closed filled conduit has been accepted as proportional to its length, diameter and fluid velocity head, and can be calculated as

$$H_L = f \frac{L}{D} \frac{V^2}{2g} \qquad (1)$$

where:

H_L = Head loss of flowing fluid (feet of fluid)
f = Friction factor
L = Length of pipe, feet
D = Diameter of pipe, feet
V = Average fluid velocity, feet/second
g = Acceleration of gravity, feet/second/second

Sometimes L/D is referred to as the "equivalent length in diameters." Valves, fittings, etc., have often been expressed in equivalent diameters. If the friction factor f is combined with L/D, a constant K is derived for any pipe, valve or fitting. If $K = f(L/D)$, then

$$H_L = K\left(\frac{V^2}{2g}\right) \qquad (2)$$

The term K is called the "head loss factor" or "pressure drop factor."

Most control valves are rated with a capacity term called C_v. C_v is the number of gallons of water per minute that will flow through the valve with a one pound per square inch pressure drop across the valve.

$$C_v = Q/\sqrt{\Delta P} \qquad (3)$$

Q = Quantity of water in gpm
ΔP = Pressure drop in psi

Rearranging Equation 3:

$$\Delta P = Q^2/(C_v)^2 \qquad (3a)$$

This equation may be rearranged so that C_v in terms of our K factor may be expressed as:

$$C_v = \frac{29.9\,(d)^2}{\sqrt{K}} \qquad (4)$$

where d = inside diameter of the pipe or fitting, inches.

The head loss or pressure drop of a fluid flowing through various pipes, valves, or fittings in series can be stated:

$$(\Delta P)_{Total} = (\Delta P)_1 + (\Delta P)_2 + (\Delta P)_3 + \text{——} \qquad (5)$$

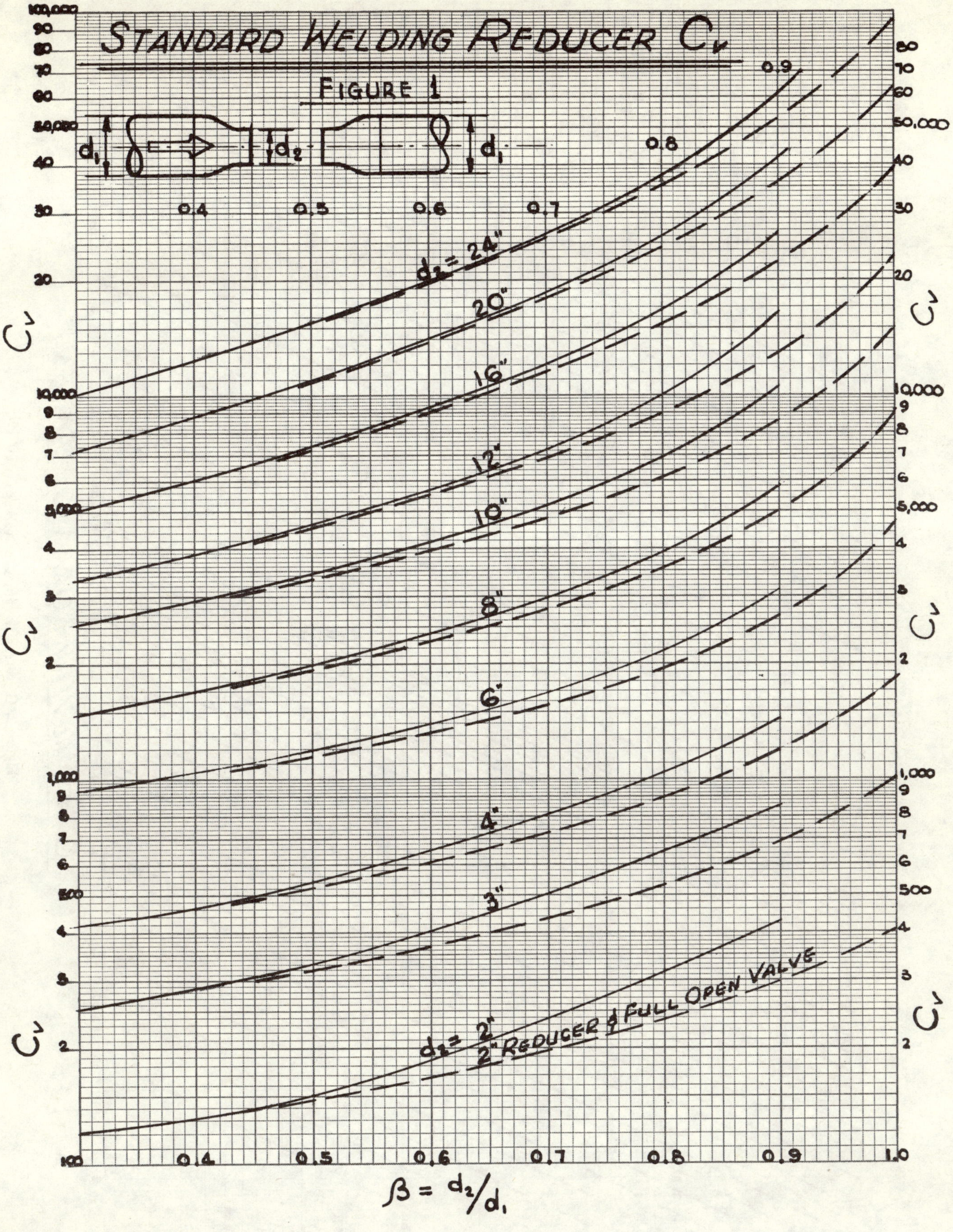

Figure 1. C_v for standard welding reducers.

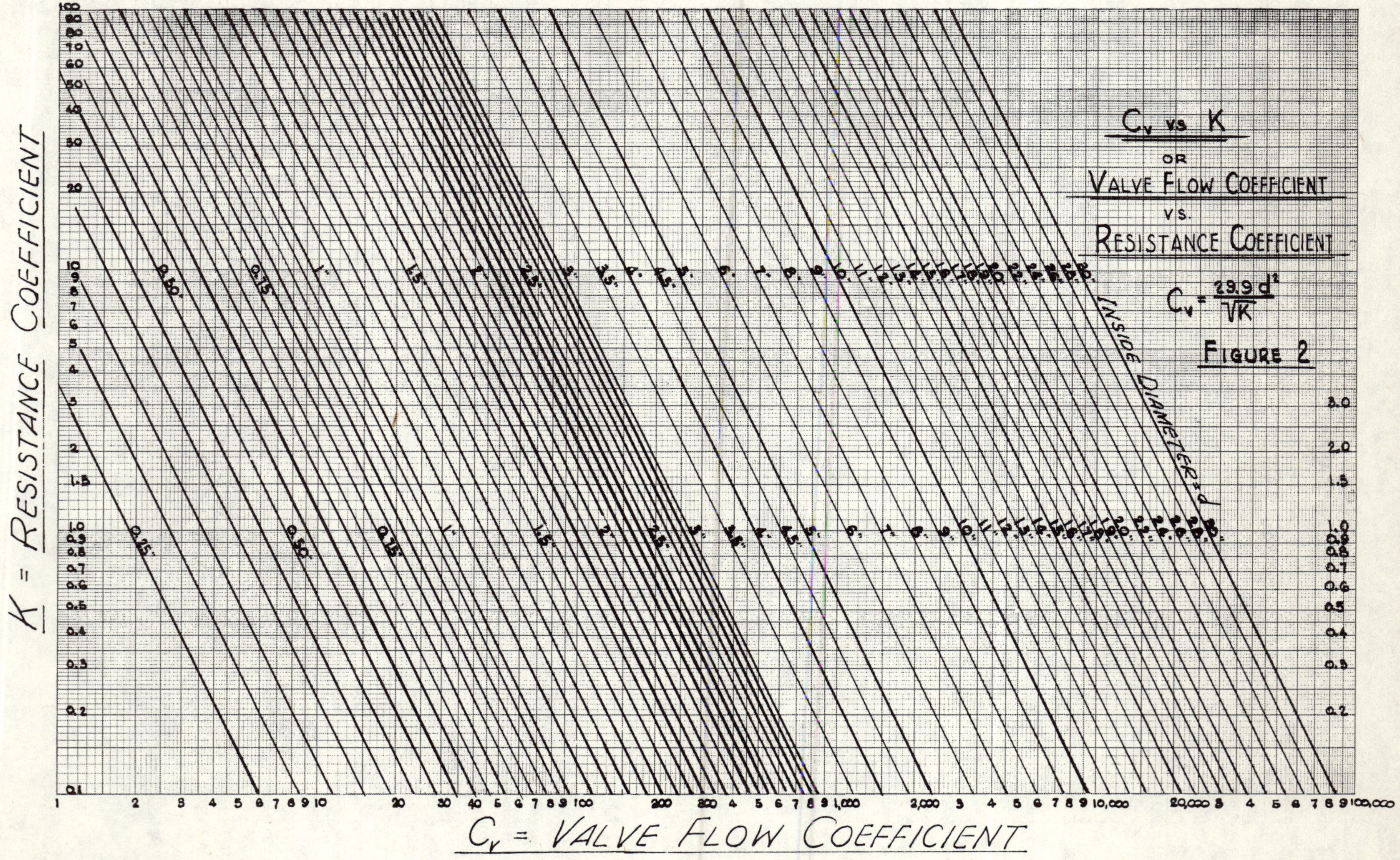

Figure 2. C_v versus K values for valves.

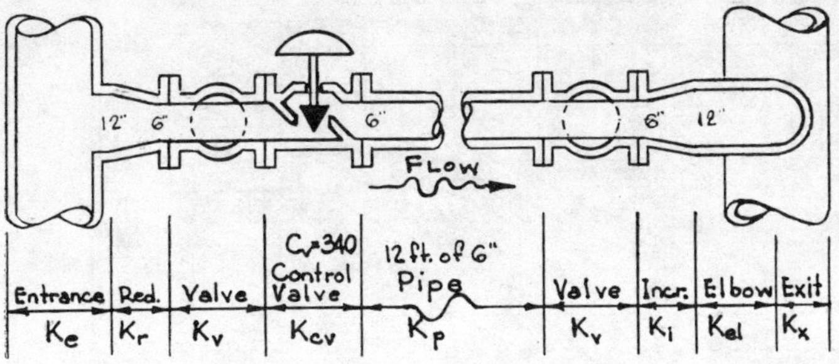

Figure 3. Piping configuration.

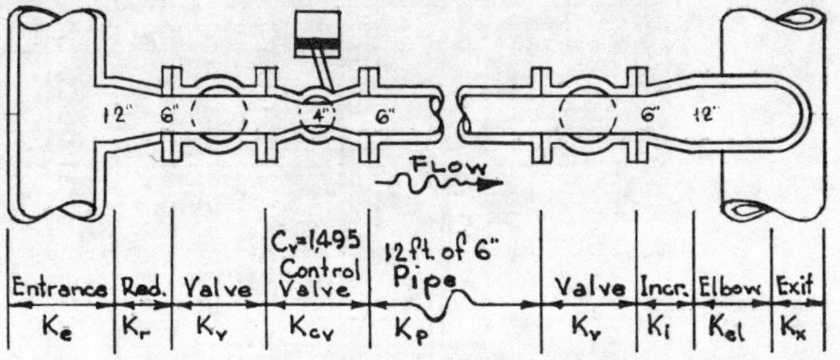

Figure 4. Piping configuration.

Substituting Equation 3a in Equation 5,

$$\frac{1}{(C_v)^2_{\text{Total}}} = \frac{1}{(C_v)_1^2} + \frac{1}{(C_v)_2^2} + \frac{1}{(C_v)_3^2} + \text{——} \qquad (6)$$

This equation is unwieldy if several fittings are used. By equating Equation 4, substituting in Equation 6 and dividing through by $[29.9(d)^2]^2$:

$$K_{\text{Total}} = K_1 + K_2 + K_3 + \text{——} = \Sigma K \qquad (7)$$

Substituting in Equation 4 for K_T an over-all C_v can be derived:

$$C_v = \frac{29.9\ (d^2)}{\sqrt{\Sigma K}}$$

K Factors for Various Valves and Fittings

For full open valves, pipes or fittings, the K factor is approximately equal to $f(L/D)$ or Equation 1.

Where

L = Equivalent length of valve or pipe or fitting in inches (or feet)

D = Diameter of the valve in inches (or feet)

f = Friction factor

Some of the recognized K factors for various welding fittings are shown in Table 1.

Other fitting K factors can be obtained from any fluid mechanics, hydraulics or piping handbook.

Low, average and high values of K given in Table 1 point out caution necessary in selecting specific types of fittings. Pressure losses in valves and fittings can be more accurately calculated by consulting the references.

All K factors, L/D's or C_v's must be converted to the same basis using the same diameter d when using an over-all C_v for a piping run. This can be accomplished with the following equations:

$$K_a = K_b\ (f_a/f_b)\ (d_a/d_b)^4 \qquad (8)$$

$$(L/D)_a = (L/D)_b\ (d_a/d_b)^1 \qquad (9)$$

$$(C_v)_a = (C_v)_b\ \ (d_b/d_a)^2 \qquad (10)$$

Table 1

K Factor			
Low	**Average**	**High**	**Type of Welding Fittings and Valves**
0.10	0.20	0.40	90° Elbow
0.05	0.12	0.25	45° Elbow
0.25	0.50	1.00	Tee (Enter run, out branch)
0.75	1.00	1.20	Tee (Enter branch, out run)
0.05	0.10	0.15	Tee (Enter run, out run)
	See Figs. 1 and 2		Standard Reducer-Increaser
0.25	0.50	1.50	Plug Valves (See manufacturer's bulletin)

EXAMPLES—Refer to Fig. 3 for a schematic of a typical control valve station using an old style control valve. Flow is from a header through a 12-inch by 6-inch reducer, a 6-inch isolating valve, 6-inch globe body control valve, 6-inch pipe spool, a second isolating valve, a 6-inch by 12-inch increaser, a 12-inch welding elbow and into a header. Note should be made of Fig. 1 in that the C_v from the solid curve is for the complete reducer-increaser combination with regard to the ratio of diameters. In this case, the Beta ratio is 6/12 or 0.5 and C_v is 1,200 for the small diameter of 6-inch. Since C_v's cannot be simply added, they must be converted to K factors. To do this, refer to Fig. 2, which converts C_v to K.

Also, to be considered are the isolating valves and control valves. Resistance coefficients, C_v's or pressure drop factors are published by most valve manufacturers and may be converted to a K factor by using Fig. 2. In this case, 6-inch K_v is 0.05.

Pipe resistance must also be considered (Fig. 5). The L/D ratio must be calculated with reference to the same units of L and D. In the example, 12 feet of 6-inch pipe has a K_p equal to 0.36.

All resistances must be converted to the same base diameter (in this case 6 inches). The entrance factor, K_e is 0.05 for 12-inch. Referring to Fig. 6, the 6-inch K_e is about 0.04; the 12-inch elbow coefficient converted to 6-inch, 6-inch K_{el} is 0.02, and the 12-inch exit coefficient, 6-inch K_x, is 0.08. A tabulation and sum of these coefficients as determined from Fig. 3 would be:

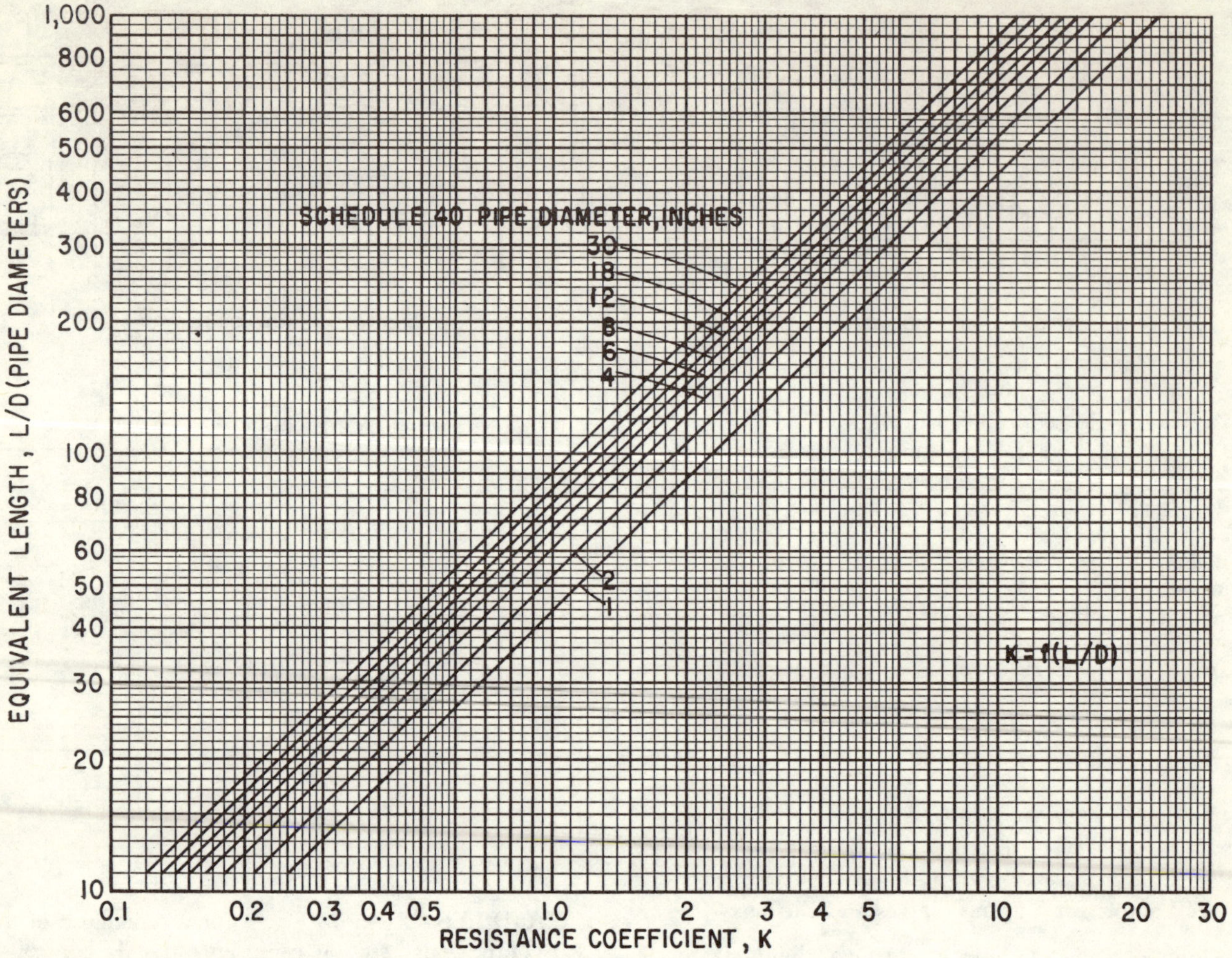

Figure 5. Resistance coefficients vs. equivalent length.

12″ Entrance	6″ K_e	=	0.04
6″ Reducer—Increaser	6″ K_{rt}	=	0.80
2—6″ Valves	6″ K_v	=	0.10
6″ Control Valve	6″ K_v	=	10.00
12-ft. spool of 6″ Pipe	6″ K_p	=	0.36
12″ Elbow	6″ K_{el}	=	0.02
12″ Exit	6″ K_x	=	0.08
	ΣK	=	11.40

Refer to Fig. 2 for a conversion of K to C_v. With the above values noted, the over-all 6-inch C_v is 320, approximately a 6-percent reduction of the original C_v of 340.

Now refer to Fig. 4. This is an identical control valve run with Fig. 3, except that a 6-inch by 4-inch by 6-inch high-capacity ball valve has been substituted for the 6-inch single-ported globe body control valve. Note that the ΣK now is 1.93, which converted to an over-all C_v is 775. This is a reduction of about 48 percent from the original 6-inch by 4-inch by 6-inch control valve C_v of 1,495.

If in Fig. 4, the 6-inch by 4-inch by 6-inch valve were substituted with a full open 6-inch ball valve, the over-all C_v is 895. This represents a reduction in C_v of about 81 percent from the original 6-inch ball valve C_v of 4,700.

The methods and graphs presented in this article are proposed for sizing and calculating pressure drops in a control station. The assumption of assigning values to certain unusual piping configuration must be recognized. The author has some field data confirmation on similar control station designs.

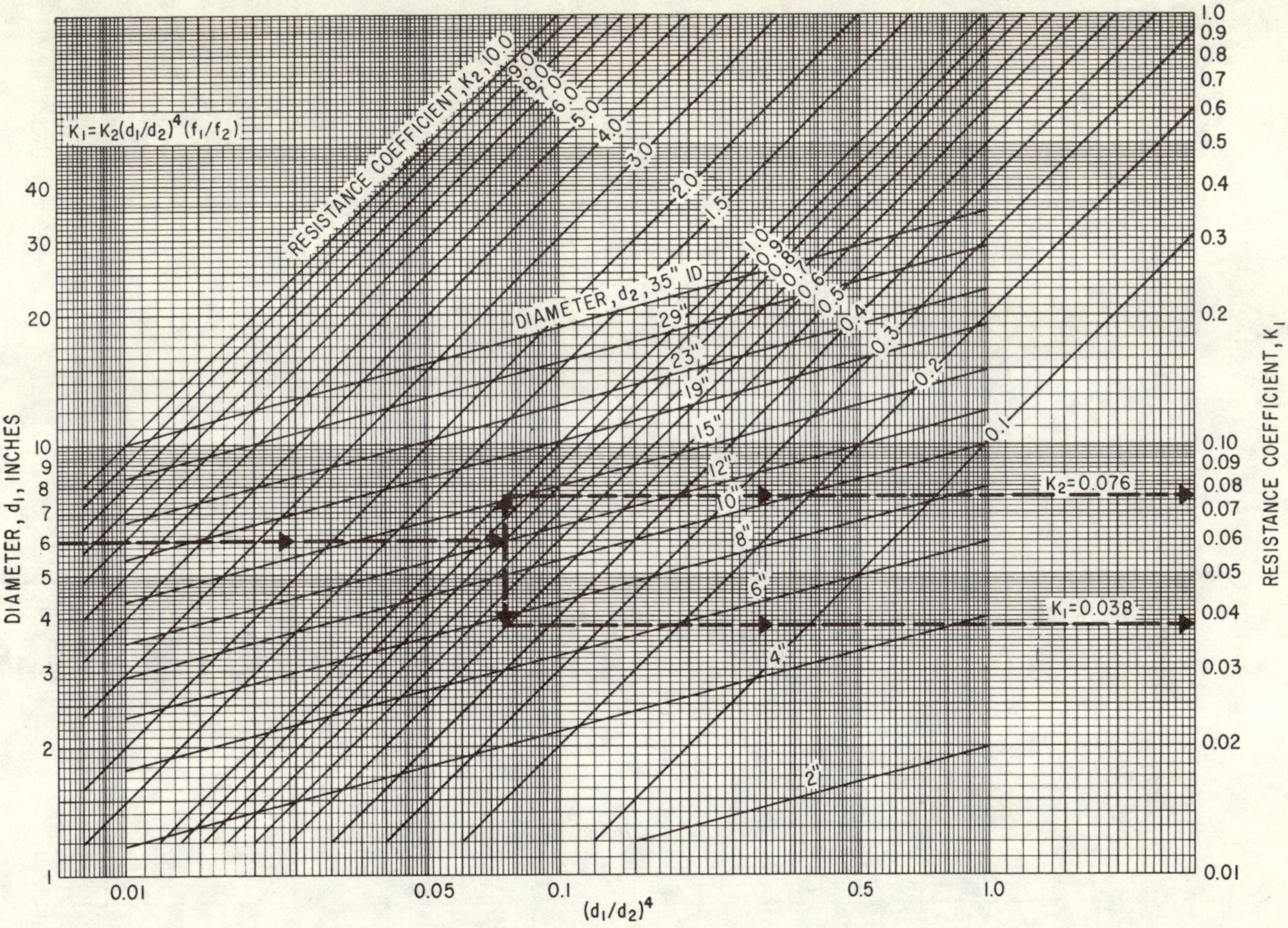

Figure 6. Equivalent resistance coefficients vs. diameter ratios.

12. Gas—Metering

Determine required orifice diameter for any required differential when the present orifice and differential are known in gas measurement

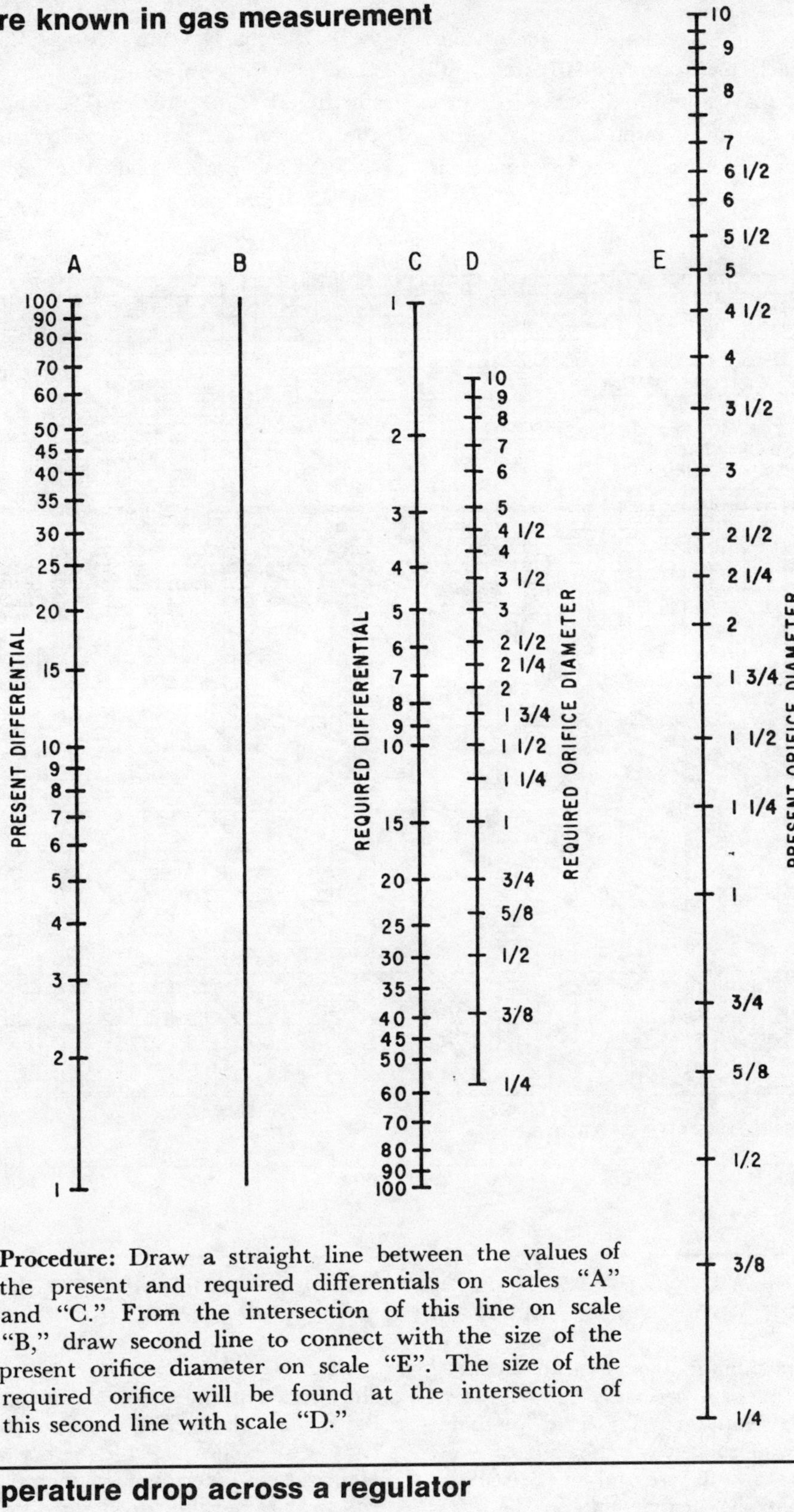

Procedure: Draw a straight line between the values of the present and required differentials on scales "A" and "C." From the intersection of this line on scale "B," draw second line to connect with the size of the present orifice diameter on scale "E". The size of the required orifice will be found at the intersection of this second line with scale "D."

Estimate the temperature drop across a regulator

The temperature drop that will be experienced when gas is passed through a regulator may be estimated as 1° drop for each atmosphere (15 psi) of pressure drop.

Example: Determine the temperature drop that will be experienced when reducing a gas from 750 psig to 600 psig.

Solution: 750 — 600 psig = 150 psi

$$\frac{150}{15} = 10 \text{ atmospheres}$$

10 times 1 degree/atm = 10 degrees of drop

Estimate natural gas flow rates

THIS CHART provides a good estimate of natural gas flow rates through 4-inch meter runs equipped with flange taps. No calculations are required to obtain a flow rate. Orifice plate size, and differential and static pressures from the orifice meter chart are used to make the estimate.

To use the Gas Flow Rate Estimator, enter the chart at the proper differential value at the indicated lower horizontal scale. Project this value vertically upward to the appropriate static pressure line, and then horizontally to the proper plate size line. Drop vertically from the plate size line to the lowermost horizontal rate scale at the base of the chart to obtain the estimated flow rate.

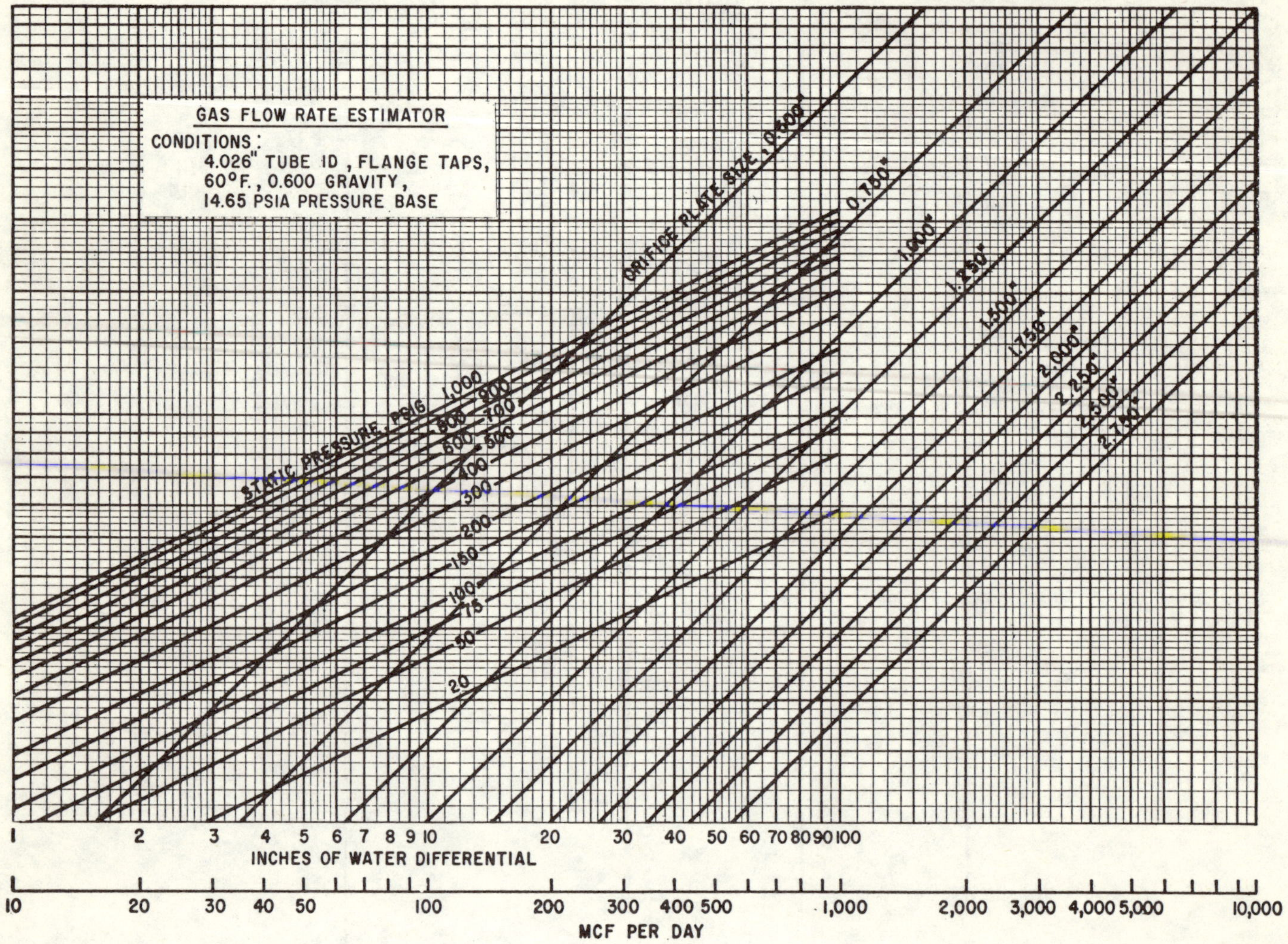

How to estimate the average pressure differential on the remaining meter runs of a parallel system when one or more runs are shut off

On gas systems where inferential meters are installed in parallel, it may at times be necessary to shut down one or more of the runs temporarily for repair or maintenance work. When this occurs, it is important to know beforehand what the pressure differential on the remaining runs will be so that the meters will not be overranged. The equation is:

Average differential for runs remaining in service (A) =

$$\left[\frac{(\text{No. of runs in service})\sqrt{(\text{Average differential})}}{(\text{No. of runs remaining in service})}\right]^2$$

Example: A measurement engineer wishes to shut off one meter run on a four-meter parallel system to inspect the orifice plate. The average differential on the four runs is 49 inches and his meters have a 100-inch range. He wishes to determine whether shutting down one meter will over-range the three remaining meters.

$$(A) = \left(\frac{4 \times \sqrt{49}}{3}\right)^2 = (9.3)^2 = 86.5 \text{ inches}$$

Since the average differential for the remaining runs is 86.5 inches, the operator may safely shut the meter down as planned to inspect its orifice plate.

Sizing a gas metering run

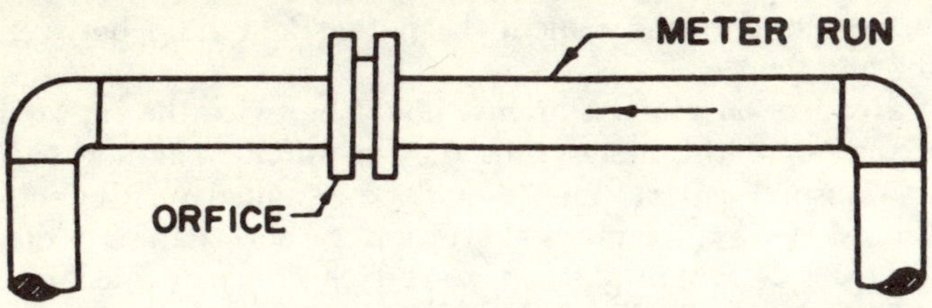

Multiply the absolute pressure by the desired differential pressure (the differential in inches of water should not be greater than the absolute pressure in psia). Take the square root of the product. Divide this figure into the maximum flow rate in cubic feet per hour to obtain the approximate coefficient. Then divide the maximum coefficient by 250 and take the square root to determine the approximate inside diameter of the orifice in inches. The metering run diameter should not be less than one-third larger than the orifice diameter.

Example: Find the orifice diameter and meter run for a line designed to carry a maximum flow of 2 Mmcf per hour at a pressure of 600 psia. Use a differential pressure of 50 inches of water.

$$\sqrt{600 \times 50} = \sqrt{30{,}000} = 173.2$$

$$\frac{2{,}000{,}000}{173.2} = 11{,}540$$

$$\frac{11{,}540}{250} = 46.2$$

$$\sqrt{46.2} = 6.8 \text{ inches, approximate orifice diameter}$$

Meter run should not be less than ten inches in diameter.

List of typical specifications for domestic and commercial natural gas

1. **Heating value:** Not less than 950 Btu's per std. cubic foot.
2. **Purity:** The gas to be delivered shall be free from dust, gums, crude oils, hydrocarbons liquefiable at temperatures in excess of 15° F at 800 psig.
3. **Sulfur content:** The gas shall contain not more than one grain of hydrogen sulfide per 100 cubic feet of gas, and not more than 20 grains of total sulfur per 100 cubic feet of gas.
4. **CO_2 content:** The gas shall contain not more than 2 percent carbon dioxide.
5. **Moisture content:** The gas shall contain not more than 4 pounds of water vapor per MMcf of gas measured at a pressure of 14.4 psi and a temperature of 60° F.
6. **Temperature:** Maximum temperature at point of delivery shall be 120° F.

Determine the number of purges for sample cylinders

To determine the proper number of times a sample cylinder should be pressured to insure removal of the original gas, multiply the reciprocal of each filling pressure (in atmospheres) by each succeeding reciprocal. The final filling will determine what fraction of the original gas is still in the cylinder.

Example: The problem is to fill a gas sample for analysis to obtain an accuracy of 0.05%. Source pressure is 450 psia.

Solution:

First filling to line pressure and emptying:

$$\frac{1}{450/15} = \frac{1}{30} \times 100 = 3.33\%$$

Second filling to line pressure and emptying:

$$\frac{1}{30} \times \frac{1}{30} \times 100 = 0.11\%$$

Third filling to line pressure:

$$\frac{1}{30} \times \frac{1}{30} \times \frac{1}{30} \times 100 = 0.0037\%$$

Estimate for variations in measurement factors

Orifice measurement:

1. **Each 10-degree change in flowing temperature** will cause a one percent change in flow calculated.
2. **Each 0.01-change in specific gravity** of the gas measured will result in a one percent change in passing flow calculated.
3. **Each 10-psi error in static pressure** will make the following error in flow calculated.

   at 100 psia—5%
   at 500 psia—1%
   at 1000 psia—0.5%

4. **Each 0.1 inch of differential pressure error** will make the following error in flow calculated.

   at 5 inches differential—1%
   at 25 inches differential—0.2%
   at 50 inches differential—0.1%
   at 100 inches differential—0.05%

5. **Supercompressibility** affects volume measured approximately 0.5% per 100 psi pressure in flowing temperature range of 60-90° F.

Find the British thermal units (Btu) when the specific gravity of a pipe line gas is known

Use the graph below prepared from the Headlee formula:

Btu @ 30 inches Hg and 60 degrees saturated = 1545 (Specific gravity) + 140 — 16.33 (% N_2)

Example: Assume a gas having a gravity of 0.670 which contains 1.0% nitrogen. From the chart, read up from 0.670 gravity on the base line to the intercept with the curve marked 1% nitrogen. Reading horizontally across, the total Btu of the gas is 1,160 Btu at 30 inches, 60 degrees saturated.

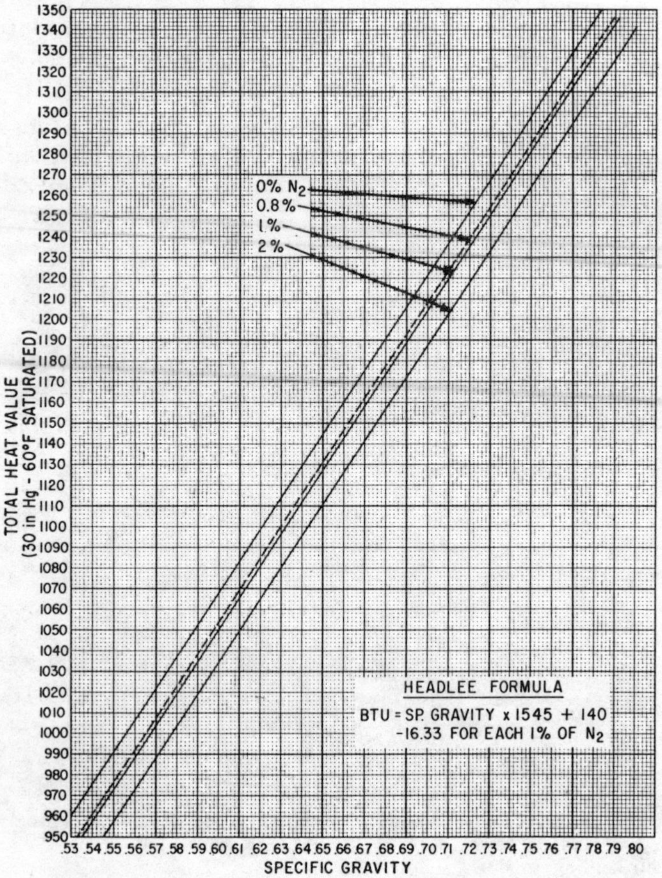

Rules for measurement of gas by orifice meter

Where the static pressure remains constant, to reduce the flow by one-half, lower the differential reading to one-fourth. For example, if the differential is reading 64 inches, to reduce the flow by one-half, reduce the differential to 16 inches.

For better measurement, keep the Beta ratio between 0.2 and 0.6.

Headers on meter stations should be twice the internal diameter of the meter runs they manifold together.

Measurement stations should be installed on the suction side of a compressor station rather than the discharge side where possible, to reduce effect of pulsation.

Use the same size orifice plate in meter runs that are manifolded together where possible; recordings can then be used to check against the other since their differentials should be approximately the same.

Make quick calculation of the gas flow thru orifice meters on meter

runs up to and including 10-in. can be accomplished by multiplying the coefficient for a one-inch orifice times the actual orifice installed squared times the square root of the extension (the product of the differential in inches of water times the static pressure in pounds per square inch absolute). AGA-3 coefficients for a one inch orifice in 2, 3, 4, 6, 8 and 10 inch standard orifice meter runs using flange taps should be committed to memory. Base conditions would be for a 0.60 specific gravity, 60 Fahrenheit and 14.73 psia pressure base. The answer would be in standard cubic feet per hour or day depending upon the coefficient used.

AGA Committee Report Number 3 coefficients at the conditions shown are:

2 inch meter run—1 inch orifice—271.4 cfh, 0.6 G, 60 F and 14.73 psia
3 inch meter run—1 inch orifice—262.8
4 inch meter run—1 inch orifice—260.8
6 inch meter run—1 inch orifice—259.3
8 inch meter run—1 inch orifice—258.7
10 inch meter run—1 inch orifice—258.5 (heavy)

Example: Find the hourly gas flow through an 8 inch meter run having a 2.00 inch orifice when the differential pen is recording 16 inches of water and the static pressure is 900 pounds per square inch absolute.

Solution: From memory, the coefficient for an 8 inch meter run, one inch orifice is 258.7. Multiplying 258.7 times 4 (or 2 squared) times 120 (the square root of the product of 900 times 16) equals 124,176 cfh, the approximate flow in cubic feet of gas per hour at 0.6 gravity, 60 F temperature, and 14.73 psia base pressure.

How to measure high pressure gas

Figure 1. Meter runs, piping, etc., should be designed as a Class III area facility. If possible, a properly vented walk-in type meter building should be used, with the total installation laid out with the safety of the operating personnel firmly in mind.

Accurate gas measurement is obtained only through a good maintenance and operating schedule

T. J. Lewis, Measurement Engineer, Columbia Gulf Transmission Co., Houston

THE RAPIDLY increasing value of natural gas, along with rising operating costs, has made accurate gas measurement vital for the natural gas industry.

Accurate gas measurement is obtained only through an efficient maintenance and operating schedule, adhered to rigidly, after a properly designed measuring station has been installed. Any error in the measurement of gas at high pressure and large volumes magnifies itself, so that there can be either a considerable gain or loss in revenue.

Field measurement at high pressure is defined as the measurement of gas at a specific point in a gas producing area at a pressure higher than that of contract base pressure.

Contracts. The gas purchase contract is of prime importance in determining the requirements for gas measurement at a given location.

Information such as gas analysis, gas/oil ratios, specific gravity determinations, flowing temperatures, pressures, etc., should be gathered before a gas purchase contract is written. This collection of data should be used to eliminate any assumptions regarding the requirements for performing accurate measurement. It also provides information for the engineering and accounting functions, so that they are made aware of their responsibilities and limitations in helping provide good measurement.

The contract should also state as clearly as possible all parameters of the gas measurement operation to prevent possible disputes by having a clear understanding between the buyer and the seller.

Design and planning. In the planning and designing of a natural gas measuring station, many items of equal importance must be considered if accurate measurement is to be obtained.

The design and fabrication of the orifice meter runs must meet the specifications of the AGA Gas Measurement Committee Report No. 3, latest revision. In addition to these requirements, consideration must be given to safety, volume and quality of gas, term of contract, flexibility of station design, and physical location of the station. The most important point in this whole procedure is remembering that the measuring station is like a large cash register with money flowing both ways for the company.

A meter tube should be designed to allow easy internal inspection. Shut-off valves should be placed on each end of the meter, so that it may depressured independently of any other piping. The instrumentation and/or control piping should be so located and supported to furnish maximum protection against accidental breakage.

Meter runs, piping, etc., should be designed as a Class III area facility. If possible, a properly vented walk-in type meter building should be used, with the total installation laid out with the safety of the operating personnel firmly in mind (Fig. 1). As periodic orifice plate inspections are necessary, provisions should be made so that these inspections can be made without interruption of the gas flow. This can be done by installing a bypass line or by using a senior type orifice fitting.

The flow and temperature recording instruments should be located in a walk-in type meter building. This arrangement provides a year-round working area for chart changing, meter tests, maintenance work, etc.

Installation. The measuring or metering components in a measuring station are divided into two elements: primary and auxiliary.

The primary element of an orifice-type measuring station consists of the meter tube, the orifice plate fitting, the orifice plate, pressure taps and straightening vanes. While the primary element is being fabricated, its components should be checked against the specifications listed in AGA Gas Measurement Committee Report No. 3. Any discrepancies found then can be corrected before the meter tube is finally assembled and shipped to the field.

When the meter tube is being installed in the field, the orifice plate is checked to make sure it is centered in the meter tube, and that it meets AGA Report No. 3 specifications. Meter tubes are normally installed in a horizontal or vertical plane. Pressure taps in a vertical plane are preferred, since this allows the gage lines to be installed from the top of the orifice fitting. The pressure taps located on the bottom of the orifice fittings then can be used for installing drips or drains.

The auxiliary element consists of the meters (recording devices) used for measuring the static pressure, differential pressure, and temperature of the gas in the meter tube. Meters are placed as close as possible to the orifice or flange fitting. Gage lines are connected from the meter to the orifice fitting in a way that will best provide a drain to the meter tube. Liquids trapped in a gage line can cause significant errors in the recorded pressures.

There are two principal types of orifice meter gages: the mercury Manometer-type and the bellows-type meters. Each of these meters is suitable for field measurement, with the bellows-type meter becoming the favorite for wet gas measurement.

Installation of these elements in a measuring station should be under the supervision of experienced measurement personnel.

Operation and maintenance. After a measuring station has been properly installed, accurate measurement is ob-

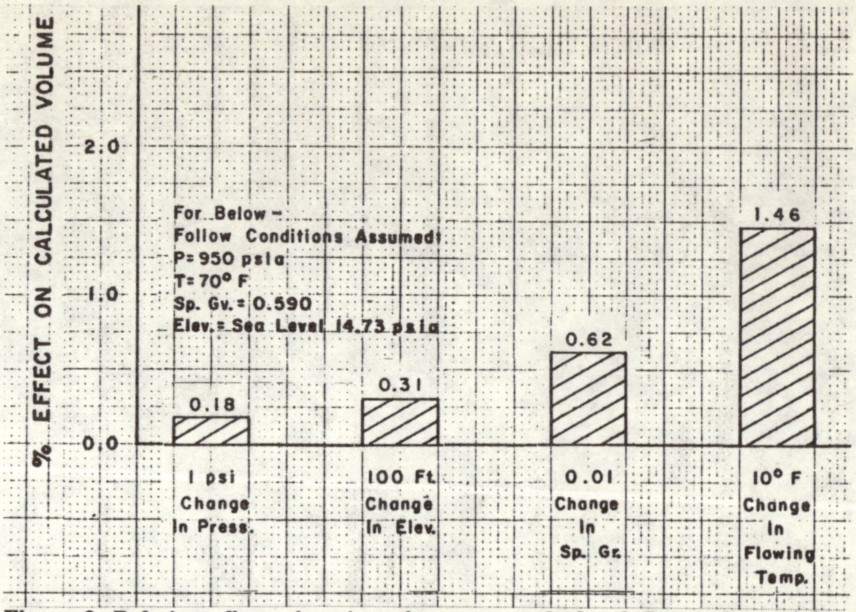

Figure 2. Relative effect of various factors on calculated flow volume (also Figure 3).

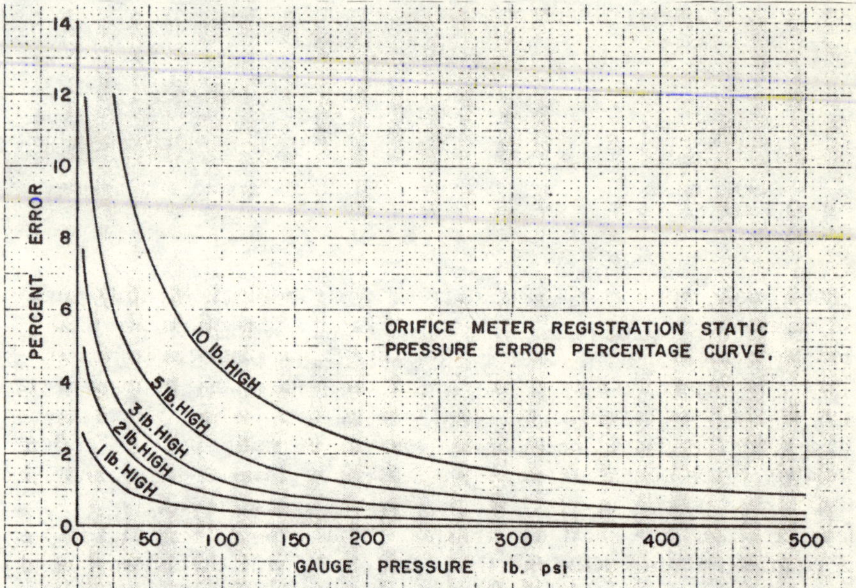

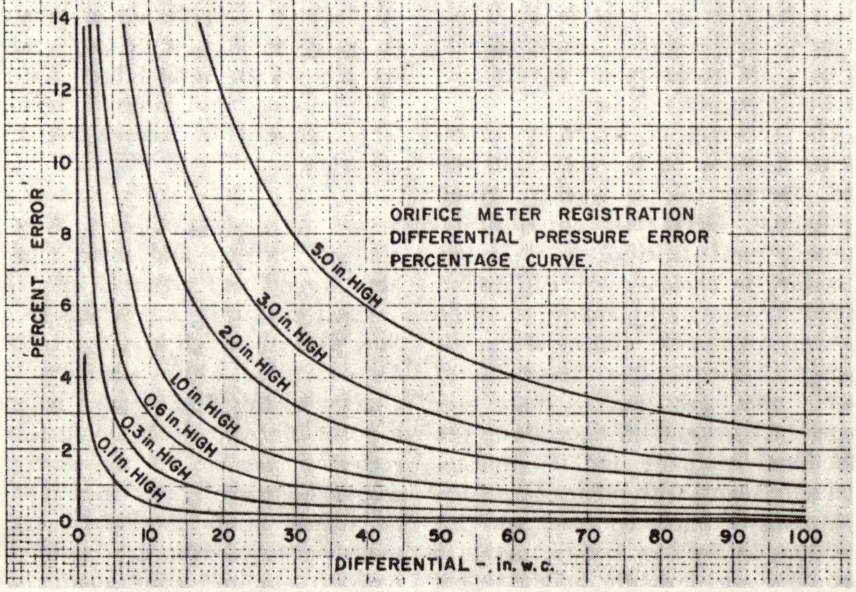

Figure 3. The above (also Figure 2) shows what happens to volume calculations when there are small errors in static pressure, differential pressure, temperature and specific gravity masurements.

Figure 4—Charts are shown being changed, and clocks wound and adjusted. Each measuring station has its own flow pattern, which is recognizable by experienced chart changers.

tainable only by conscientious care and maintenance by the field measurement technicians.

Any discrepancy noticed on the chart should be explained on the chart and a correction should be made immediately. Since immediate correction is not always possible, a report should then accompany the chart describing the condition of the meter and when the correction was completed. These reports, along with test and inspection reports, are necessary for the calculation of accurate volumes.

The accuracy of recording instruments depends on the men who test and operate the measuring station equipment. Good test equipment, properly maintained, is absolutely necessary in checking the calibration of temperature and pressure recording instruments and the quality of the gas being measured. The frequency of meter tests, gas quality tests, and meter tube inspections depends on the individual company and its contractual agreements with other companies.

Monthly calibration tests and inspections on the flow and temperature recorders are normally adequate for accurate measurement. In special cases, however, a weekly inspection may be required. Monthly calibration tests and inspections are vital to accurate measurement practices. Figs. 2 and 3 show what happens to volume calculations when there are small errors in static pressure, differential pressure, temperature and specific gravity measurements.

Since external tests and inspections do not reveal the internal condition of a meter tube, the meter tube and its components must be inspected internally on an annual basis for possible buildups and internal corrosion. This inspection insures continuing accuracy.

Chart changing. Charts must be changed on time and clocks must be kept wound and adjusted (Fig. 4).

Each measuring station has its own flow pattern, which is recognizable by experienced chart changers. Any change in this flow pattern indicates a possible error, which should be corrected if possible. If the correction cannot be made by the chart changer, he should notify the field measurement personnel who are qualified to perform required adjustment and/or repairs.

Each chart must be properly dated and identified along with field information necessary for calculating the required volumes. Automatic chart changers have become increasingly popular, especially in remote areas such as offshore Louisiana.

Periodic pulsation test should be made at each measuring station, particularly where a measuring station is located on the suction or discharge side of a compressor station. In this instance, pulsation dampeners must be installed between the compressor and the measuring station. Errors caused by pulsation can be very large, but they will not necessarily show in calibration tests.

A comprehensive maintenance program, faithfully followed, can detect and prevent costly errors before they occur. Leaking valves, gage lines, etc., will cause errors in the chart records, which are reflected in an increase or decrease in company revenue. Such leaks should be corrected as soon as possible.

Wet gas measurement. The problems encountered in measuring wet gas are considerably different from dry gas measurement.

Wet gas contains entrained liquids consisting of water and hydrocarbon condensates which can cause hydrate formations in the orifice plate fitting, flow recorders and instrument lines. These hydrates are formed with the aid of free water under certain pressure and temperature conditions. Formation of hydrates can be reduced or eliminated in several different ways.

One successful solution to prevent hydrate formation is to use drip or condensate bottles in instrument and gage lines. Infrared gas heaters can be used in meter buildings for heating gage lines and flow recorders. This type of heater has become very popular lately, since it does not produce an open flame.

Errors caused by hydrate formations in the meter tube and around the orifice plate are common in wet gas measurement. The formation of these hydrates can be prevented by continuously injecting methanol into the gas stream ahead of the meter run. Experience and experimentation will determine the amount of methanol to be injected.

The use of methanol-containing inhibitors is not recommended as the inhibitor often solidifies in the meter tubes.

Four ways to calculate orifice flow in field

- **A.G.A. formula**
- **Computers and slide rules**
- **Flow graphs**
- **Quick method**

Rogers G. Thompson,
Measurement Technician,
Texas Eastern Transmission Corp.,
Shreveport, La.

SEVERAL METHODS are available for orifice flow field calculations.

The particular method used is determined by such factors as company policy, advance warning, time available, individual preference, accuracy required and the information and equipment available at the time of need. A good rule to follow is to calculate as precise a volume as possible at all times.

Any volume calculated manually will be considered a field calculation.

Most companies now use some type of computer system and rarely will a manually calculated volume match a computer volume. Field calculations are usually made under conditions which eliminate the use of adding machines and calculators. Accuracies can vary from an extremely accurate spot check to a very basic calculation. External factors usually determine what procedure to use.

Uses for the calculations determine the degree of accuracy required for each particular instance. Since time is usually at a premium it would be absurd to calculate to the cubic foot when calculating to the Mcf would suffice. A good example is automatic control of a compressor station where rates are given in terms of MMcf.

Field calculations are used for such things as determining purge gas volumes, relief valve losses, set point determination and sizing of plates. In general, volume, rate, differential, and plate size are the form variables that are determined by field calculations.

This discussion will be based on the AGA Committee Report No. 3 as it is the most commonly used. It is also the most complicated due to the large number of corrections applied.

Four categories have been devised to aid in discussing field calculations: the A.G.A formula, graphs, computers, and the quick method.

A.G.A. Formula. In making field calculations a good rule to follow is—the more simple the better. Detailed calculations are difficult to perform under usual field conditions.

Another good rule to follow is cool and check. Always allow some time to elapse between calculating and checking. This can be from five minutes to five hours. Even better is to have someone else check your work. Following these rules will eliminate most errors.

The gas flow equation as stated in the AGA Committee Report No. 3 is:

$$Qh = Fb\ Fr\ Y\ Fm\ Fpb\ Ftb\ Fm\ Fa\ Fl\ Ftf\ Fg\ Fpv\ \sqrt{h_w pf}$$

Obviously the most accurate method

Table 1

Combined specific gravity and temperature correction factors—orifice meters combined factor $= \left(\sqrt{\frac{1}{G}}\sqrt{\frac{520}{460° F}}\right)$

Flow Temp. °F	Specific Gravity .590	.591	.592	.593	.594	.595	.596	.597	.598
35	1.3343	1.3332	1.3321	1.3309	1.3298	1.3287	1.3276	1.3264	1.3254
36	1.3330	1.3319	1.3208	1.3296	1.3285	1.3274	1.3263	1.3251	1.3241
37	1.3317	1.3306	1.3295	1.3283	1.3272	1.3261	1.3250	1.3238	1.3228
38	1.3303	1.3292	1.3280	1.3269	1.3258	1.3247	1.3235	1.3224	1.3214
39	1.3290	1.3279	1.3267	1.3256	1.3245	1.3234	1.3222	1.3211	1.3201
40	1.3277	1.3266	1.3254	1.3243	1.3232	1.3221	1.3209	1.3198	1.3188
41	1.3264	1.3253	1.3241	1.3230	1.3219	1.3208	1.3197	1.3185	1.3175
42	1.3251	1.3240	1.3228	1.3217	1.3206	1.3195	1.3184	1.3172	1.3162
43	1.3238	1.3227	1.3215	1.3204	1.3193	1.3182	1.3171	1.3159	1.3149
44	1.3223	1.3212	1.3201	1.3190	1.3179	1.3168	1.3156	1.3145	1.3135
45	1.3210	1.3199	1.3188	1.3177	1.3166	1.3155	1.3143	1.3132	1.3122
46	1.3197	1.3186	1.3175	1.3164	1.3153	1.3142	1.3130	1.3119	1.3109
47	1.3184	1.3173	1.3162	1.3151	1.3140	1.3129	1.3118	1.3106	1.3096
48	1.3171	1.3160	1.3149	1.3138	1.3127	1.3116	1.3105	1.3093	1.3083
49	1.3158	1.3147	1.3136	1.3125	1.3114	1.3103	1.3092	1.3080	1.3070
50	1.3147	1.3135	1.3124	1.3113	1.3102	1.3091	1.3080	1.3069	1.3059
51	1.3134	1.3122	1.3111	1.3100	1.3089	1.3078	1.3067	1.3056	1.3046
52	1.3121	1.3109	1.3098	1.3087	1.3076	1.3065	1.3054	1.3043	1.3033
53	1.3108	1.3096	1.3085	1.3074	1.3063	1.3052	1.3041	1.3030	1.3020
54	1.3095	1.3083	1.3072	1.3061	1.3050	1.3039	1.3028	1.3017	1.3007
55	1.3081	1.3070	1.3059	1.3048	1.3037	1.3026	1.3015	1.3004	1.2994
56	1.3070	1.3059	1.3048	1.3037	1.3026	1.3015	1.3004	1.2992	1.2982
57	1.3057	1.3046	1.3035	1.3024	1.3013	1.3002	1.2991	1.2980	1.2970
58	1.3044	1.3033	1.3022	1.3011	1.3000	1.2989	1.2978	1.2967	1.2957
59	1.3032	1.3021	1.3010	1.2999	1.2988	1.2977	1.2966	1.2955	1.2945
60	1.3019	1.3008	1.2997	1.2986	1.2975	1.2964	1.2953	1.2942	1.2932
61	1.3006	1.2995	1.2984	1.2973	1.2962	1.2951	1.2940	1.2929	1.2919
62	1.2994	1.2983	1.2972	1.2961	1.2950	1.2939	1.2928	1.2917	1.2907
63	1.2981	1.2970	1.2959	1.2948	1.2937	1.2926	1.2915	1.2904	1.2894
64	1.2970	1.2959	1.2948	1.2937	1.2926	1.2915	1.2904	1.2893	1.2883
65	1.2957	1.2946	1.2935	1.2924	1.2913	1.2902	1.2891	1.2880	1.2870
66	1.2945	1.2934	1.2923	1.2912	1.2901	1.2890	1.2879	1.2868	1.2858
67	1.2932	1.2921	1.2910	1.2899	1.2888	1.2877	1.2866	1.2855	1.2845
68	1.2920	1.2909	1.2898	1.2887	1.2876	1.2865	1.2855	1.2844	1.2834
69	1.2908	1.2897	1.2887	1.2876	1.2865	1.2854	1.2843	1.2832	1.2822
70	1.2895	1.2884	1.2874	1.2863	1.2852	1.2841	1.2830	1.2819	1.2809
71	1.2884	1.2873	1.2862	1.2851	1.2840	1.2829	1.2818	1.2807	1.2798
72	1.2872	1.2861	1.2850	1.2839	1.2828	1.2818	1.2807	1.2796	1.2786
73	1.2859	1.2848	1.2837	1.2826	1.2815	1.2805	1.2794	1.2783	1.2773
74	1.2847	1.2836	1.2825	1.2815	1.2804	1.2793	1.2782	1.2771	1.2761
75	1.2835	1.2825	1.2814	1.2803	1.2792	1.2781	1.2770	1.2760	1.2750
76	1.2824	1.2813	1.2802	1.2791	1.2780	1.2770	1.2759	1.2748	1.2738
77	1.2811	1.2800	1.2789	1.2778	1.2767	1.2757	1.2746	1.2735	1.2725
78	1.2799	1.2788	1.2777	1.2767	1.2756	1.2745	1.2734	1.2723	1.2713
79	1.2787	1.2776	1.2766	1.2755	1.2744	1.2733	1.2722	1.2712	1.2702

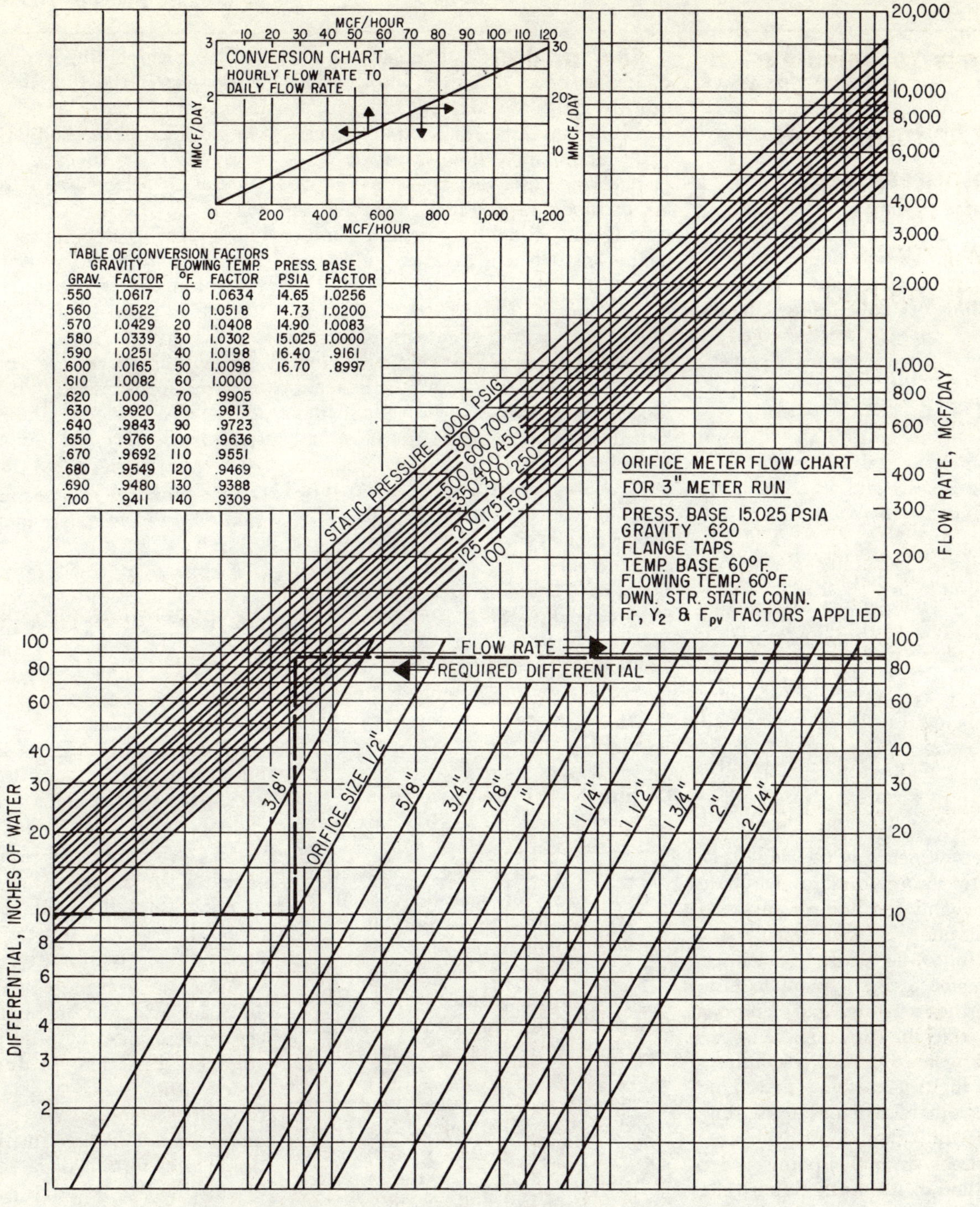

GRAVITY		FLOWING TEMP.		PRESS. BASE	
GRAV.	FACTOR	°F.	FACTOR	PSIA	FACTOR
.550	1.0617	0	1.0634	14.65	1.0256
.560	1.0522	10	1.0518	14.73	1.0200
.570	1.0429	20	1.0408	14.90	1.0083
.580	1.0339	30	1.0302	15.025	1.0000
.590	1.0251	40	1.0198	16.40	.9161
.600	1.0165	50	1.0098	16.70	.8997
.610	1.0082	60	1.0000		
.620	1.000	70	.9905		
.630	.9920	80	.9813		
.640	.9843	90	.9723		
.650	.9766	100	.9636		
.670	.9692	110	.9551		
.680	.9549	120	.9469		
.690	.9480	130	.9388		
.700	.9411	140	.9309		

Figure 1. Example of orifice meter flow graph used by Texas Eastern.

of volume calculation would use all of the above factors. Most field calculations will use some modified form of the AGA equation.

In order to present a logical discussion these correction factors are broken down into three groups—major, minor and intermediate. This grouping is based upon the effect each factor has on the volume.

The major factors are the basic orifice factor *Fb* and the specific gravity factor *Fg*. These two factors have to be included in all computations.

The intermediate factors are the pressure base correction *Fpb,* flowing temperature correction *Ftf* and supercompressibility correction *Fpv.* The value of these factors is variable. Usually they are determined by operating conditons at each specific location or are set by contract as in the case of pressure base.

Supercompressibility can become a large correction with transmission companies and producers where gas is measured at elevated pressures. For field calculation purposes, intermediate factors will be considered as either major or minor.

The *Fr, Y, Fm, Ftb, Fa,* and *Fl* corrections are minor factors for field calculation because of their minimal effect upon the final volume. Most of these factors are usually ignored.

Condensed Formula. As previously stated, the most accurate method of field calculation is the AGA formula using all corrections. However, this is usually not possible because of insufficient time, advance warning or operating information.

If a little work is done prior to the

time of use, the formula can be condensed to three or four calculations. However, some operating information must be available. The static pressure, differential, run and plate size, specific gravity, and gas composition must either be known or assumed.

Assumptions should be based on knowledge of normal operating procedures and characteristics of similar M&R stations. As an example, the inert composition of gas is assumed to be 0 percent nitrogen and 0 percent carbon dioxide until an analysis is performed. This procedure permits volume calculations until more definite information is available.

By using known and assumed data, an hourly coefficient composed of *Fb, Fr, Y, Fm, Ftb, Fpb, Fa* and *Fl* can be computed. The only corrections left to be made are *Fg, Ftf* and *Fpv.* By giving some thought, time and prior work the formula can be reduced to $Qh = C\ F_g\ Ftf\ F_g\ \sqrt{h_w pf}$.

Tables are available which further reduce the amount of calculation that has to be done. Combined specific gravity and temperature tables are available from different sources. Also available are extension values for $\sqrt{h_w pf}$ or the individual values of $\sqrt{h_w}$ and $\sqrt{pf}$. These tables eliminate what is probably the hardest part of a field calculation.

When following these guidelines, calculations become fairly simple. One note of caution: always make computations in a legible and logical order. This makes everything fairly simple to check. As many errors are caused by illogical layout of calculations as any other factor.

All methods from this point on are an adaptation of the AGA Report No. 3 equation.

Flow Graphs. Usually the next most accurate method is the company developed flow graphs.

Although the graphs necessarily have to be based on average data, they are tailored to each particular company's operating conditions. Thus, you get more accurate results for your particular requirements. Corrections such as gravity can be made to the final figure if desired, but usually are not.

The graph has the advantage of giving considerable information quickly. All plate size and differential combinations are shown at a glance. Daily and hourly flow rates are also given. This is the fastest and easiest of all methods to use and accuracy is usually good enough to use under most conditions. No calculations are necessary under most conditions.

Most paper graphs are bulky, requiring a straight edge and space to use. Humidity will destroy them. However, replacements usually are available upon request. Oddly enough this is probably the least used of all methods.

Most companies use the AGA equation to develop their graphs. The newer factors *Fa* and *Fl* probably are not included in current data. Most factors are based on a particular company's average data.

Some factors are given a value of 1.000 and ignored. Supercompressibility corrections are usually by average specific gravity and inert composition.

Computers. Flow computers and slide rules are probably the most widely used method of field calculation. This is due to their compactness and simplicity of operation. Accuracy is generally in the acceptable range for field calculations.

Most manufacturers of measurement equipment furnish some type of flow computer either free or at a nominal cost. Construction usually is of paper and plastic.

All of these flow computers are based on the AGA Report No. 3 flow equation. Most of the minor corrections are not used, with the omissions listed in the instructions. Since computers are designed for use throughout the industry, they must be kept standard. Consequently they will not fit a particular set of conditions without making additional corrections.

Most flow computers incorporate *Fb, Fg, Ftf, Ftb* and sometimes a generalized form of *Fpv.* The minor factors are usually ignored. The pressure base is usually set at 14.7 or 14.73, but some computers do incorporate a pressure base correction feature.

The biggest single drawback to computers is the fact that they must be general in scope and not slanted to one particular method of calculating volumes. Corrections are kept too general to fully correct for a specific set of conditions. There must be enough use of a computer for a person to maintain speed and accuracy of operation. If not, the instruction booklet will have to be studied each time the computer is used.

Flow computers are compact, easy to carry and store, and are very quick to use. As long as their limits are recognized they can do an excellent job.

Quick Method. Primarily, the quick method uses the major factors *Fb* and *Fg* only.

Inclusion of the intermediate factors is desirable if sufficient information is available. These intermediate corrections are made in a general way only. For example, if the pressure base is 15.025, subtract two percent of the volume. Algebraically, add one percent of the volume for each 10° variation from the temperature base. For each 100 psi of static pressure, add eight percent of the volume.

This procedure will suffice for a lot of field calculation requirements. Keep in mind that this method is a rough calculation and is not intended to be used as a precise quantity.

Some mention must be made of the extension which probably causes more trouble than any other part of a volume calculation. In determining average differential and pressure, spot readings are the easiest to get and the most commonly used method. Sometimes predetermined values are used. Quite often differential and pressure are averaged over several hours. When averaging, care should be exercised for differentials under 25 inches and especially for differentials under 10 inches. Due to the square root effect, errors are magnified in these ranges.

Extension books are available either in the form $\sqrt{h_w pf}$ or individual $\sqrt{h_w}$ and $\sqrt{pf}$ tables. Although the combined tables are bulky, the individual tables can be limited to a few small sheets. Use of these tables will improve accuracy and speed.

Practical maintenance tips for positive displacement meters

Here are the most common malfunction causes and points to watch on the widely used meters

J. A. Willson, Meter Shop Manager, Northern District, Skelly Oil Co., Pampa, Texas

THOROUGH UNDERSTANDING of a meter's function is necessary to properly repair any mechanism within it.

There is a definite reason for each meter part, and each part has a definite relation to the other. The alignment of any one part with another has been precisely established for each type and size of meter.

The functions of a positive displacement meter are similar to a four-cylinder engine. For one revolution of the tangent, each chamber must fill and exhaust once.

One revolution of the tangent is 360°. Therefore, to make a smooth running meter with uniform displacement, one of the chambers must start to exhaust every 90°. Since the valves control the time at which the chamber starts to fill and exhaust, they must be coordinated, one with the other, so that this 90° relationship is maintained.

Similarly, the diaphragms must also be connected to assure this relationship, at least in their extreme positions. Both valves and diaphragms must be connected to coordinate both 90° relationships.

Following are practical tips for PD meter repair and testing.

PDM REPAIRS

The repairman cannot always carry parts for every type meter. Also, testers who only do an occasional repair job are not as efficient as men performing repairs daily. Therefore, most major repairs and all major overhauls are more efficiently handled in the meter shop.

Operating conditions in the field are various, and not always the best. Some of the operating problems in the field are gas saturated with water and hydrocarbons, iron oxide, sour gas and lubrication.

Shop Repair. Meter shops, no matter how well equipped, cannot perform satisfactory meter repair without qualified and dedicated personnel.

When an experienced man disassembles a meter for repair, he inspects each part as he tears it down, considering each as a possible cause of the malfunction. When the trouble is found it is corrected before proceeding, if possible.

Some of the most common causes of malfunction and points to watch on the more widely used meters are:

► Exhaust port gaskets should be inspected to make sure the seal is good. Some meters have metal spacers under the seal and must be in place when reassembled.

► The clearance of the flag rods at the top of the meter should be checked. After a meter has been time checked and balanced, it will sometimes run rough. Look for a bright spot right over the top of the flag rods in the top cover. This is not a common problem, but is difficult to locate, and a few strokes with a file will take care of it.

► The reassembly of the tangent gear between the valve gears of meters, as well as other gear-driven valves, should be centered carefully. There is usually a small amount of side play as well as end play on the tangent shaft bracket. A little extra care here will contribute to a quiet operating meter.

► The body seal O-rings on the later models should be placed carefully in the slot when the top cover is put into place.

Flag Arms. Because new materials are being used in meters and the new aluminum-case meters are designed differently, there have been changes

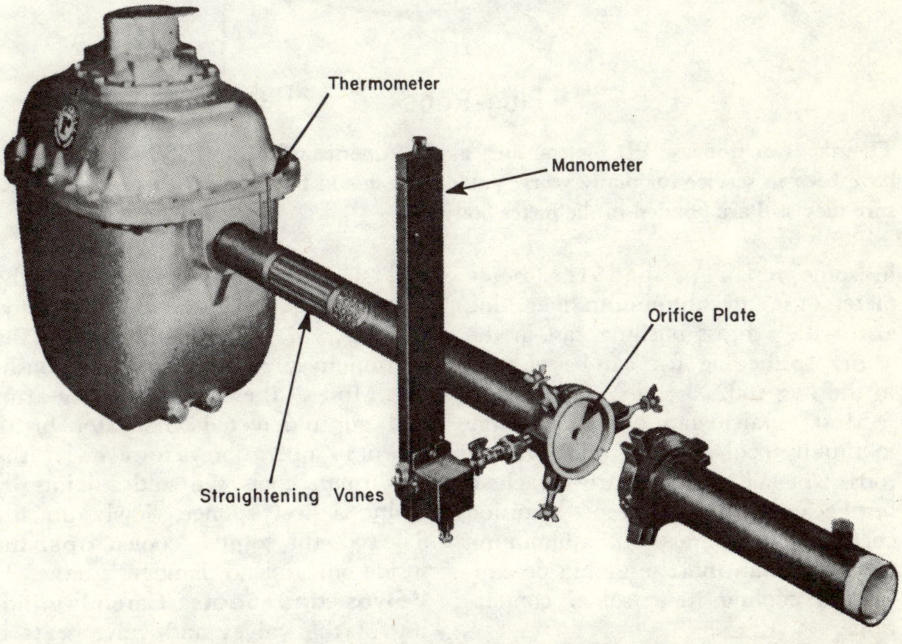

Low pressure flow provers for PD meters can be used in the field or shop. The principle is essentially that of the orifice meter, except the orifice discs are calibrated for different flow rates instead of using various coefficients. Shown above is nomenclature for the Rockwell prover.

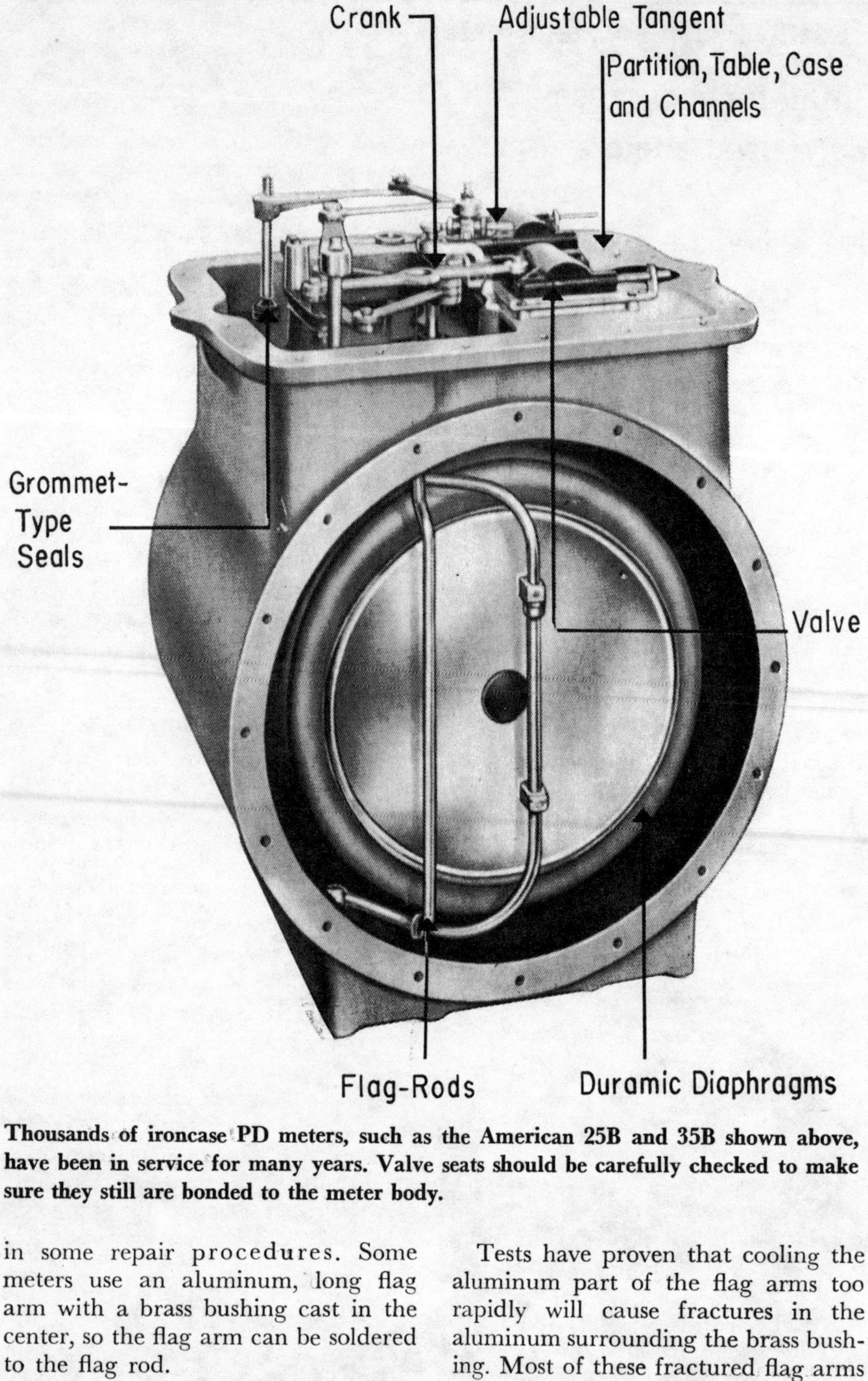

Thousands of ironcase PD meters, such as the American 25B and 35B shown above, have been in service for many years. Valve seats should be carefully checked to make sure they still are bonded to the meter body.

in some repair procedures. Some meters use an aluminum, long flag arm with a brass bushing cast in the center, so the flag arm can be soldered to the flag rod.

Most repairmen use a wet sponge to rapidly cool a solder joint on flag rods when flag arms are attached, but because of the different expansion coefficients of brass and aluminum, the repairman must use extra precaution in cooling these solder connections.

Tests have proven that cooling the aluminum part of the flag arms too rapidly will cause fractures in the aluminum surrounding the brass bushing. Most of these fractured flag arms will go undetected and later break down in operation. However, if the repairman cools the solder joints by easing a wet sponge slowly up the flag rod, the joint is cooled from the inside out and no damage is done.

Valves and Seats. Careful grinding of the valves and valve seats is very important. Number 80 grit may be used to clean up bad valves and seats, finishing with number 120 grit. Most repairmen use the figure eight motion in grinding, letting the weight of the glass block control the pressure on the valve seat. Use a very light touch on the valves. Where it is practical, both valve seats should be ground together. Grinding paper that is self adhesive is certainly recommended over grinding paper that has to be cemented to a glass block.

There are thousands of ironcase meters that have been in service for a great number of years. The valve seats on these meters should be carefully checked to make sure they are still bonded to the meter body. Remove and reseal the seats if there is any doubt about the strength of the bond. Use an old grinding block to true up the bottom of the valve seats and buff the meter case clean to insure a good bond.

Clean all surplus cement from the intersurfaces of the valve seat opening. Be careful not to use screws that are longer than standard in valve guides that screw into the valve seat. An overlength screw can jack the valve seat off the body and break the seal.

Watch the clearance on the new plastic parts when they are installed in old style meters. The raised reinforcing rib on some of these parts may pose clearance problems in many meters. Soldering acid will dissolve some plastic so if you must use acid core solder, be careful to keep it off the plastic.

Balance and Time Check. Take plenty of time to balance and time the valves and diaphragms to insure a smooth operating meter. Use care in setting the tangent in proper relation to the valves. Time spent here will be regained manyfold when the meter is calibrated.

Most shops have a utility-type vacuum sweeper. This is an excellent means of rotating a meter before the top cover is installed. This gives the repairman a good chance to watch the valve action and make a final check on timing and balance, gage and depth of the undergear driving arm in relation to the tangent. Make sure the tangent wrist is engaged, but do not touch the short flag arm.

Installing a long pointer on the index driver dog will help proof test the meter since the circumference of the circle will be much longer than the test-hand on the index. It is a good practice to rotate the meter at least two rotations of the test-hand after the index is installed. This gives a final check to make certain the meter is operating smoothly. A vacuum cleaner can also be used to good advantage here.

Pan-type diaphragms can be rebuilt with a minimum amount of investment in jigs and other equipment. Many parts, such as crank brackets and flag arms which have pressed-in bushings, are easily repaired. Rebuilding parts can save up to 75 percent of the cost of a new item.

When it comes to repairing meters, "about right" isn't good enough. The meter repairman who wants to make his company money will strive to make all checks and adjustments as accurately as possible.

PDM TESTING

Positive displacement meters may be tested by any of the following methods: (1) bell type meter prover, (2) low pressure flow prover, (3) critical flow prover, (4) by test meter. All of these methods when correctly used under proper conditions will give similar results.

Bell Prover. Meter shop personnel are more familiar with the bell prover than with the other three types.

When a bell of sufficient size is available, large-capacity meters can be calibrated very rapidly in the shop. The use of a slope gage will certainly reveal if the repairman has assembled a well balanced, timed meter with a minimum of internal friction.

Bell provers should be in a temperature controlled area, and all meters should be at room temperature before testing. Normally both a 20 percent rate and capacity rate are run when proof testing flow rate on the meter. Two inches of water pressure on the bell is used instead of the 1½ inches used to test small domestic meters.

Low Pressure Flow Prover. The low pressure flow prover is used both in the field and in the shop. It is very compact and portable, consisting of two pieces of aluminum pipe with straightening vanes in the upstream sections.

The principle is essentially that of the orifice meter, except the orifice discs are calibrated for different flow rates instead of using various coefficients. Many shops use the low pressure flow prover to test large capacity meters when a bell prover of sufficient size is not available. One or two blowers may be used to supply the necessary volume of air, or low pressure gas may be used where safe.

Critical Flow Prover. Critical flow is a condition occurring in the gas flow when the downstream pressure is approximately one-half the upstream pressure.

The critical flow prover consists of a piece of 2-inch, extra heavy pipe approximately 18 inches long. One end is threaded to screw into a test connection placed as close to the meter outlet as possible. The other end has a special orifice plate holder. The critical flow prover can be used to test positive displacement meters operating on pressure in excess of 15 psi by passing gas or air through the meter and the prover, and discharging it into the air.

The orifice plates of a critical flow prover are stamped on the down stream side with the time (in seconds) required for one cubic foot of air to pass through the orifice at critical flow conditions. Plates used in critical flow provers have rounded edges on the upstream side, whereas orifice plates in a low pressure flow prover are sharp on the upstream edge.

It is vitally important that both the critical flow and low pressure orifice plate be handled with care to prevent damage, which would change the time stamped on the plate.

The stopwatch, which is important in timing prover tests, should be serviced and checked at regular intervals. Like many pieces of mechanical equipment, the stopwatch will operate for a long time before it completely gives out, and there is a very good possibility that during this time it will be operating inaccurately. Remember that tests taken with a defective timer are inaccurate.

Test Meter Method. Test meters have become popular in recent years with larger companies and contract testers. Continued development of rotary displacement has made the test meter a practical method for field use.

Although the shop tester normally runs only two rates—check rate and capacity rate—on a bell prover, most testers run three rates on the critical flow, low pressure and test meter methods. These rates are usually about 20, 50 and 100 percent.

Testing frequency is a much discussed subject. Sales through a large capacity meter represent a great deal of money; thus, it is important to regularly check the meter's accuracy. Some of the larger meters should be tested at 90-day intervals. Others of intermediate size could perhaps operate for two, five or seven years without being tested and at the end of the scheduled time, simply be changed out.

Many factors affect a testing and change-out schedule. The meter's mileage, its type of service, the type and amount of gas the meter must measure, and the length of time the meter is in service are important factors in setting up a testing and change-out schedule. Only after all of these factors have been considered, can an adequate testing and change-out program be established.

Sizing headers for meter stations

Calculate the velocity of gas in meter headers for proper sizing to achieve good measurement

Glenn E. Peck, Senior Engineer
Pipe Line Division
Bechtel Corp., San Francisco

METER AND REGULATOR station design for natural gas transmission systems continues to grow more complex and demand more of the design engineer. Basic elements of today's station remains about the same but higher pressures and ever increasing capacities have replaced the positive displacement meter with multi-tube orifice meters and automatic chart changers. Buildings are often pre-fabricated structures designed specifically to house telemetering transmitters, supervisory and communications equipment in addition to basic recording instruments. The multi-run setting may have a computer giving the dispatcher instantaneous rate and totalized flow for the station. Electrical equipment may cost several times that of the entire station of a few years ago.

There is a sparsity of published criteria for sizing of orifice meter headers in the multi-run station. A discussion of factors that affect the sizing of the header and arrangement of component parts for satisfactory operating conditions should be in order.

We shall assume that the orifice tube and its component parts are designed in accordance with AGA Gas Measurement Report No. 3, that piping meets all the requirements of USASI B31.8 for pressure, materials, and class of construction, that all valves and fittings meet appropriate codes or standards, and that the differential pressure across the orifice plate will not exceed 100 inches of water.

The design goals may be listed as follows:

1. Commercially available equipment will be used that is specified to the closest reproducible economic tolerances.
2. The arrangement of the equipment shall conform to accepted practice for good measurement and allow fullest use of the available meter tubes.
3. Fittings and connections to the header shall minimize the flow turbulence in so far as is practical.

Meeting the above goals allows considerable latitude in the choice of components according to the judgement and experience of the designer without contradicting code, contract, or economic requirements.

In sizing the header a number of factors must be considered, namely:

- The number of orifice tubes which will eventually be connected to the header, or the ultimate flow capacity.
- The setting layout, or arrangement, which governs the resistance flow paths the gas must take from entering the inlet header and dividing through the meter tubes before recombining in the outlet header and exiting.
- The type of connections which the tubes make with the header.

Station Layout. There are as many opinions of what constitutes good meter station layout as there are engineers working with gas measurement. Four of the more common arrangements are illustrated below and will be discussed relative to the flow-resistance paths offered:

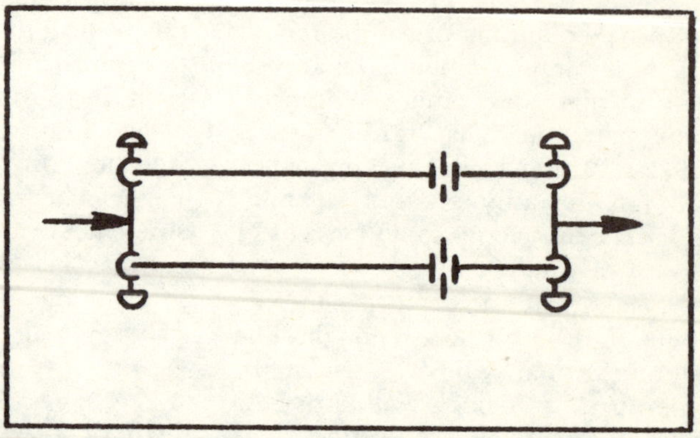

Figure 1. The flow-resistant paths are identical for this layout and each of the two runs will furnish its full capacity with near identical differential readings on the meters. However, when additional capacity is required, an additional run will have a higher or longer resistance-flow path and will furnish a lower differential reading and capacity as compared with the first two runs.

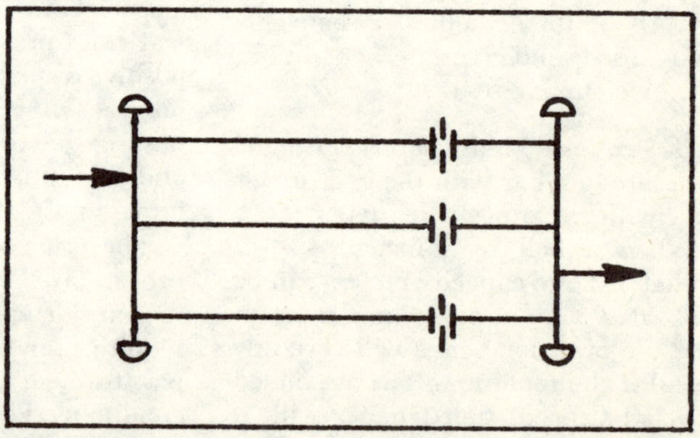

Figure 2. In this arrangement the middle tube has the lower resistance-flow path compared with the outer tubes and will affect the meter reading accordingly. The difference between meter readings will be more noticeable if the headers are undersized.

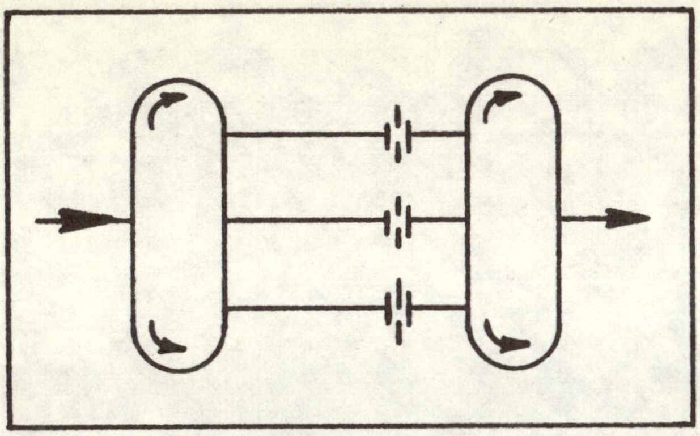

Figure 3. The "donut" header has been used frequently when large headers could be utilized. At higher pressures and capacities this layout places a high economic burden on the station and requires more space which can be a problem when a limited site is available.

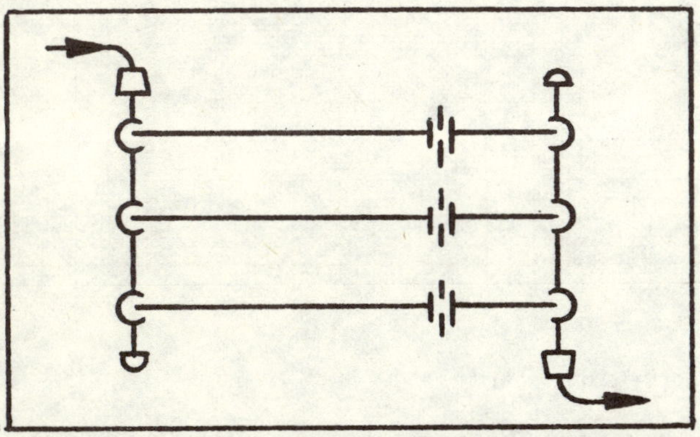

Figure 4. This layout has been a favorite for many since the advent of the extruded or drawn nozzle header and because the resistance-flow paths are as nearly identical as is possible using commercially available equipment. It is limited to three or four meter tubes before the header begins to be "out-sized."

Header Connections. Not many years ago, the header was a common piece of pipe with "orange-peel" ends and notched-mitre nipple connections. The next step was a shaped nipple reinforced by a welding saddle or pad. Higher pressures brought forged steel tees into use. On larger sized tees the spacing between runs or tubes depended on the measurement for the tee and required a number of girth welds. With the introduction of the extruded nozzle header, the limitation of tees were eliminated for a generally more economical header.

The radius of curvature or flaring of the nozzle as extruded has superior flow characteristics over the reduced branch of a forged tee and both are vast improvements over the square cut opening for a shaped nipple. Some designers have gone a step further and reduced the 90° header nozzle by the use of a 45° or 60° lateral nipple to the axis of the header. This reduces the turbulence entering or leaving the header and could reduce the diameter of the header. This method can be much more costly for fabrication, is suspect for structural integrity at higher pressures and extreme ambient conditions, and goes beyond the configurations used in establishing the proportions, beta ratios, and coefficients for the orifice meter.

Runs Per Header. The number of runs, or meter tubes, for a given header is limited by the velocity criterion selected, the operating pressure, the size of fittings obtainable, and the relationship of the peak load to the normal load or capacity for which the meter setting is designed. The operating pressure has an important bearing on the number of orifice tubes used with a given header because the capacity of the tube is proportional to the square root of the pressure while the capacity (velocity) of the header is proportional to the pressure in atmospheres. At lower pressures the number of runs might be limited to three while at higher pressure the limit would be four runs. Size of fittings is generally less limiting than other factors unless the meter tube is greater than 12" in diameter. For the very large tube, the extruded nozzle header is required since reducing branch tees are not generally available. The relationship of the peakload to the sizing of the header will be discussed under the velocity sizing criterion.

VELOCITY SIZING

The following velocity criterion is suggested for sizing of headers for the sales orifice meter setting (expressed as feet per minute):

Target	Normal Limit	Maximum Limit
1500	2000	2500

The relationship of the peak hour load to the average hour for the peak day designed for the station determines the range of velocity used in sizing the header. For example, the peak hour which is 10% of the peak day and one that is only 1% higher than the average hour for the peak day obviously demand that different limits be used for economic justification. The 10% peak hour suggests that for higher fluctuation in demand, a higher velocity be allowed in the header for the short time involved.

Keeping the velocity as low as possible within this criterion will provide the basis for good measurement and minimize the difference between adjacent meter readings. The way that gas enters and leaves the header

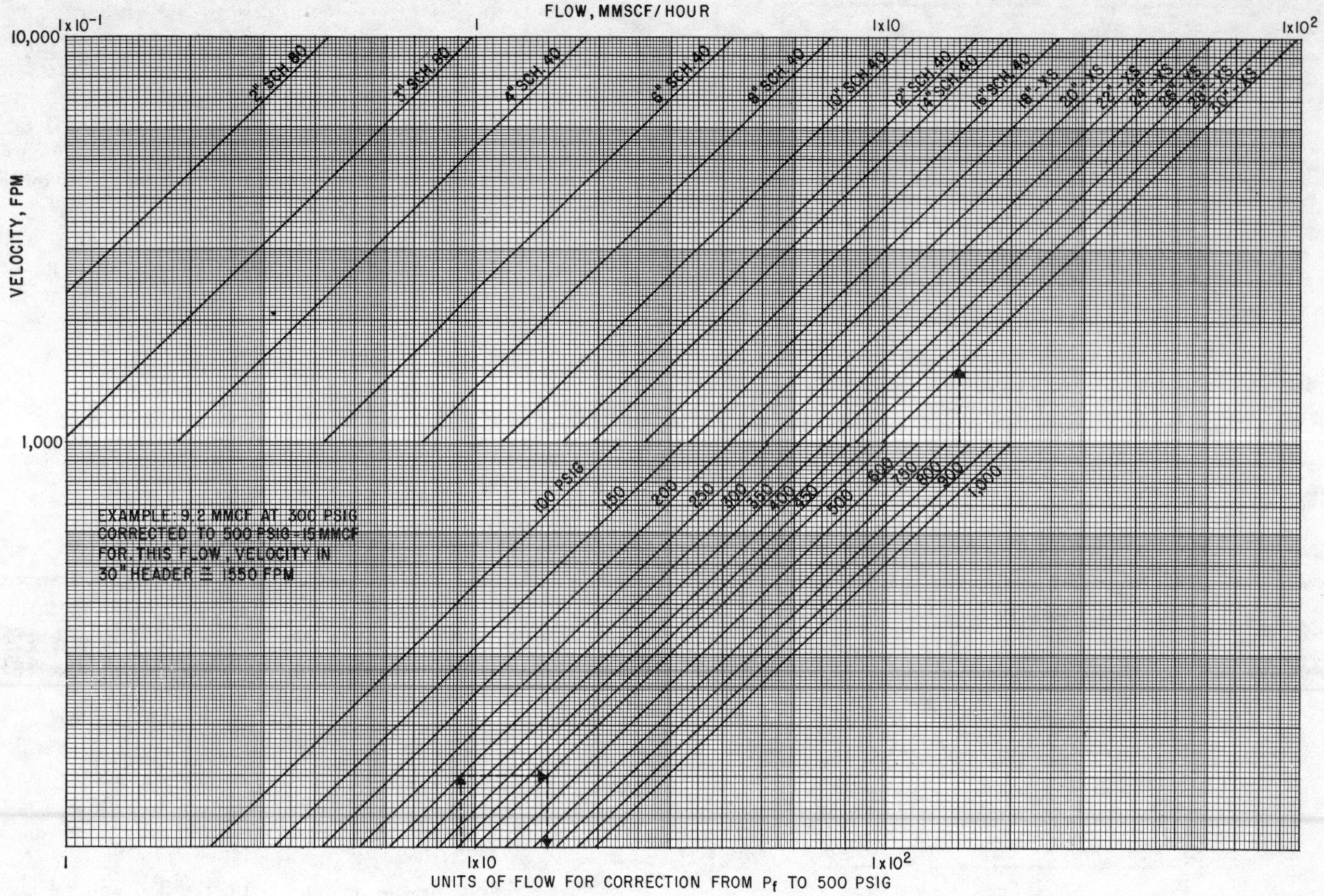

Figure 5. Pipe and header velocity sizing chart.

connections to the meter tube is important in considering the velocity criterion which is chosen.

Sizing Procedure. The normal procedure for sizing the station follows:

1. Take the peak day forecast or contract load and calculate the appropriate peak hour figure—provided the peak hour figure has not been fixed as a part of the forecast or contract. (Suggested period: 10 years, minimum)
2. Divide the peak hour by the number of parallel tubes desired.
3. Use a flow rule or chart to size the orifice tube. Check the initial minimum hour against the minimum beta ratio and differential desirable since this may affect the size and number of tubes.
4. Use a chart, such as Fig. 5, to calculate the velocity in the header for the proper criterion. Fig. 5 has been found useful since the capacity can be traced up to the desirable velocity and the nearest appropriate size of header chosen.

Fig. 5 was prepared for 500 psig or 514.65 psia at a temperature of 60° F in order that sizing of pipe or headers be made against velocity and capacity in Mscf/h on the upper part of the chart. The lower part of the chart is used to correct the flow at other pressures to the proportionate flow at 500 psig. Since the sizing procedure is an approximate one, it is felt that corrections for temperatures other than the standard condition and for compressibility are not necessary. Use of Fig. 5 or a chart prepared to your own physical conditions will allow sizing of headers for the orifice meter setting quickly and accurately within the present state of the art.

13. Oil & Products Pipe Lines —Design & Engineering

Determining the viscosity of crude

If the viscosity of a gas-saturated crude oil at the saturation (bubble-point) pressure is known, using this nomograph you can quickly estimate the viscosities at higher pressures.

For Example: Find the viscosity at 4200 psia for a crude oil when its viscosity is 30 cp at the saturation pressure of 1200 psia. Notice that 4200 psia is 3000 psi above the saturation pressure. Connecting 3000 on the pressure difference scale (left) with 30 on the curved scale for viscosity at the bubble-point pressure. The intersection with the scale on the right at 48 cp is the desired value.

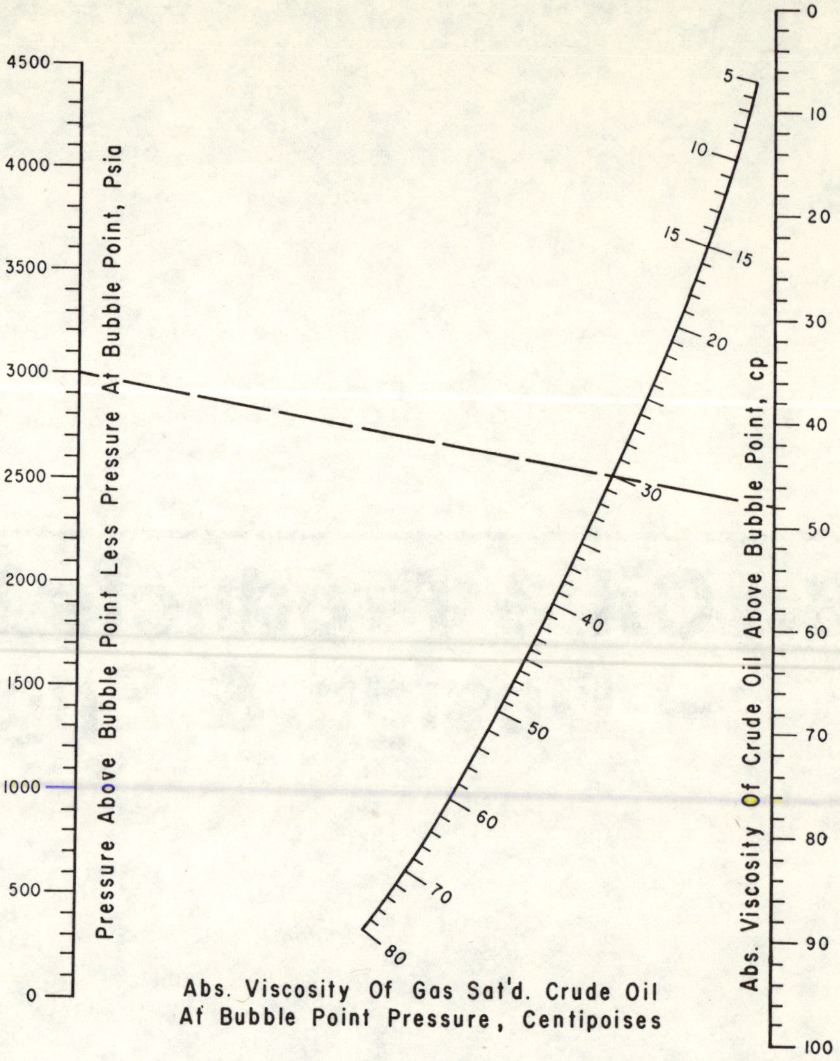

Sizing pipe lines for water flow

This form gives a reasonable degree of accuracy.

$$d^5 = \frac{(GPM)^2 (D)}{(10)\ (H)\ (100)}$$

d = inside diameter of pipe in inches

GPM = gallons per minute

D = equivalent distance (in feet) from standpipe (or pump) to outlet

H = total head lost due to friction in feet

Example: What size line should be run from a water tower to a cooling tower for a minimum of 160 gpm if the equivalent distance (actual distance in feet plus allowances for valves and fittings) is 1000 feet and the minimum head supplied by the tower will always be at least 40 feet?

Solution:

$$d^5 = \frac{(160)^2 (1000)}{(10)\ (40)\ (100)} = 640 \text{ inches}$$

Rather than extract the fifth root of 640, simply raise inside diameters of logical pipe sizes to the fifth power and compare them with 640.

If three-inch ID pipe were used

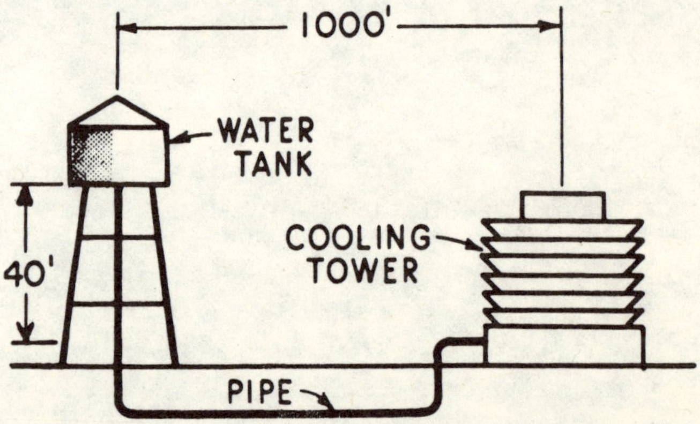

$$d^5 = (3)\ (3)\ (3)\ (3)\ (3) = 243$$

This is obviously too small.

Trying four-inch pipe

$$d^5 = (4)\ (4)\ (4)\ (4)\ (4) = 1024$$

Liquid gravity and density conversion chart

This line chart provides an easy method for converting units of liquid gravity and density. Draw a horizontal line perpendicular to the scale line through a known value and read the equivalent value on any other scale.

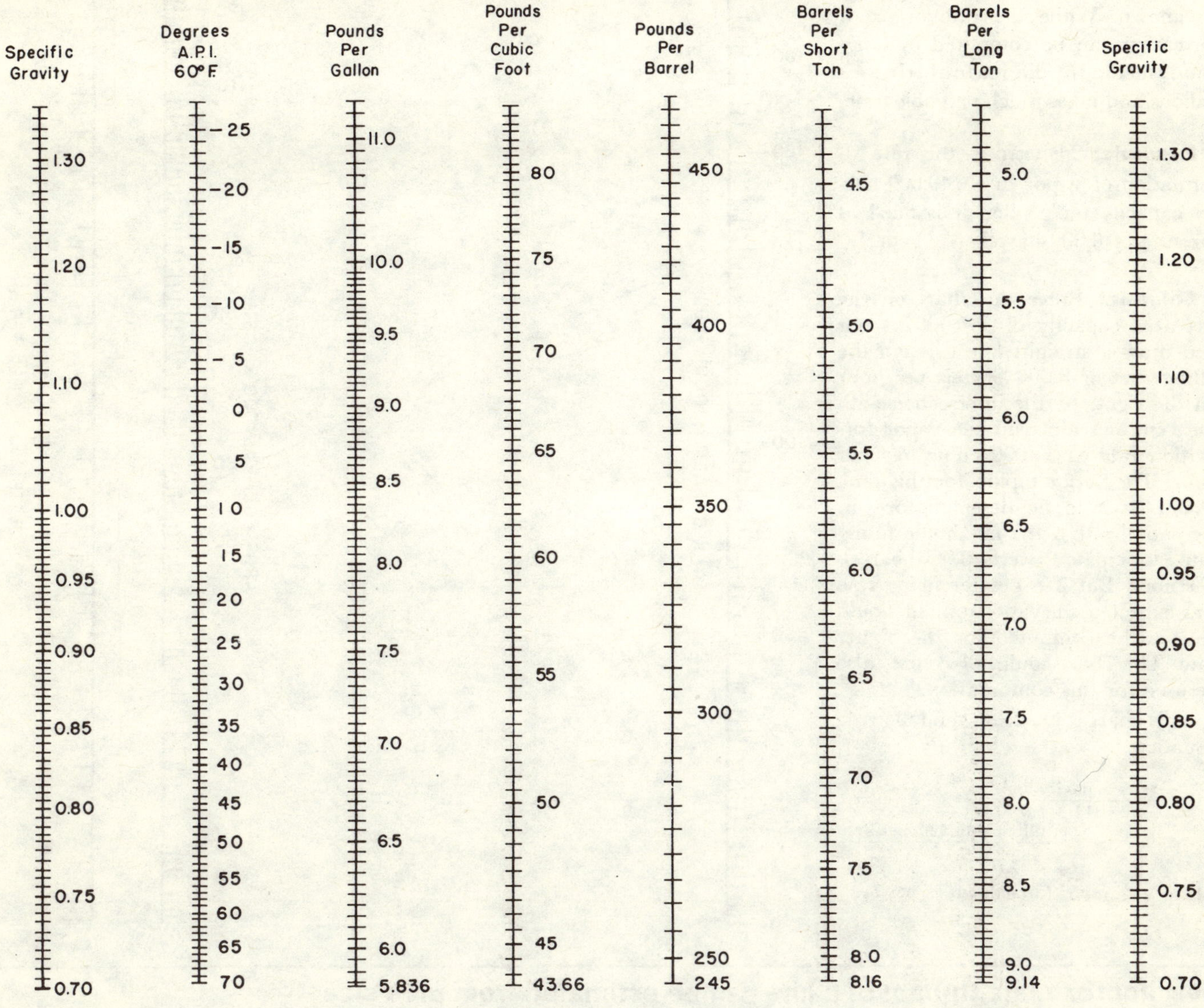

LIQUID GRAVITY AND DENSITY

Estimate the volume of a pipe line per linear foot using the inside diameter

To find the volume of a pipe line in gallons per foot square the inside diameter in inches and multiply by 4 percent. A more accurate answer may be obtained by adding 2 percent of this result to the first answer.

Example: Find the volume of a pipeline having an ID of 7.981 inch expressed in gallons per foot.

Volume = $(0.04) \times (7.981)^2 = 2.548$ gallons per foot

For greater accuracy add 2 percent of 2.548.

$2.548 + 0.051 = 2.599$ gallons per foot.

Actual volume = 2.599 gallons per foot.

To find the volume of the pipe line in barrels per foot, square the inside diameter in inches and divide by 1000. For greater accuracy subtract 3 percent of the first answer.

Example: Find the volume of a pipe line with an I.D. of 7.981 inches in barrels per foot.

$$\text{Volume} = \frac{(7.981)^2}{1000} = 0.0637 \text{ barrels per foot.}$$

For greater accuracy subtract 3 percent of 0.0637

$0.0637 - 0.0019 = 0.0618$ barrels per foot

Actual volume = 0.0619 barrels per foot.

Chart gives tank's vapor formation rate

When sizing the vapor piping for a manifolded expansion-roof tank system, the rate of vapor formation must be known. While the rate of vapor formation can be computed by longhand methods, the calculation is tedious and takes much valuable time.

Example: Determine the rate of formation of vapor in a 140,000 barrel capacity tank when it is filled at the rate of 8,000 barrels per hour.

Solution: Enter the chart on the left at a capacity of 140,000 barrels and draw a straight line through the filling rate of 8,000 barrels per hour on the right. At the intersection with the central scale read the vapor formation rate as 55,000 cubic feet per hour. The vapor piping for this tank would have to be designed for this formation rate if the maximum filling rate anticipated were 8,000 barrels per hour. But if a greater filling rate was expected, the vapor formed would have to be computed for the higher rate. The chart could, of course, also be used for this computation.

This chart is based on the following equation:

$$V = \frac{\text{tank capacity, bbl.}}{14.3} + \frac{\text{filling rate, bbl. per hr.}}{0.178}$$

where V = vapor formed, cu. ft. per hr.

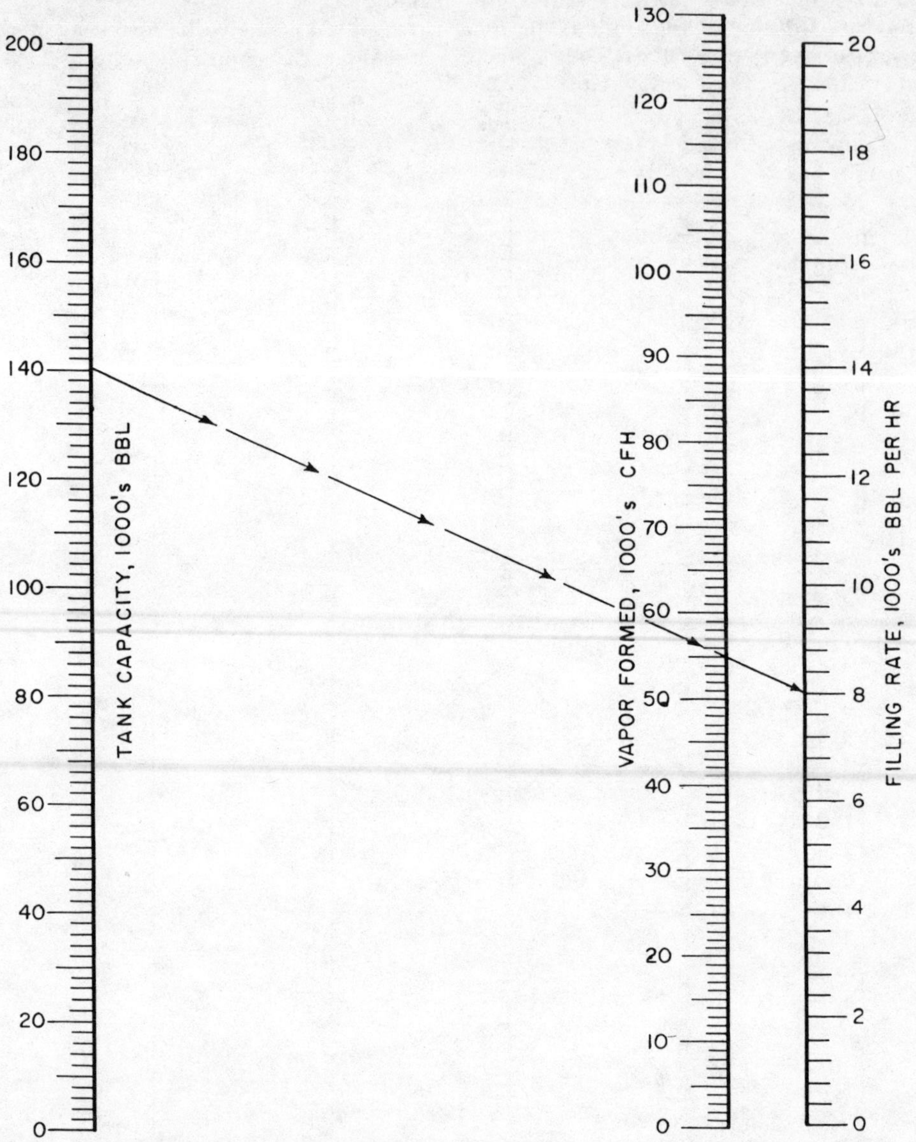

How approximate thruput of a line can be estimated from pipe size

Multiply the square of the nominal pipe size by 500.
Example: The throughput of a 12-inch line is approximately 144 by 500 = 72,000 barrels per day.

Naturally, the results of this rule are at best only a very rough approximation, since no account is taken of gravity, viscosity, pressures, or even of spacing between pump stations. Nevertheless, almost all of the oil and products lines now in operation have throughput capacities which will be given by the above rule, with the constant (given as 500) ranging between 300 and 600. The rule does *not* apply to short lines, such as loading lines, or runs within a refinery yard, but only to cross-country pipe lines.

Gage liquid flow where no weir or meter is available

Rule: Measure number of seconds required to fill five gallon pail and divide this number into the factor 10,000 and get the answer in barrels per day.

Actually the factor is 10,286, but it is so close to 10,000 that 10,000 can be used for a rule of thumb number since it is easy to remember.

Example: It requires 100 seconds to fill a 5-gallon bucket. 10,286 divided by 100 equals 102.86 bbls. per day, or for practical purposes: 10,000 ÷ 100 = 100 b.p.d.

Proof: 100 seconds to fill 5-gallon pail equals 20 seconds per gallon equals .05 gallons per second.

$60 \times 60 \times 24 = 86{,}400$ seconds in a day.
$86{,}400 \times .05 = 4320$ gallons per day.
$4320 \div 42$ gal. per bbl. $= 102.86$.

Use nomograph to find tank capacity

This simple nomograph can be used to find the capacity of your vertical cylindrical tanks. Here's how it works:

Draw a straight line from the "height" scale through the "diameter" scale and to the first "capacity, barrels" scale. Read directly the capacity of the tank in barrels. (Note: The "height" scale may be used to indicate the over-all height of the tank or the depth of liquid in the tank.)

Draw a second straight line connecting the two "capacity, barrels" scales at the same reading on each scale. Read the capacity of the tank in gallons and cubic feet on the proper scales.

The nomograph was constructed as follows:

1. The "height" scale is based on two log cycles per ten inches with a range of 1 to 60 feet.
2. The "capacity, barrels" scale is based on four log cycles per ten inches with a range of 20 to 150,000 barrels.
3. The "diameter" scale is based on three log cycles per ten inches with a range of 4 to 150 feet.
4. The distance between the height and diameter scales is exactly two-thirds the distance between the height and "capacity, barrels" scale.
5. Determine points to locate the diameter scale from the following equation: Capacity, barrels $=0.1399$ (diameter)^2 (height), units in feet.
6. The "capacity, gallons" scale is based on four log cycles per ten inches. The initial point on the scale is determined as follows: 20 barrels x 42 gallons per barrel — 840 gallons. The range of the scale is 900 to 6 million gal.
7. The "capacity, cubic feet" scale is based on four log cycles per ten inches. The initial point on the scale is determined as follows: 20 barrels x 5.6146 cubic feet per barrel = 112.292 cubic feet. The range of the scale is 120 to 800,000 cubic feet.

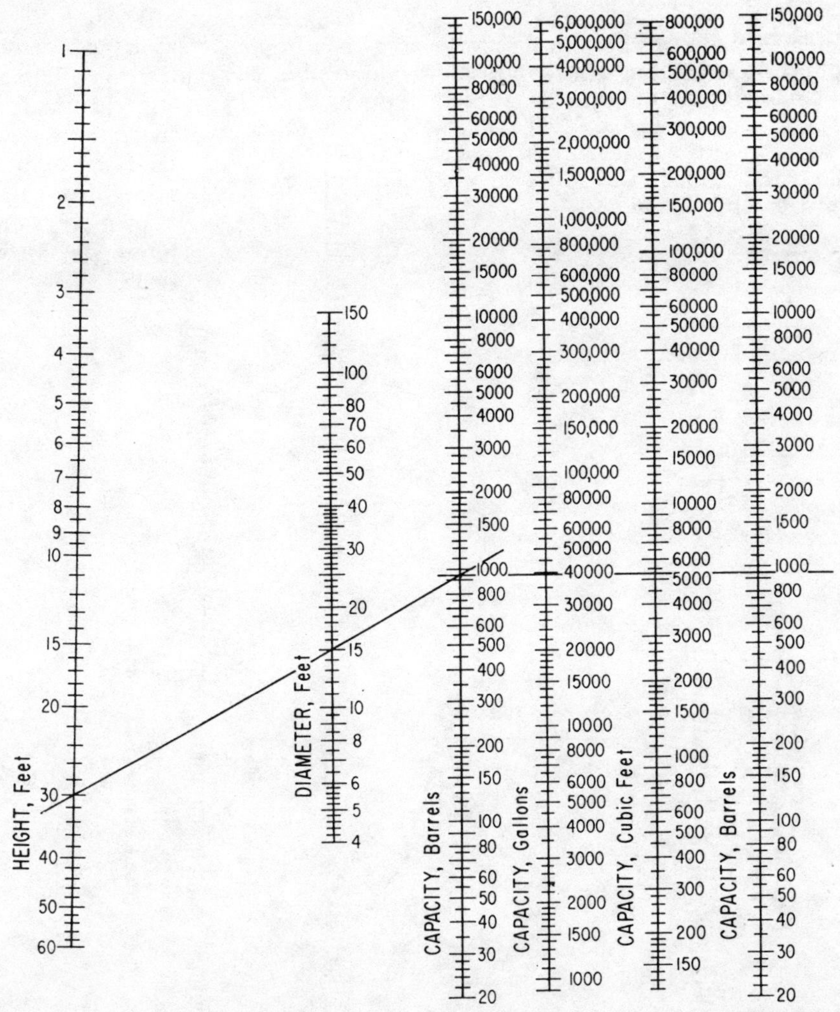

Estimate crude gathering line thruput for a given pipe diameter

To select the proper size line or estimate thruput for crude oil gathering lines, an engineer with a leading oil company has devised this handy rule: **Thruput in barrels per hour is roughly equal to a constant times the square root of pressure drop per mile (psi/mile).** Constants were derived from field data for various diameter lines between 2 and 10 inches.

Pipe diameter (inches)	Constant (K)
2	7.8
3	19.2
4	37.8
6	105
8	209
10	385

Example: What size gathering line would be required to move 240 barrels per hour of crude through a 10-mile line using a pump discharge pressure of 400 psi? Assuming no significant changes in head due to elevation and a pressure drop of 40 psi per mile, for the first trial calculation a 4-inch line appears to be the best choice so try the constant 37.8.

Thruput (4-inch) $= 37.8\sqrt{40}$
$= (37.8)\ (6.4) = 241$ barrels per hour. So a 4-inch gathering line will be adequate.

Estimating the contents of horizontal cylindrical tanks

Horizontal cylindrical tanks are frequently used for water and fuel storage and in many cases it is important to be able to gage these vessels to determine the volume of liquid contained in them. However, it is normally much more difficult to establish a volume per inch scale for a horizontal tank than for one which is in a vertical position. The accompanying nomograph simplifies this problem.

To use the nomograph, it is necessary to gage the tank and determine the ratio of the depth of liquid in the tank to tank diameter. After this is found, draw a straight line from the point on the "Ratio" scale through the known point on the "Diameter of Tank" scale and read the intercept on the "Gallons per Foot of Length" scale. From this point, draw a second line through the known point on the "Length of Tank" scale and read the intercept on the "Gallons (or Barrels) in Total Length" scale.

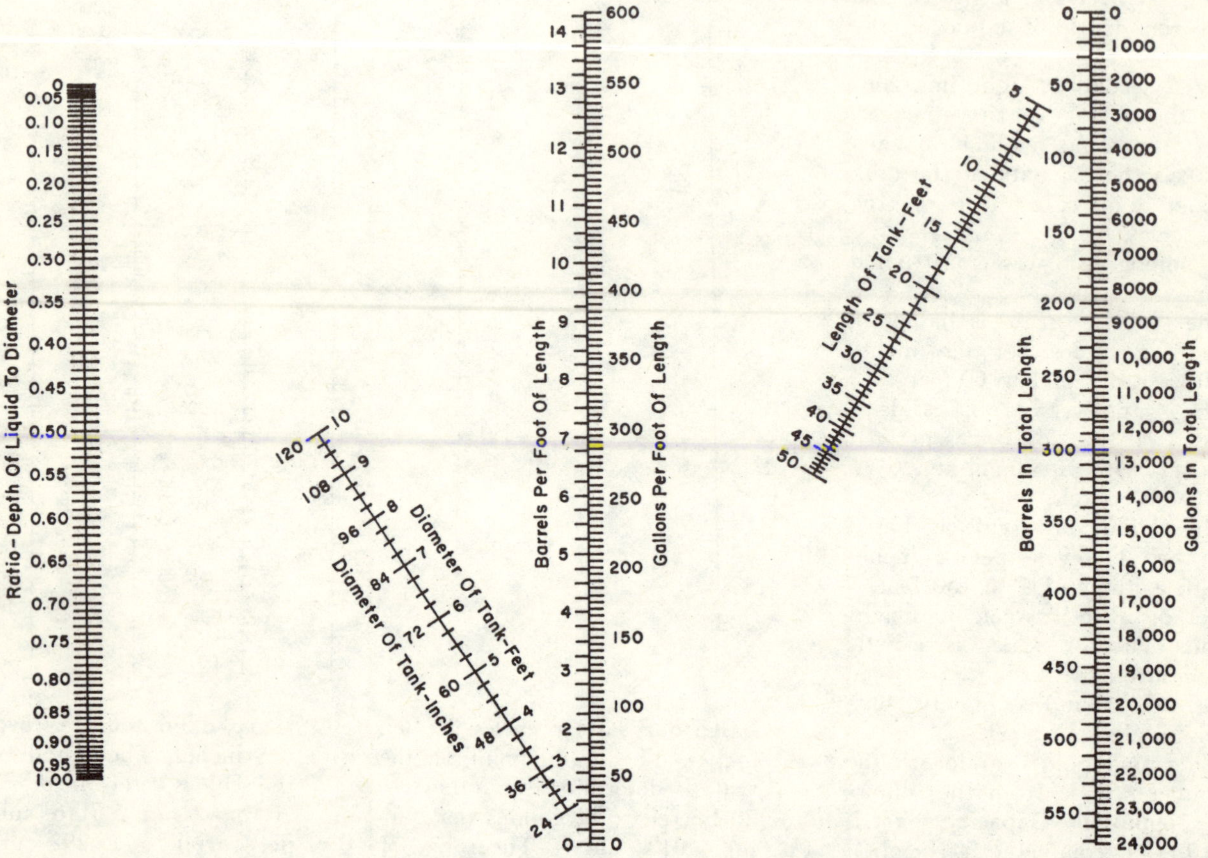

Find the volume of liquid contained in a horizontal cylindrical tank 7 feet in diameter and 20 feet long when the depth of the liquid is 4 feet, 10.8 inches.

The ratio of depth of liquid to tank diameter is:

$$58.8/84 = 0.70$$

Connect 0.70 on the Ratio scale with 7 feet on the Diameter scale and continue the straight line to obtain the intercept 215 on the Gallons per Foot of Length scale. Draw a second line from the point 215 through the 20 on the Length of Tank scale and continue the line to obtain the intercept 4,300 on the Gallons in Total Length scale. Total liquid content of the tank is 4,300 gallons.

How to gage a horizontal cylindrical tank

Express the depth in % of the diameter; then the result will be given in % of total capacity of the tank.

Rule (1), for depth up to 30; multiply the square root of the depth by the depth, and then by 0.155.

Example: liquid depth is 16% of tank diameter

$\sqrt{16} \times 16 \times 0.155 = 4 \times 16 \times 0.155 = 9.9\%$

(The correct answer is 10.3%; error is about .4%)

Rule (2), for depth between 30 and 50; subtract 10 from the depth, multiply by 1.25.

Example: liquid depth is 44% of tank diameter

$(44 - 10) \times 1.25 = 34 \times 1.25 = 42.5\%$

(The correct answer is 42.4%)

The maximum error, for depths less than 5% may be as great as 10%; gauging in this range is always difficult, and a very small slope can introduce a much larger error than this. When the depth is greater than 50%, apply the same rule to get the volume of the empty space above the fluid, and subtract.

Chart gives API gravity of blends quickly

API gravity of a crude oil blend may be readily estimated from the gravity of the components and their percentage composition. It is a simple procedure to use these curves to find the resulting API gravity of the blend.

Example: If you blend a 14 degree API pitch (60 percent) with a 36 degree API cutting stock (40 percent), the resulting fuel oil has an API gravity of 22 degrees as read from the nomograph. Calculated result from gravity tables would be 22.1 degrees API.

Results found using this nomograph checked out within one degree API over the range of gravities and percentage of components (1) and (2) in the below nomograph. Estimates from the nomograph are based on the assumption that volumes of blends are additive and that no light components flash off in blending.

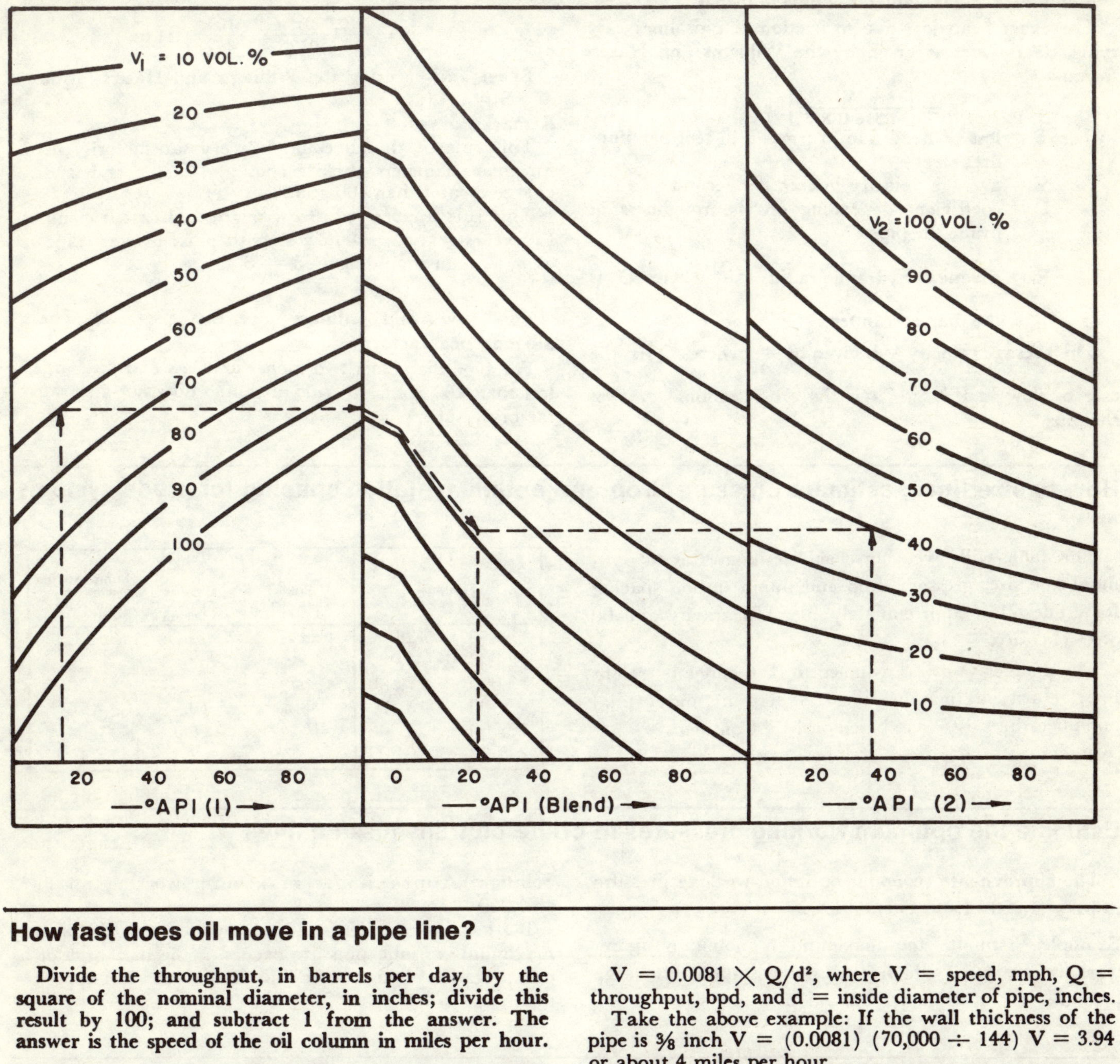

How fast does oil move in a pipe line?

Divide the throughput, in barrels per day, by the square of the nominal diameter, in inches; divide this result by 100; and subtract 1 from the answer. The answer is the speed of the oil column in miles per hour.

Example: A 12-inch line has a throughput of 70,000 bpd; 70,000 ÷ 144 = 486; 486 ÷ 100 = 4.86; 4.86 — 1.00 = 3.86. The oil moves about 4 miles per hour.

This rule is rough, but seldom is an accurate answer needed for this problem. Greater accuracy can be had by using the true inside diameter of the line in the following formula:

$V = 0.0081 \times Q/d^2$, where V = speed, mph, Q = throughput, bpd, and d = inside diameter of pipe, inches.

Take the above example: If the wall thickness of the pipe is ⅜ inch V = (0.0081) (70,000 ÷ 144) V = 3.94 or about 4 miles per hour.

Example: A 20-inch pipe line has a throughput of 200,000 barrels per day. How fast does the oil move inside the pipe line?

By the first method: 200,000 ÷ 400 = 500; 500 ÷ 100 —1 = 4 miles per hour.

By the second method: $V = (.0081)(200{,}000 \div 19.25^2)$ V = 4.37 or about 4 miles per hour.

How to determine head loss due to friction in ordinary iron pipe line carrying clear water

The form given below gives a reasonable degree of accuracy.

$$S = \frac{G^2}{10\ d^5}$$

where:

S is the head loss due to friction in feet per 100 feet length of pipe line
G is the rate of flow in gallons per minute
d is the nominal diameter of pipe in inches

The exact head loss due to friction in pipe lines carrying clear water is given by the Williams and Hazen formula

$$S = \left[\frac{V}{1.318\ C\ R^{.63}}\right]^{1.85}$$

where S = loss of head due to friction in feet per hundred feet
V = average velocity in feet per second
C = coefficient depending on the roughness of inside the pipe

R = the mean hydraulic radius = $\frac{D}{4}$ where D is the diameter in feet.

The average velocity V is given by $\frac{Q}{a}$ where Q is the rate of flow in ft.³/sec. a is the cross sectional area of the pipe.

C is a function of the material and age of the pipe and the type of water. For smooth new iron pipe C = 120. For 15-year-old ordinary pipe C = 100. For 25-year-old ordinary pipe C = 90.

Problem: What will be the head loss per 100 feet if a 30-inch line flows 10,000 gpm?

Solution:

$$S = \frac{(10{,}000)^2}{10\ (30)^5} = \frac{100}{243} = .41 \text{ ft.}$$

Exact answer using the Williams and Hazen formula S = .34 ft.

Remarks:

This rule of thumb could be very satisfactorily used for pipe diameters three inches and larger and a discharge greater than 400 gpm.

This rule applies only to wrought and cast iron pipes (20 years service). If new smooth brass or steel pipe is used, a factor should be used = .6.

For 30-year-old ordinary pipe factor = 1.2. These are empirical factors.

For pipe diameters two inches and less and discharges less than 200 gpm, this rule of thumb does not give satisfactory results.

How to size lines, estimate pressure drop and optimum station spacing for crude systems

This table will give "horseback" estimates as to optimum line size, pressure drop and pump station spacing for crude oil of approximately 40° API gravity and 60 SSU viscosity.

The depreciation was assumed to be 6 percent straight line; return on the investment 8 percent, and capital and operating costs were based on Mid-Continent averages:

FOR	USE		
Thruputs (Bbls. per day)	Pipe Diameter (inches O.D.)	Pressure Drop (psi per mile)	Pump Station Spacing (miles) (For 800 psi disch. Press.)
0 to 2,000	3½	. .	. . .
2,000 to 3,000	4½	32	25
3,000 to 7,500	6⅝	16	50
7,500 to 16,500	8⅝	10.5	76
16,500 to 23,500	10¾	8.5	94
23,500 to 40,000	12¾	7	112

Estimate the optimum working pressures in crude oil transmission lines

The approximate economic optimum working pressure in psi = 1,250 - 15 × O.D. of line.

Example: Estimate the maximum economic optimum working pressure—which is usually station discharge pressure—for a 20-inch crude oil pipe line.

Solution: Approximate maximum working pressure = 1,250 - 15 x 20 = 950 psi.

If the line profile is approximately horizontal, the maximum working pressure occurs at the discharge of a pump station. This method is useful only for approximations.

What is the linefill of a given pipe in barrels per mile?

Multiply the inside diameter of the pipe, in inches, by itself, and multiply the result by 5.13; the answer is the linefill in barrels per mile.

Example:

Six-inch schedule 40 pipe; inside diameter is 6.065 inch; 6.065 × 6.065 × 5.13 = 188.7 barrels per mile.

If the correct inside diameter is used, the rule gives the correct linefill in standard 42-gallon barrels per mile of pipe.

How to size crude and products lines for capacity increases

Any new pipe line for crude or products should be sized to carry 2/3 of its ultimate capacity.

Example: Here's the way it works out.

Design a line to take care of an initial capacity of 100,000 barrels per day. This could be done by laying a 16-inch line and spacing some 14 pump stations along its length. But to increase it 50 percent or to 150,000 barrels per day, it would be necessary to add intermediate stations—14 in all. Excessive operating costs would likely result.

However, if the line were designed so that the 100,000 barrels per day tallied up to only 2/3 its ultimate capacity, a 20-inch line with five original stations would be installed. To increase this capacity to 150,000 bpd, it would be necessary to add only five stations.

This is based on an estimate that **when expanding the capacity of a line, loop it if the increase is to be greater than 50 percent of the original capacity. If it is to be less, simply add horsepower.** This, too, often holds true.

Both of these approaches are based on statistical results, rather than on economic studies, etc.

How to determine the maximum surge pressure in liquid-filled pipe line when a main valve is closed suddenly

The maximum surge pressure that occurs in a crude or products line when a main line valve is suddenly closed may be quickly determined using the simple rule of thumb:

Line pressure equals .8 times the weight of liquid per cubic foot times its velocity in feet per second.

This saves considerable calculation required in using the standard equation:

$$P = \frac{12\,w\,V\sqrt{\dfrac{32.2}{w}\left(\dfrac{1}{\dfrac{1}{K}+\dfrac{D}{t\,E}}\right)}}{144 \times 32.2}$$

where P = surge pressure, psi
w = weight of liquid per cubic foot, lbs.
V = velocity change, fps
K = bulk modulus of the liquid, psi
D = outside diameter of pipe, inches
t = pipe wall thickness, inches
E = Young's modulus of the pipe material, psi

Example: A products line transports gasoline at 4.5 feet per second in an 8-inch line. Suddenly the main line valve is closed. Assuming the weight of gasoline to be 45 pounds per cubic foot, what would be the maximum surge pressure to be expected in the system?

Using the simplified equation, $P = .8wV = .8 \times 45 \times 4.5 = 162$ psi. Using the standard equation, P = 159 psi.

Using the Rule of Thumb the error is less than 2 percent.

Correct the volume of light fuels from actual temperature to a base of 60° F.

To approximate quickly the volume of gasoline or other light liquid fuel at 60°F. from a known volume at any temperature in the atmospheric range use the formula:

$$V_a - V_{60} = 0.0006\,(T_a - 60)\,V_{60}$$

where V_a = Volume at actual temperature
V_{60} = Volume corrected to 60°F.
T_a = Actual temperature of fuel

Example: A tank contains 5500 gallons of gasoline at 46°F. Correct the volume to a base of 60°F.

$$(5500 - V_{60}) = 0.0006\,(46 - 60)\,V_{60}$$
$$(5500 - V_{60}) = 0.0006\,(-14)\,V_{60}$$
$$5500 = V_{60} - 0.0084V_{60}$$
$$5500 = 0.9916V_{60}$$

Volume at 60°F. = 5546.6 gallons

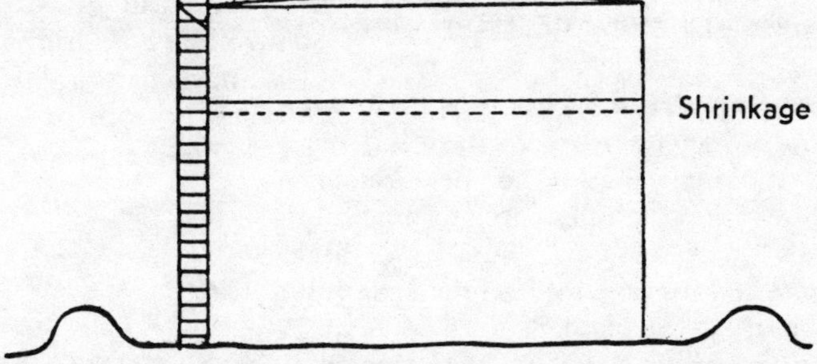

To approximate the shrinkage or expansion obtain the difference between the actual volume measured and the corrected volume. In this case:

Shrinkage = 5546.6 — 5500 = 46.6 gallons

Volume of liquid in vertical cylindrical tanks

Measure the depth of the liquid and either the diameter or circumference of the tank; then the volume in

Gallons $= 0.0034\ d^2h$ or $0.00034\ c^2h$
Barrels $= 0.000081\ d^2h$ or $0.00082\ c^2h$
Gallons $= 5.88\ D^2H$ or $0.595\ C^2H$
Barrels $= 0.140\ D^2H$ or $0.0142\ C^2H$

Where d = diameter in inches D = diameter in feet
c = circumference in inches C = circumference in feet
h = depth in inches H = depth in feet

If the circumference is measured on the outside, then three times the thickness of the tank wall should be subtracted before using the formula. Naturally, these rules cannot supplant the results of accurate tank strapping, which take many other factors into account.

Example: How many gallons will a tank 12 feet in diameter and 16 feet high hold when full?

Gallons $= 5.88\ D^2H$
$= (5.88)\ (144)\ (16)$
$= 13{,}548$ gallons

Example: How many barrels will a tank 8 feet in diameter and 16 feet high hold when full?

Barrels $= 0.140\ D^2H$
$= (0.140)\ (64)\ (16)$
$= 143$ barrels

What is the hydrostatic pressure due to a column of oil H feet in height?

Multiply H by 61.41; divide this by 131.5 + degrees API.

Example: A column of 37-degree oil is 20 feet in depth; what is the pressure at the base?

$P = 20 \times 61.41 / 131.5 + 37$
$= 1228.2 / 168.5$
$= 7.88$ pounds per square inch.

Example: What is the pressure on a closed block gate, if the highest point on the line is 1380 feet above the gate, and the line is standing full of 36-degree oil?

$P = 1380 \times 61.41 / 131.5 + 36$
$= 84{,}750 / 167.5$
$= 506$ pounds per square inch.

This rule is not an approximation, but is as accurate as the determination of specific gravity.

How engineers make pipe fit the ditch

By R. J. BROWN

THE FIVE-MEN crews of engineers that keep track of the pipe and calculate side bends, overbends, sags and combination bends can make money—or lose money—for any contractor crossing rough terrain. If they function smoothly, the bending crew and pipe gang have easy going and the lowering-in can be a breeze. But when they make mistakes, the welded-up pipe may not fit the ditch and this spells trouble for the contractor.

Bending Crew. As a rule there are two rod men, two chairmen and one instrument man in the measurement crew. Their duties are to measure each joint of pipe, number it and stake its position in the line. In addition they calculate degrees of bends, etc. Here is the way they perform these duties:

Measuring and Numbering Joints. (See Figure 1.) First, the crew measures all joints of pipe and numbers them from 0 to 100. (When 100 is reached, the numbering starts over.) Then, they stake the position of each joint of pipe in the line. They use stakes which are numbered according to the pipe, with the number of the joint behind on front of the stake and the joint ahead on back of the stake.

Stakes Set Back. Because the setup gang will move the pipe back half a joint's length and the pipe gang back another half, the engineers set the stakes from one to two joints length behind the corresponding numbered pipe.

Notes on Figure 1. Note that joint 25 was measured 39.9 feet, and the gap between stakes 24-25 and 25-26 was set at 29.9 feet to allow a 10-foot lap for tie-in. This leaves a pup long enough to be carried ahead and worked into the line. If the pups are distributed over sections ot bends, there is a possibility of saving a cut in a full joint. In event the stakes catch up with the pipe, one joint is carried ahead of the stringing.

Calculating Bends. The method used to determine the size of overbend and sags is illustrated in Figure 2. Note that at 0+00 the instrument is set up on the ditch bottom at a height of 5.0 feet. The rod is set on the ditch bottom ahead of the radius of overbend. (Care should be taken to set neither transit nor rod on the arc of the change in slope.) The transit is moved behind the radius of overbend at 1+00 and the height of the instrument is measured to be 4.90 feet.

The rod is set just ahead of the change in radius of overbend and is

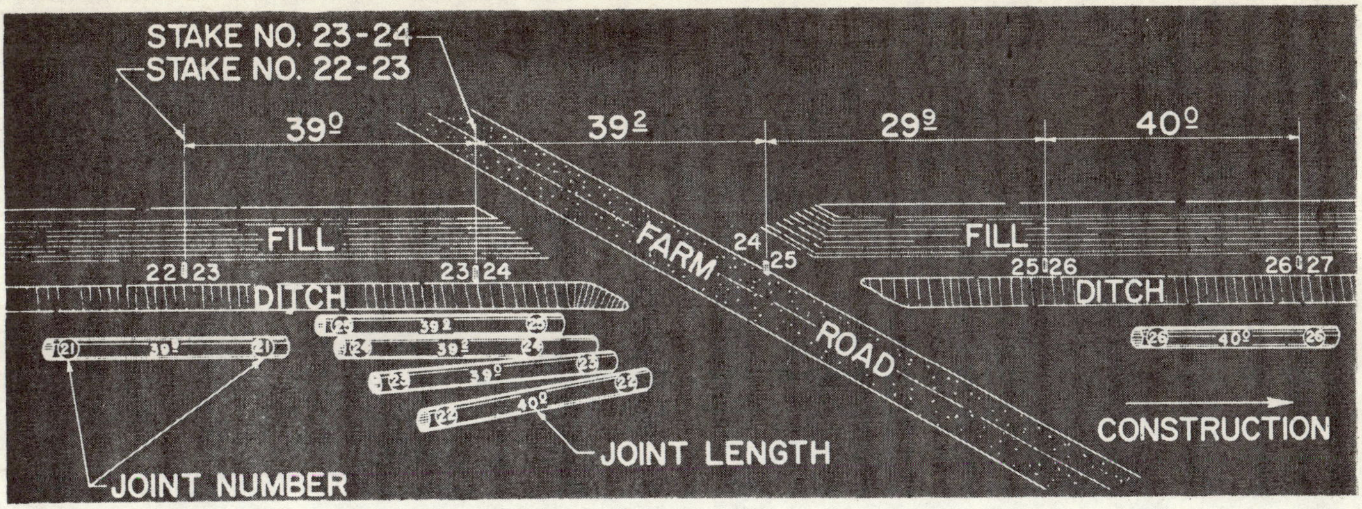

Figure 1.

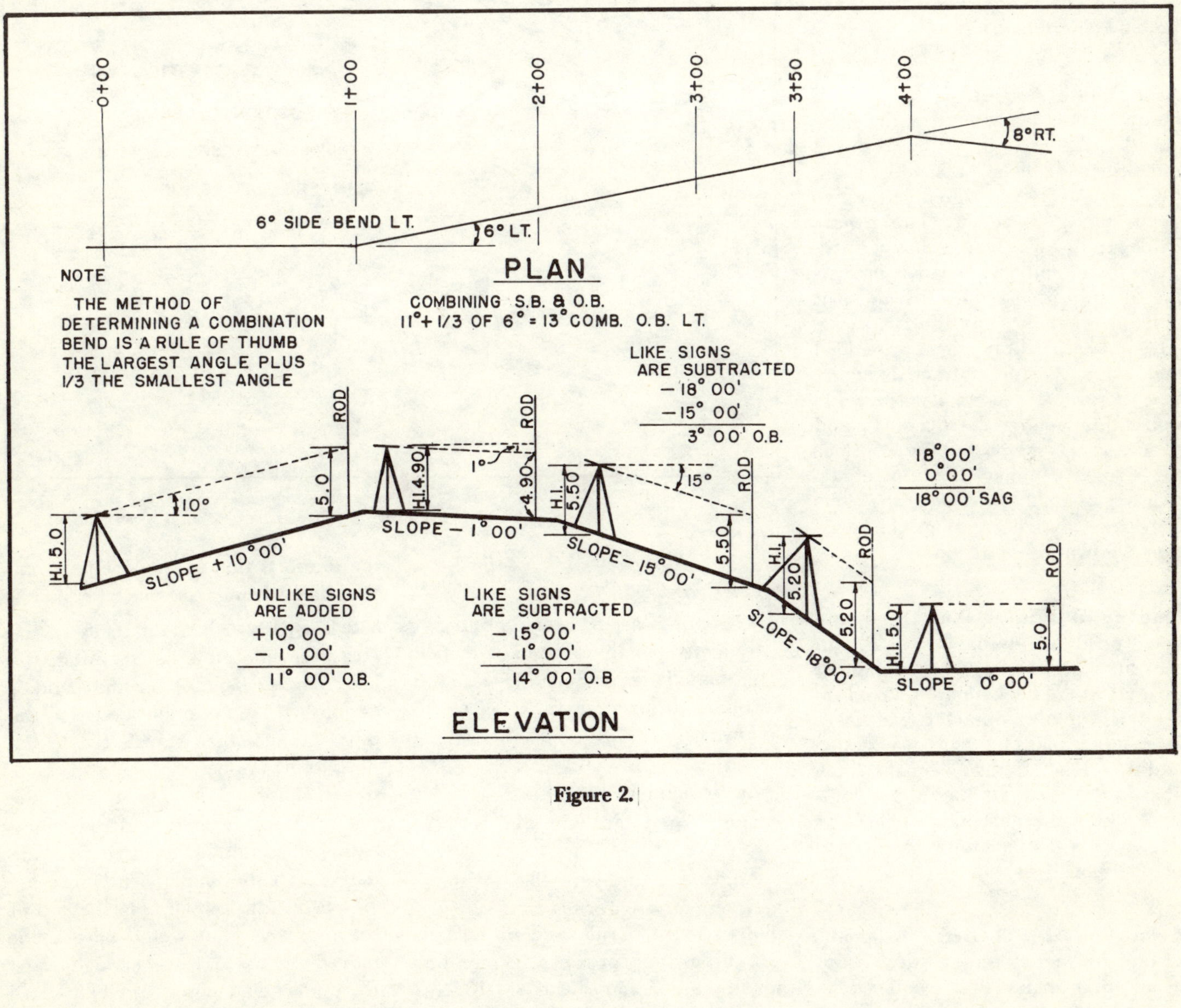

Figure 2.

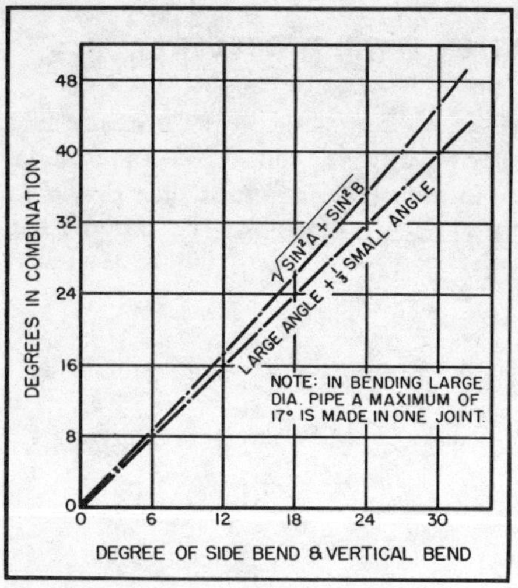

Figure 3.

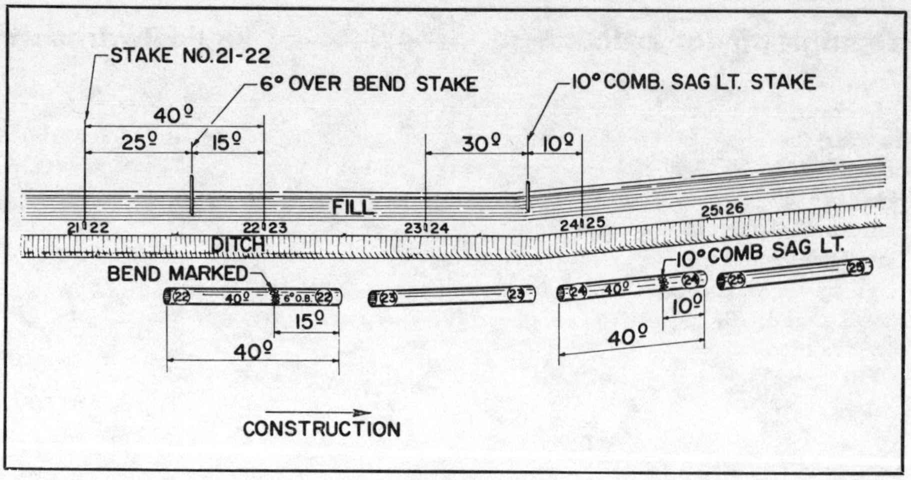

Figure 4.

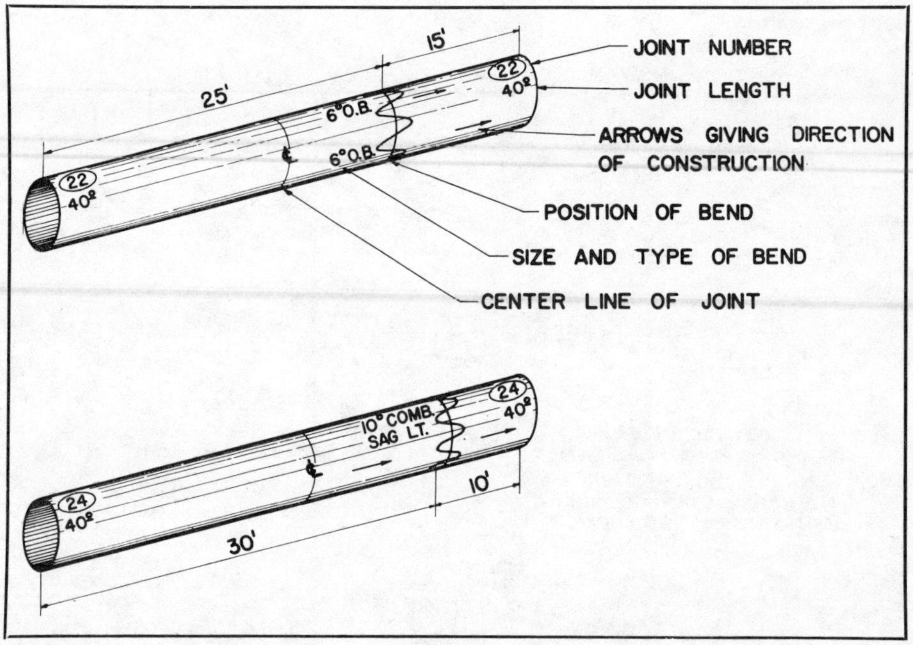

Figure 5.

read at 4.90 feet, establishing a slope of —1°. The magnitude of bend can now be determined.

Add Unlike Angles. If the slope is positive and the tangent of the slope ahead is decreasing, an overbend is established. If the slope is positive and the tangent ahead is greater positive slope, a sag is noted. The reverse holds true for change in negative slope, if the tangent ahead is increasing the transition is an overbend; if it is decreasing, the transition is a sag. The unlike algebraic signs are added and like signs are subtracted. Figure 2, station 2+00, shows a positive 10° slope approaching the overbend and a negative 1° slope leaving it. Since the signs are unlike, the sum of the two angles is 11° and since the slope is positive and decreasing, the notation overbend is used.

Side Bend Angles. Side bends are determined by "eyeballing" a P.I. in the ditch and setting the instrument up over it. A back sight is taken on the ditch or preceding P.I. whichever is easier read. The scope is flipped and a foresight is taken on the ditch ahead. Referring to Figure 2 at station 1+00, the deflection angle is 6° left; station 4+00 shows a deflection of 8° right.

Combination Bends. At points where side bends and vertical bends occur the angles are combined to make one angle rolled. The method of calculating is a rule of thumb: *the larger angle plus one-third the smaller angle.*

This rule is accurate enough to give satisfactory results in actual practice. The actual derivation of this rule comes from an approximation of the following:

$$\text{Sin (combination angle)} = \sqrt{\text{Sin}^2 A + \text{Sin}^2 B} \quad (1)$$

Angle A and B are the horizontal and vertical angles. Figure 3 shows a plot of the 1/3 rule and of formula (1) From the plot it can be seen that for the maximum bend of 18.5 an error of slightly over 1° is introduced. The maximum error evidenced on the graph is due to the side and vertical bands being equal.

Marking Pipe. The final step of the measurement crew is marking the joints to be bent. After the bend is established, its position between the pipe stakes is then transferred to the pipe joint number matching the pipe stakes number. The bend size, the centerline of bend and arrows indicating direction of construction are marked on the joint. (See Figure 5.) If a bend falls too close to the end of the pipe the joint would have to be moved, so as to leave enough room for the bend. A pup could be inserted immediately behind the joint in question to give better bend positioning.

Nomograph for calculating viscosities of liquid hydrocarbons at high pressure

Lockhart and Lenoir developed a graphical correlation showing the effect of pressure on viscosity of liquid hydrocarbons. This correlation is shown in Figure 1 which is based primarily on data of Hersey and Hopkins,[2] including pure hydrocarbons, lubricating oils, bright stocks, and distillates. Data from Reference 1 have also been included.

To use the nomograph, the characterization factor of Watson, K_w, and the viscosity of the liquid at atmospheric pressure are required.

The accuracy of the correlation decreases as pressure increases.

Example

What is the viscosity of an oil at 5,400 psia, if its characterization factor is 11.8 and its viscosity at atmospheric conditions is 90 centipoises?

Enter 5,400 psia in the pressure scale to the viscosity line of 90 and proceed horizontally to middle reference scale. Follow the curve lines to intersect the vertical line drawn at $K_w = 11.8$ and read the ratio of viscosity on the extreme left scale at 2.6. The viscosity of the oil is (2.6)(90) = 234 centipoises.

Source

Lockhart, F. J. and Lenoir, J. M., *Petroleum Refiner,* 40, No. 3, 209 (1961).

References

1. Griest, and others, *J. Chem. Physics,* 29, 711 (1958).
2. Hersey, M. D., and Hopkins, R. F., *Viscosity of Lubricants Under Pressure,* A.S.M.E., 1954.

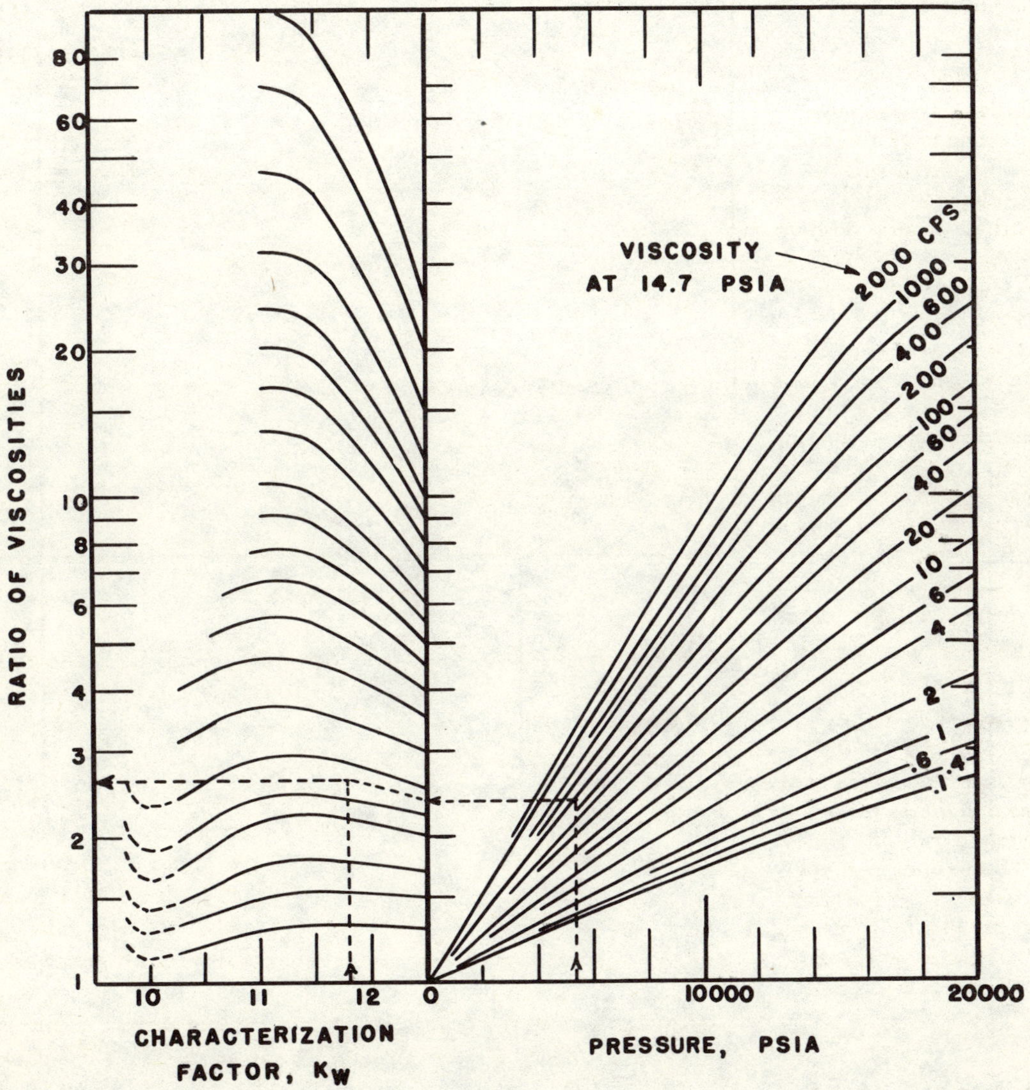

Figure 1. Isothermal effect of pressure on viscosities of liquid hydrocarbons at low reduced temperatures. (Reproduced by permission, *Petroleum Refiner,* Vol. 40, No. 3, page 209, copyright 1961, Gulf Publishing Co., Houston.)

Nomograph for calculating viscosity of blends

Viscosity is not an additive function of percentage composition. Kendall and Monroe[2] showed that for a wide variety of binary liquid mixtures, the cube roots of the viscosities are linear functions of the composition. Epperson and Dunlap[1] found that this relationship applies to lubricating oils as well and proposed the use of the Kendall and Monroe equation for this industrially important case. This equation is

$$(\mu)^{1/3} = \frac{x(\mu_A)^{1/3}}{(1-x)\,(\mu_B)^{1/3}}$$

where

μ = viscosity of the blend, centipoises
μ_A = viscosity of component A, centipoises
μ_B = viscosity of component B, centipoises
x = mole fraction of component A in the blend.

The equation is conveniently handled by the nomograph, Figure 1.

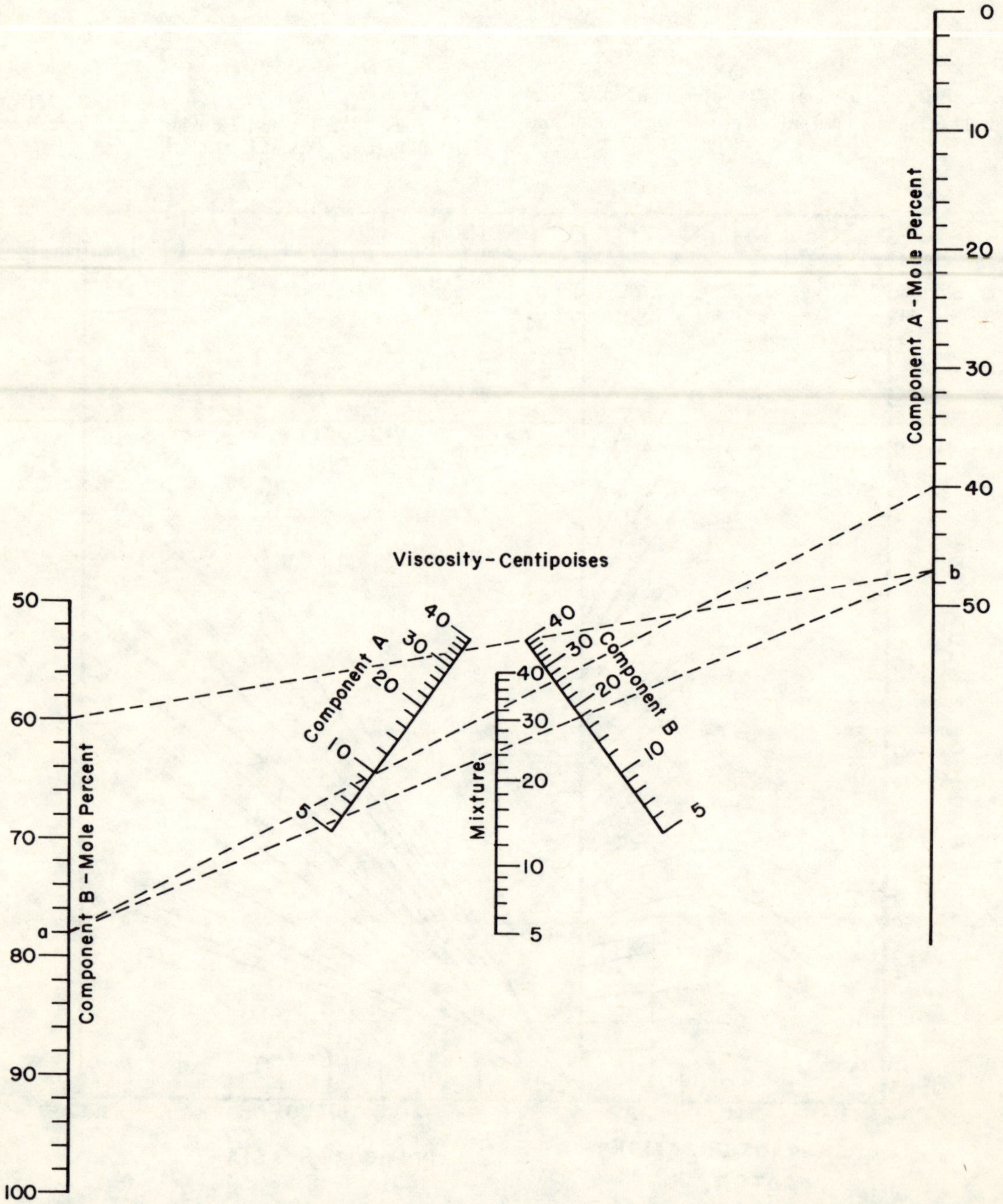

Figure 1. Viscosity of blends. (Reproduced by permission *Petroleum Refiner,* Vol. 38, No. 1, page 216, copyright 1959, Gulf Publishing Co., Houston.)

Example

Find the viscosity of a blend that contains 40 mole percent of an oil with a viscosity of 10 centipoises and 60 mole percent of an oil with a viscosity of 40 centipoises.

Connect 40 mole percent on the component A scale and 10 centipoises on the component A viscosity scale with a straight line and draw this line to meet the left hand vertical axis at the point "a." Then connect 60 mole percent on the component B scale and 40 centipoises on the component B viscosity scale with a straight line and extend this line to meet the right hand vertical axis at the point "b." A straight line connecting points "a" and "b" intersects the mixture viscosity scale at the desired value, 24.8 centipoises.

In most instances the equation and chart are good to within 2 percent, although for blends of light Pennsylvania oil with heavy mid-continent oil the uncertainty may amount to 3 percent. Throughout, percentage compositions are expressed in mole percent and not in weight or volume percentages. Mole percentages are readily calculated by means of the formula:

$$x = \frac{100\ w/m}{w/m + W/M}$$

where:

x = mole percentage of component A
w = weight of component A
m = molecular weight of component A
W = weight of component B
M = molecular weight of component B

Source

Davis, D. S., *Petroleum Refiner,* 38, No. 1 (1959).

References

1. Epperson, E. R., and Dunlap, H. L., *Ind. Eng. Chem.,* 24, 1369 (1932).

2. Kendall, J., and Monroe, K. D., *J. Am. Chem. Soc.,* 39, 1787 (1917).

3. Wilson, R. E., and Wylde, E. P., *Ind. Eng. Chem.,* 15, 801 (1923).

Nomograph for calculating viscosity of slurries

The nomogram is based on the Hatschek equation for estimating the viscosity of slurries in an aqueous suspension.

$$\mu = \frac{\mu_w}{1 - x^{0.333}}$$

where:

μ = viscosity of slurry, centipoises
μ_w = viscosity of water at the temperature of the slurry, centipoises
x = volume fraction of dry solids in the slurry

The nomogram (Figure 1) shows a scale calibrated in terms of the temperature of the water instead of its viscosity. Thus, the actual water viscosity value is not needed.

Example

What is the viscosity of a slurry having a volume fraction of solids of 0.06 at a temperature of 39° C?

Connect T = 39° C with x = 0.06 and read μ_w = 1.1 centipoises on the middle scale.

Source

Davis, D. S., *Brit. Chem. Eng.,* 4, 9, 478 (1959).

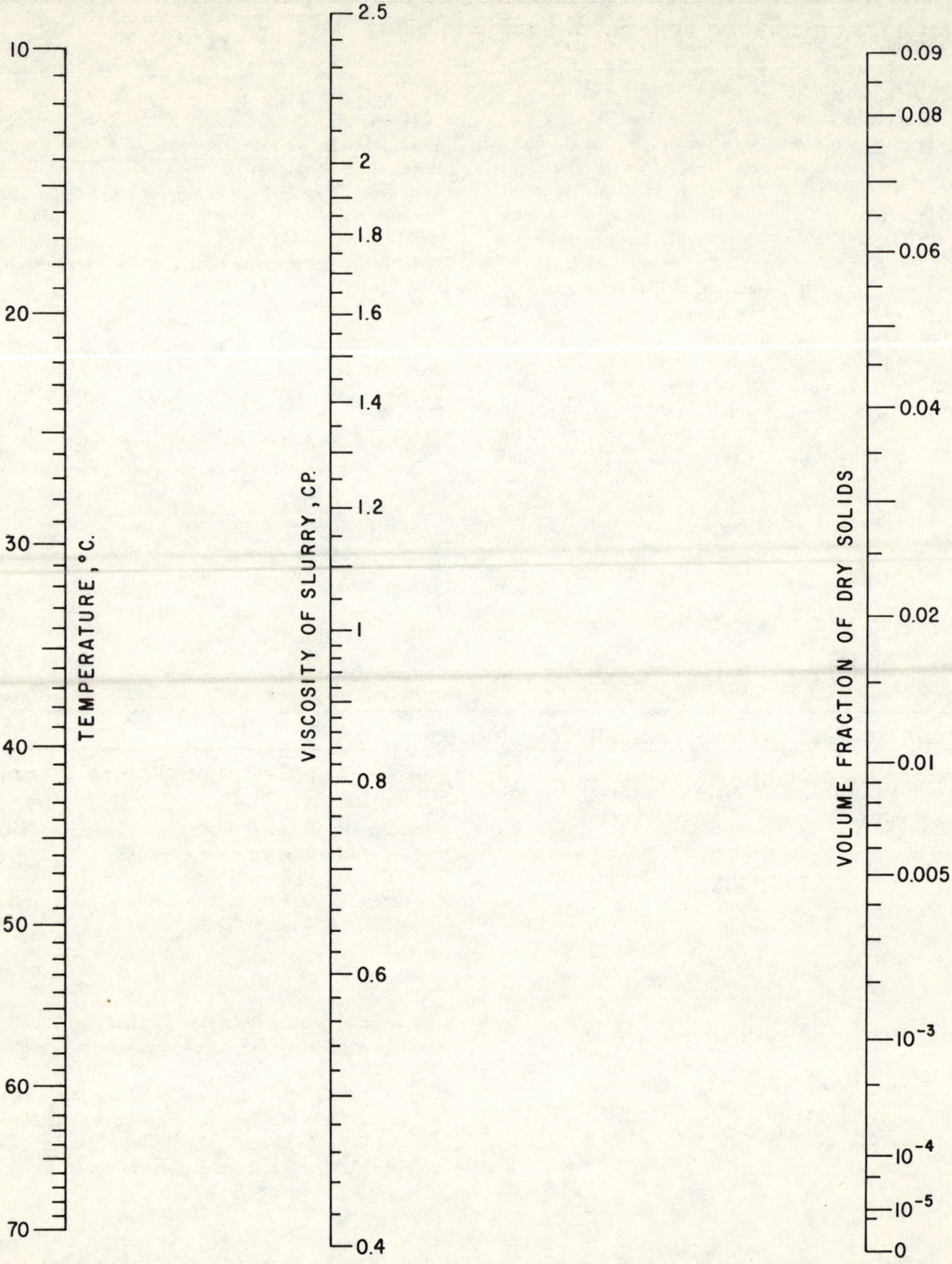

Figure 1. Viscosity of slurries. (Reproduced by permission, *British Chemical Engineering,* Vol. 4, page 478, copyright 1959.)

Nomograph for calculating velocity of liquids in pipes

The mean velocity of any liquid flowing in a pipe of circular cross-section can be calculated from the following formula:

$$v = 183.3 \frac{q}{d^2} = 0.408 \frac{Q}{d^2} = 0.0509 \frac{W}{d^2 \rho}$$

where:

v = average fluid velocity, ft./sec.
d = inside diameter of pipe, inches
q = rate of flow, cu. ft./sec.
Q = rate of flow, gals./min.
W = rate of flow, lbs./hr.
ρ = fluid density, lbs./cu. ft.

The nomograph (Figure 1, page 128) can be used to calculate the liquid velocity when the rate of flow is in cubic feet, gallons or thousands of pounds. Conversely, knowing the flow rate and velocity, the pipe diameter may be calculated.

Example

What is the velocity of fuel oil at 60° F flowing through a 2-inch schedule 40 pipe at a rate of 45,000 lbs./hr. The oil density is 56.02 lbs./gal.

Connect	With	Mark or Read
W = 45	ρ = 56.02	Q = 100 cu. ft./sec.
Q = 100	d = 2-inch schedule 40	v = 10 ft./sec.

Source

"Flow of Fluids Through Valves, Fittings, and Pipe," 3-7 *Technical Paper No. 410,* Crane Company, Chicago, Illinois (1957).

Nomograph for calculating velocity of compressible fluids in pipe (Figure 2, page 129)

The mean velocity of a compressible fluid in a pipe can be computed by the following formula, which is obtained after dividing the rate of flow in appropriate units by the cross-sectional area of the pipe:

$$V = \frac{3.06\, W\overline{V}}{d^2} = \frac{3.06\, W}{d^2 \rho}$$

where:

V = mean velocity of flow, ft./min.
W = rate of flow, in lbs./hr.
$\overline{V}$ = specific volume of fluid, cu. ft./lb.
d = internal diameter of pipe, in.
ρ = density of fluid, lbs./cu. ft.

Example

Steam at 600 psig and 850° F is to flow through a schedule 80 pipe at a rate of 30,000 lbs./hr. Find the pipe size if the velocity is to be limited to 8,000 ft./min.

Connect	With	Mark or Read
850° F	600 psig line (vertically)	600 psig
600 psig	Horizontally to	$\overline{V}$ = 1.21
$\overline{V}$ = 1.21	W = 30,000	Index
Index	V = 8,000	d = 3.7, use 4-inch schedule 80 pipe

If a 4-inch schedule 80 pipe is used, the actual velocity is found by connecting the Index with 4-inch schedule 80 to get V = 7,600 ft./min.
Note: If a different fluid is involved, the value of the density (or specific volume) needed to make the calculation can be obtained from Section 10 of this book, under "Density and specific volume of gases and vapors," page 69.

Source

"Flow of Fluids Through Valves, Fittings, and Pipes," *Technical Paper No. 410,* 3-16, Crane Company, Chicago, Illinois (1957).

Nomograph for calculating velocity of liquids in pipes

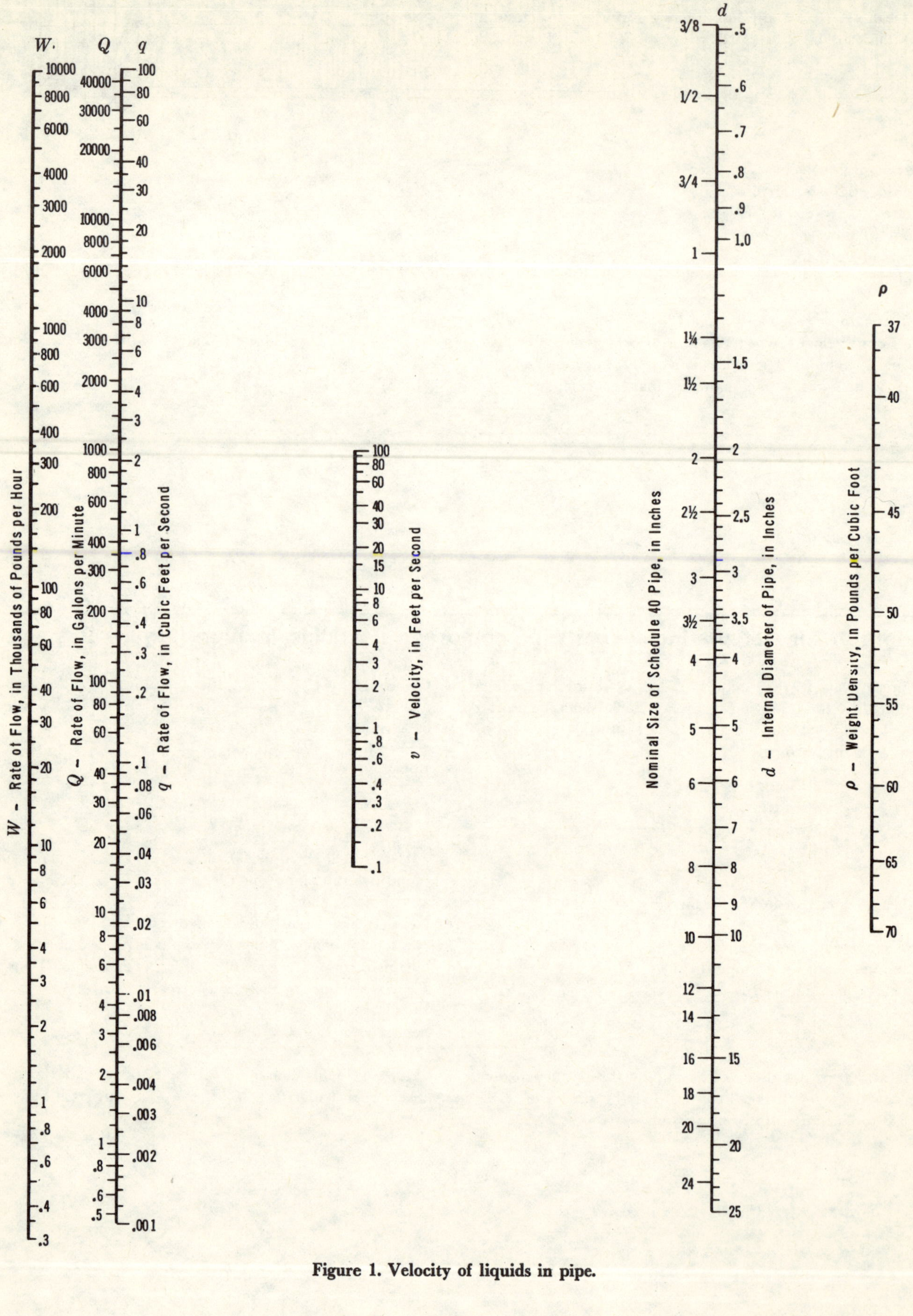

Figure 1. Velocity of liquids in pipe.

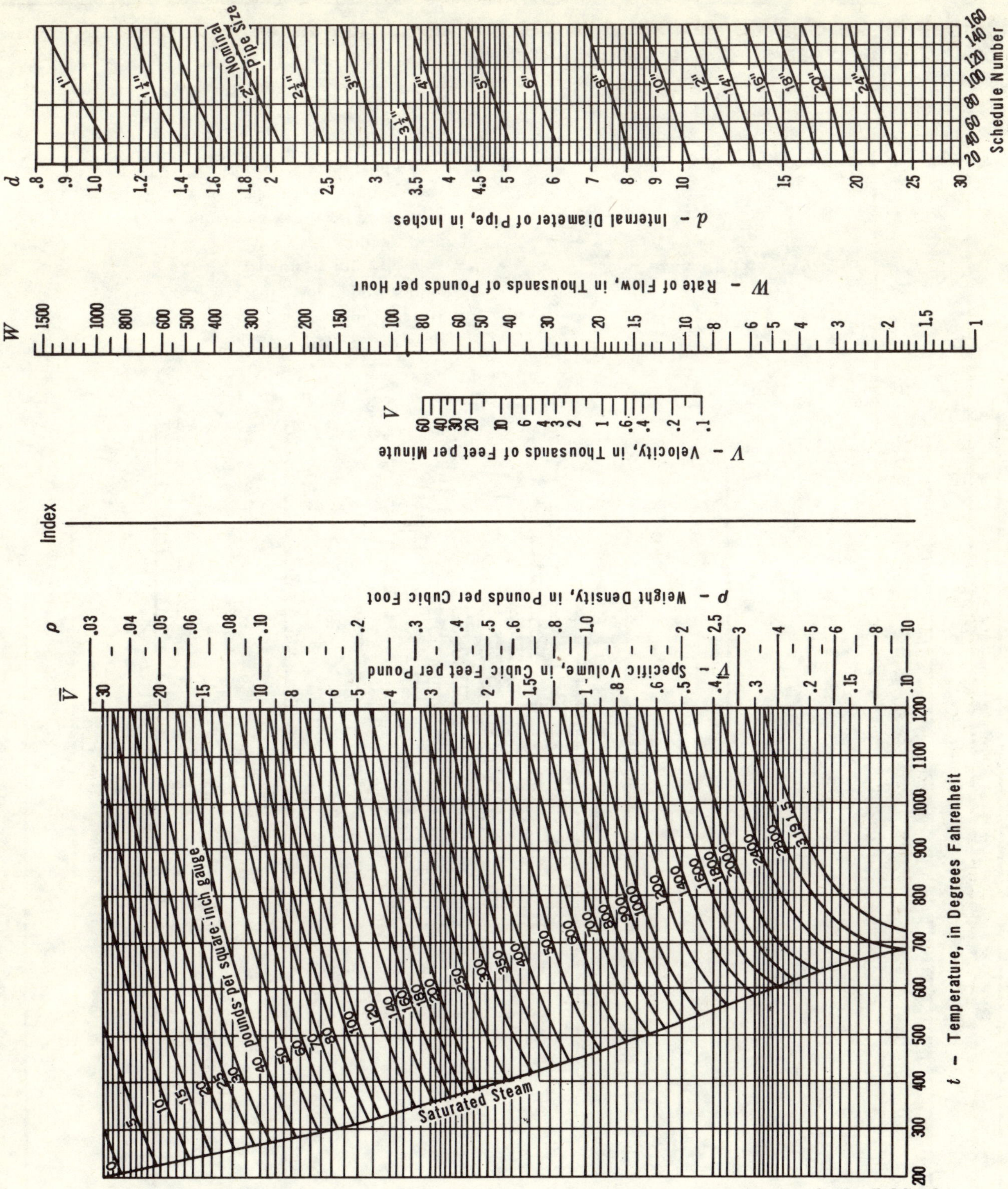

Figure 2. Velocity of compressible fluids in pipes.

Nomograph for calculating Reynolds number for flow of liquids and friction factor for clean steel and wrought iron pipe

This nomograph (Figure 1) permits calculation of the Reynolds number for liquids and the corresponding friction factor for clean steel and wrought iron pipe.

The Reynolds number is defined as:

$$Re = 22{,}700 \frac{q\rho}{d\mu} = 50.6 \frac{Q\rho}{d\mu} = 6.31 \frac{W}{d\mu} = \frac{dv\rho}{\mu a}$$

where:

Re = Reynolds number
d = inside pipe diameter, in.
q = volumetric flow rate, cu. ft./sec.
Q = volumetric flow rate, gals./min.
ρ = fluid density, lbs./cu. ft.
μ = fluid viscosity, centipoise
μa = fluid viscosity, lbs./(ft.)(sec.)
v = average fluid velocity, ft./sec.
w = mass rate of flow, lbs./hr.

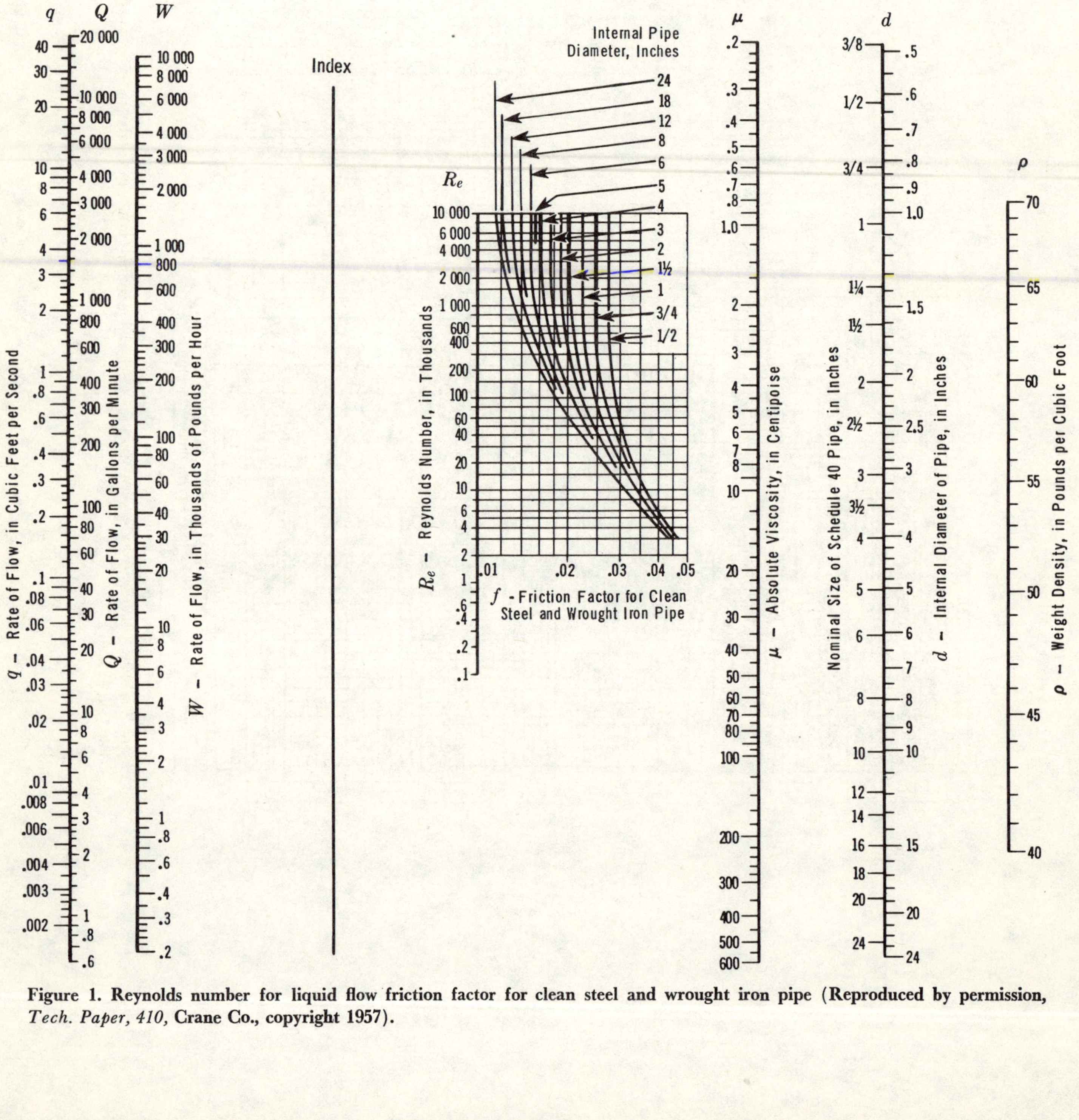

Figure 1. Reynolds number for liquid flow friction factor for clean steel and wrought iron pipe (Reproduced by permission, *Tech. Paper, 410,* Crane Co., copyright 1957).

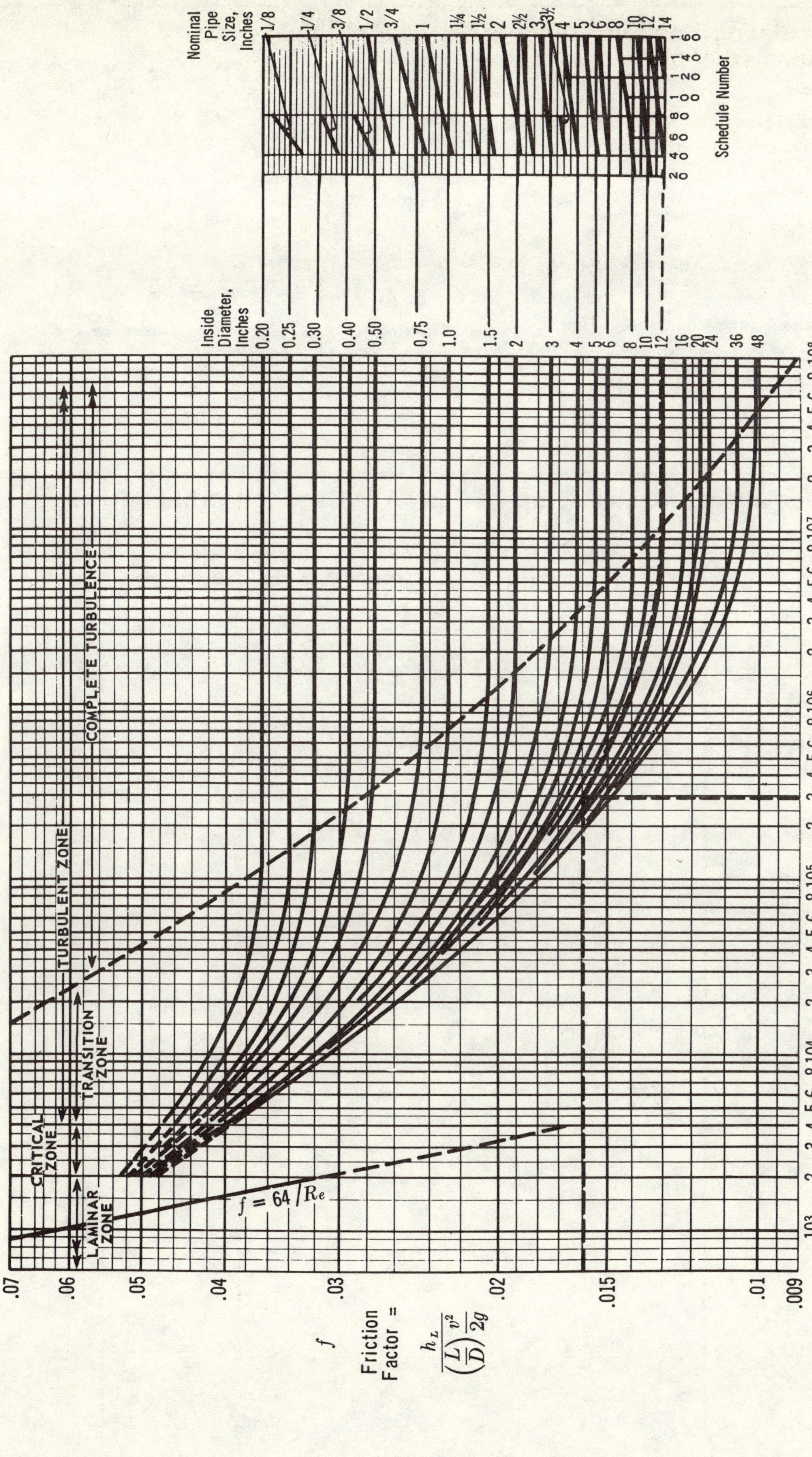

Figure 2. Friction factors for clean commercial steel and wrought iron pipe. (Reproduced by permission, *Tech. Paper*, *410*, Crane Co., 1957.)

Example

Given:

water at 200 degrees F.
d = 4-in. sch. 40 steel pipe
Q = 415 gallons/min.
ρ = 60.13 lbs./cu. ft.
μ = 0.3 centipoises

Obtain the Reynolds number and the friction factor.

Connect	With	Mark or Read
Q = 415	ρ = 60.13	W = 200,000
W = 200,000	d = 4-inch schedule 40	Index
Index mark with	μ = 0.3	Re = 1,000,000

At a Re = 1,000,000 read f = 0.017 from the graph on the center of the nomograph.

Figure 2 is an enlarged version of the friction factor chart, based on Moody's data[1] for commercial Steel and Wrought iron. This chart can be used instead of the smaller version shown in Figure 1.

References

1. Moody, L. F., *Transactions American Society of Mechanical Engineers,* 66, 671 (1944).

Nomograph for calculating pressure drop of liquids in lines for turbulent flow

The nomogram (Figure 1) is based on a modified version of the Fanning equation. It gives the pressure drop per 100 feet of pipe for liquids flowing through a circular pipe. The basic equation is:

$$P_{100} = \frac{0.0216 \cdot f\rho Q^2}{d^5} = 4350\frac{f\rho q^2}{d^5} = 0.000336\frac{fW^2}{d^5\rho} = 0.1294\frac{fv^2\rho}{d}$$

where:

P_{100} = pressure loss per 100 feet of pipe, psi
f = friction factor (4 x f' in Fanning equation)
ρ = fluid density, lb./cu. ft.
d = tube internal diameter, in.
W = flow rate, lbs./hr.
Q = flow rate, gals./min.
q = flow rate, cu. ft./sec.

The Moody[1] friction factor and relative roughness generalized charts (Figure 2 and Figure 3) are also included for use with the nomogram.

Example

The Reynolds number, estimated from the nomogram on page 130, is 100,000 when water is flowing at a rate of 100 gals./min. and a temperature of 70° F. Find the pressure drop per 100 feet of 2-inch diameter pipe, the relative roughness from Figure 3 is 0.0009.

At e/d = 0.0009 and Re = 100,000, the friction factor, f, is 0.022 from Figure 2.

Using Figure 1:

Connect	With	Mark or Read
f = 0.022	ρ = 62.3 lb./cu. ft.	Intersection at Index 1
Index 1	Q = 100 gals./min.	Intersection at Index 2
Index 2	d = 2-inches	P = 8 psi per 100 feet of pipe

Source

"Flow of Fluids Through Valves, Fittings and Pipe", *Technical Paper No. 410,* 3-11, Crane Company, Chicago, Illinois (1957).

Anon., *British Chemical Engineering,* 2, 3, 152 (1957).

Reference

1. Moody, D. F. *Transactions, American Society of Mechanical Engineers,* 66, 671, (1944).

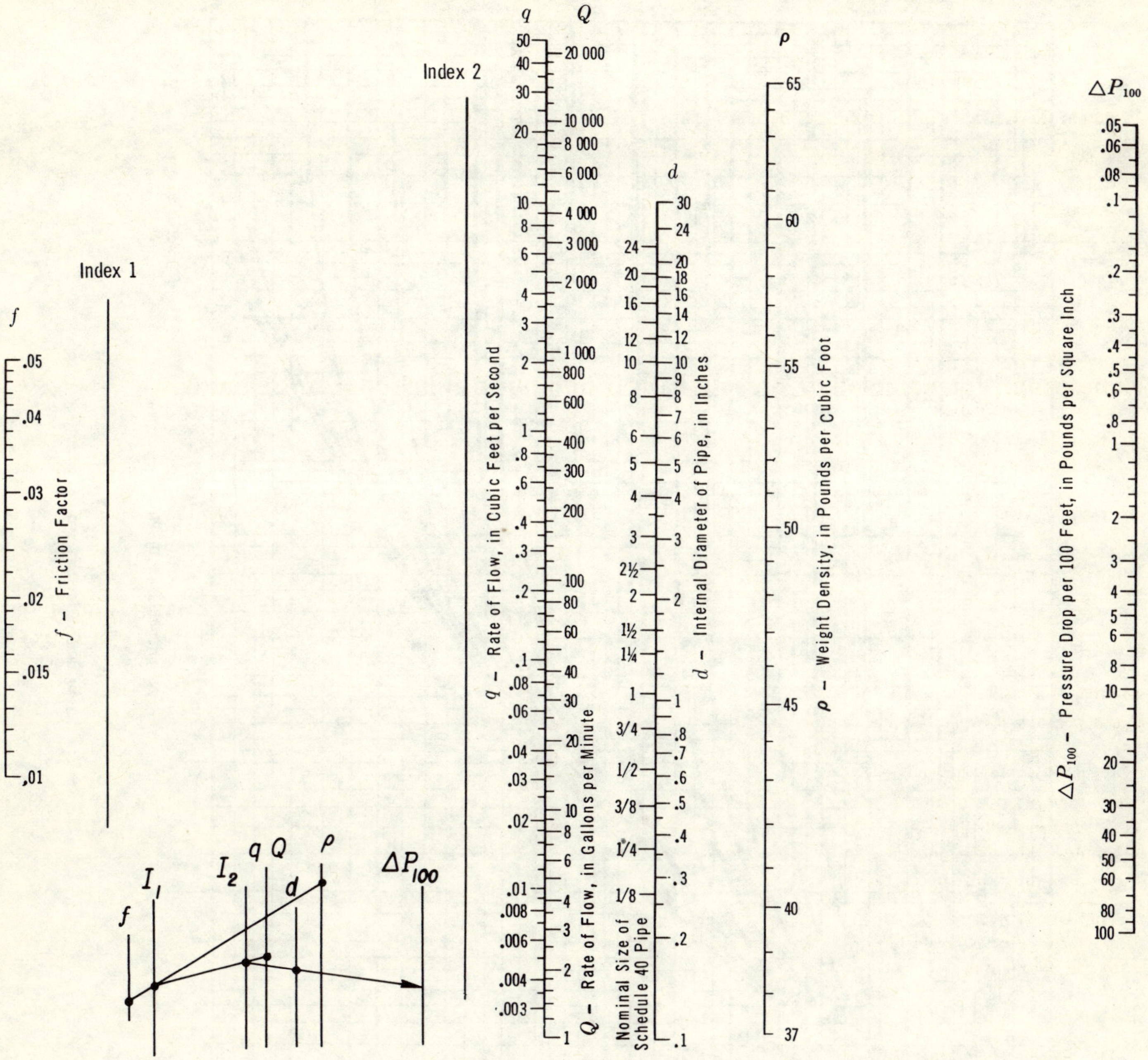

Figure 1. Pressure drop in liquid lines for turbulent flow.

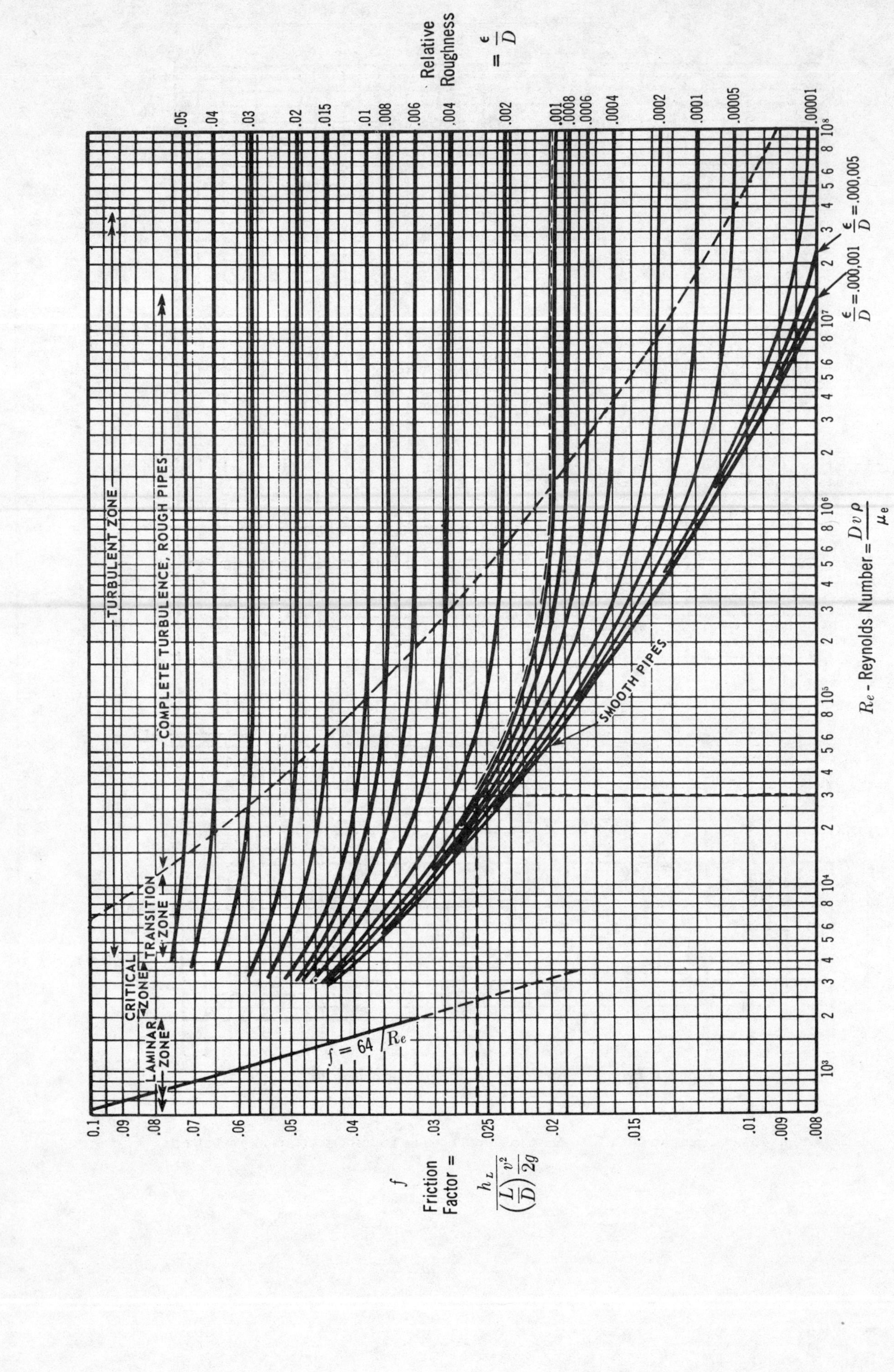

Figure 2. Friction factors for any type of commercial pipe.

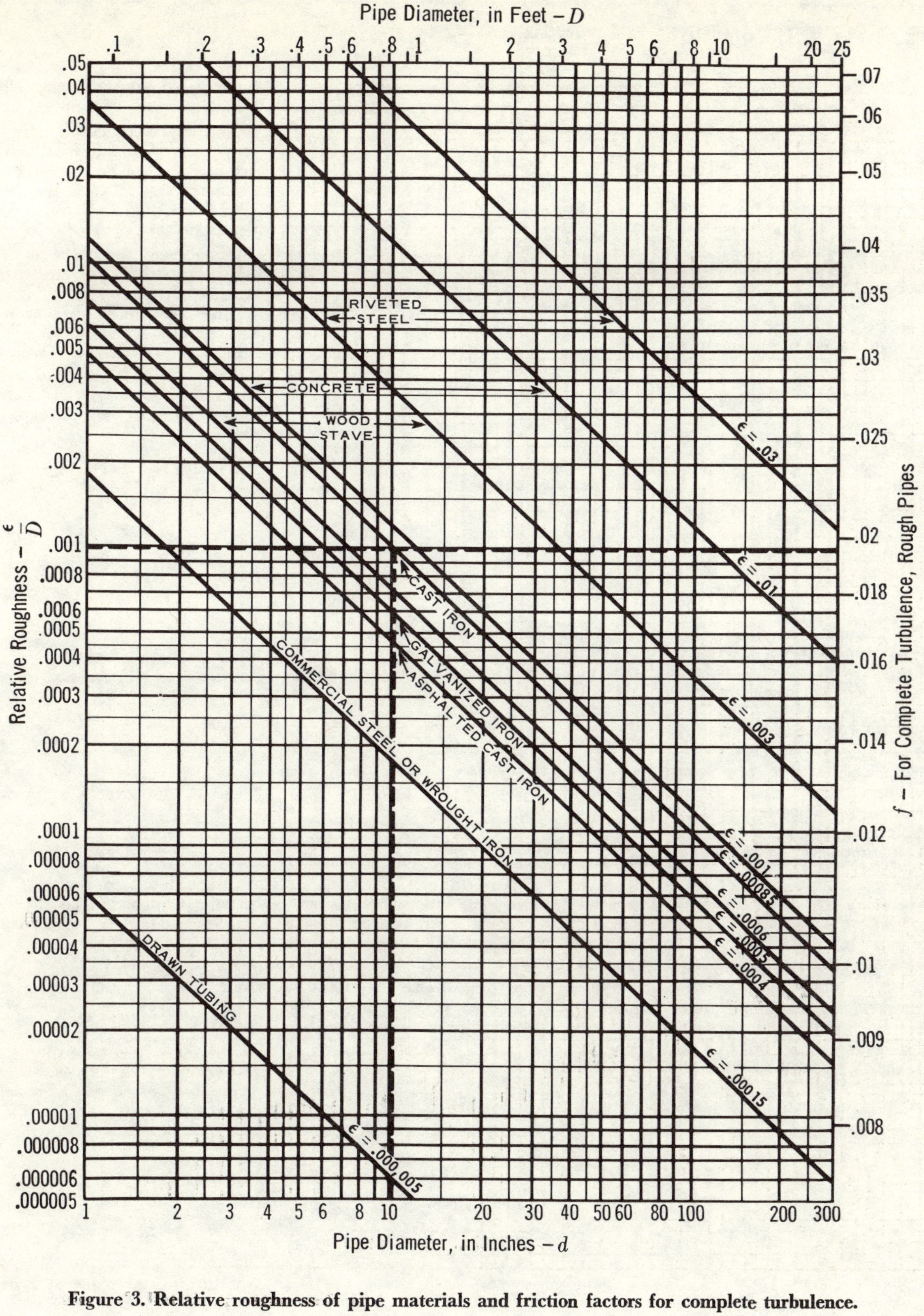

Figure 3. Relative roughness of pipe materials and friction factors for complete turbulence.

Derivation of basic ultrasonic flow equations

The actual measurements performed by ultrasonic flowmeters are of flow velocity, with all other variable effects due to temperature, pressure, density, viscosity, etc., being cancelled via a differential sensing technique.

Multiplication by the cross-sectional pipe area readily enables the output to be conditioned to read in volumetric flow rate scaled to engineering units. Refinement of the technique enables density and other physical properties to be measured as well, so that mass flow rates and other parameters pertaining to various process variables can be determined.

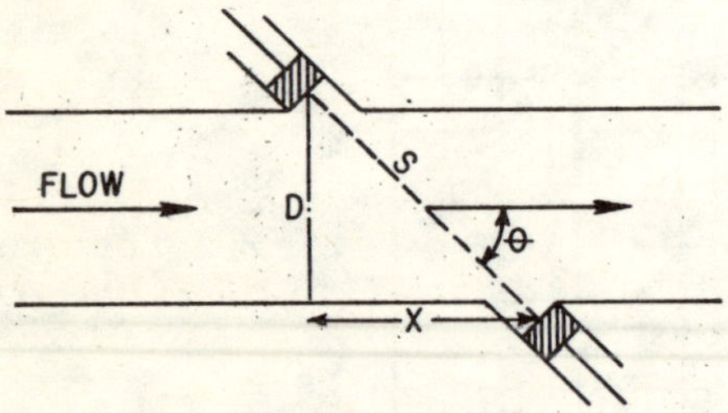

Let: D = Pipe diameter
X = Axial displacement of transducers
S = Path length between transducers
Θ = Angle of transducers to flow stream
c = Velocity of sound in the flowing medium
v = Velocity of flowing medium
t_1 = Flight time of upstream pulse
t_2 = Flight time of downstream pulse

then:

$$t_2 = \frac{S}{c + v \cos \Theta} \quad (1)$$

$$t_1 = \frac{S}{c - v \cos \Theta} \quad (2)$$

To cancel the effects of variations in the velocity of sound c in the medium itself, the circuitry is arranged to compute the reciprocal of each flight time and to subtract one from the other. This difference of the flight time reciprocals leads to the relationship:

$$\frac{1}{t_2} - \frac{1}{t_1} = \frac{2 v \cos \Theta}{S} \quad (3)$$

Solving for v

$$= \frac{S}{2 \cos \Theta}\left(\frac{1}{t_2} - \frac{1}{t_1}\right) \quad (4)$$

but

$$\cos \Theta = \frac{X}{S} \quad (5)$$

therefore

$$v = \frac{S^2}{2X}\left(\frac{1}{t_2} - \frac{1}{t}\right) \quad (6)$$

The circuitry which performs the reciprocal time computation produces an output frequency since

$$\frac{1}{t} = f \quad (7)$$

Therefore

$$\frac{1}{t_2} = f_2 \quad (8)$$

and

$$\frac{1}{t_1} = f_1 \quad (9)$$

and

$$\frac{1}{t_2} - \frac{1}{t_1} = f_2 - f_1 = \Delta f \quad (10)$$

Substituting in Equation 6 we have the basic velocity equation:

$$v = \frac{S^2}{2X} \Delta f \quad (11)$$

or

$$v = k_1 \Delta f \quad (11a)$$

The relationship

$$k_1 = \frac{S^2}{2X} \quad (12)$$

is the flow velocity scaling factor, and is usually expressed in feet per second or feet per minute. Note that it is a function of the square of the distance between the transducers. This is an important consideration when providing for re-zeroing of the meter calibration each time the transducers are removed and re-inserted.

For volumetric flow, the velocity must be multiplied by the cross-sectional area of the pipe, and the following considerations are involved:

$$A = \pi \left(\frac{D}{2}\right)^2 \quad (13)$$

Volumetric flow

$$V - v A \quad (14)$$

Therefore, volumetric flow

$$V\pi = \frac{S^2 D^2 \Delta f}{8X} \quad (15)$$

The relationship

$$k_2 = \frac{S^2 D^2}{8X} \pi \quad (16)$$

is the volumetric flow scaling factor, and is usually expressed in cubic feet per minute.

Further information on the physical properties of the flowing medium is also acquired in the process of the basic measurement. Specifically, the velocity of sound in the medium itself can be measured independent of the velocity of its motion.

Equation 3 computed the difference of the reciprocals of the upstream and downstream flight times. In the instrument this is performed in the delta computing channel. Supposing an addition of the reciprocals of the respective upstream and downstream flight times is performed in a separate computing channel, called the sigma channel. The following relationships ensue:

$$\frac{1}{t_2} + \frac{1}{t_1} = \frac{c + v \cos \Theta}{S} + \frac{c - v \cos \Theta}{S} = \frac{2c}{S} \quad (17)$$

then

$$c = \frac{S}{2} \Sigma f \quad (18)$$

Let

$$k_3 \frac{S}{2} \quad (19)$$

be the sound velocity scaling factor.

Then

$$c = k_3 \Sigma f \quad (20)$$

The speed of sound c in a medium is a function of its bulk modulus and its density, and is expressed by the following equation:

$$c = \sqrt{\frac{B}{\rho}} \quad (21)$$

where B = bulk modulus of elasticity
ρ = Density

From this, the equation for density may be derived thus:

$$c^2 = \frac{B}{\rho} \quad (22)$$

and

$$\rho = \frac{B}{c_2} \quad (23)$$

Substituting the value of c from Equation 20 into 23 gives

$$\rho = \frac{B}{(k_3 \Sigma f)^2} \quad (24)$$

For a given gas, B is essentially constant.

Therefore,

$$\frac{B}{k_3 2} = k_4 \text{ is a fair representation} \quad (24a)$$

Then

$$\rho \text{ (density)} = k_4 \, \Sigma \, f^{-2} \quad (25)$$

An interesting relationship may be derived at this point, namely, an equation for MASS FLOW. Since

$$\text{MASS FLOW} = \rho \, V \quad (26)$$

and since

$$V = k_1 \, \Delta \, f \quad (11a)$$

and

$$\rho = \frac{k_4}{\Sigma \, f^2} \quad (25)$$

then

$$M.F. = k_1 \, k_4 \frac{\Delta \, f}{\Sigma \, f^2} \quad (27)$$

or, letting

$$k_1 \, k_4 = k_5$$

$$\boxed{M.F. = k_5 \frac{\Delta \, f}{\Sigma \, f^2}} \quad (28)$$

Further, the density of a gas is related by the equation

$$\rho = \frac{P}{T} \quad (29)$$

where P = Pressure

T = Temperature (degrees Kelvin)

Therefore, if the pressure is known, the temperature may be extracted from the sigma channel thusly:

From Equation 25

$$\frac{P}{T} = k_4 \, \Sigma \, f^{-2} \quad (30)$$

and

$$T = k_4 \, P^{-1} \, \Sigma \, f^{-2} \quad (31)$$

Let us go back to Equation 24. If the chemical properties of the gas or the ratio of the mixture between two or more gases changes, then B is not a constant. If separate transducers are used to measure density, say via the pressure/temperature relationship, then the variation of B can also be measured.

Thus

$$\rho = B \, (k_3 \, \Sigma \, f)^{-2} \quad (32)$$

and

$$B = \rho (k_3 \, \Sigma \, f)^2 \quad (33)$$

The above relationship holds for liquids or gases.

In the case of a gas, we may further say

$$B = \frac{P}{T} \, (k_3 \, \Sigma \, f)^2 \quad (34)$$

(Contributed by T. J. Scarpa, Scarpa Laboratories, Inc.)

How to calculate stress in above/below ground transition

Design of gas gathering/gas reinjection project in southern Iran provides solutions for deflection and anchor block forces in long underground lines

P. J. Schnackenberg, Design Engineer,
Iran-Texas Engineering Co., Tehran, Iran

STRESSES AND DEFLECTIONS occur in pipe lines at the transition from the below ground (fully restrained) to the above ground (unrestrained) condition.

Analysis of the stresses and deflections in transition areas, resulting from internal pressure/temperature change, is necessary in determining anchor block requirements and design. Longitudinal deflections are used to determine whether an anchor block is required. Anchor block forces required to maintain the pipe in a fully constrained condition are then determined.

Iran-Texas Engineering Co. recently completed an analysis for a gas gathering/compression/reinjection project in southern Iran under contract to Oil Service Co. of Iran (OSCO). Each of the two fields, Bibi-Hakimeh and Rag-E-Safid have two gas/oil separation plants, and at present the gas is being flared. The broad scope of the project is as follows:

- ► Provision of gas gathering lines from existing and new wellhead separation to a production unit
- ► Design of four production units (each adjacent to the GOSP) comprising knockout facilities, compressor station and dehydration plant
- ► Provision of transfer pipe lines to send gas plus condensate as a two-phase flow to a natural gas liquid separation plant (under design by Foster Wheeler)
- ► Provision of reinjection system to inject lean gas back into a number of wells
- ► Some 9.4 MM m^3/d (330 MM scfd) of gas plus 15,000 bpd condensate will be going to the NGL plant. LPG will be used as feedstock and for export.

Following is a brief review of the analysis that resulted in more accurate solutions for deflection and anchor block forces. Sample calculations are for line sizes up to 41-cm (16-in.) OD, pressure to 193 bars (2,800 psig) and temperatures to 72° C (162° F).

Case I—no anchor. Considering the pipe line as shown in Fig. 1, the section up to point A is fully restrained, the

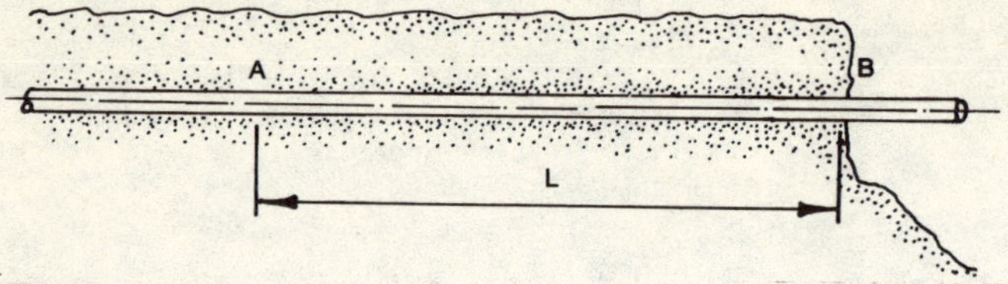

Figure 1. Area A-B depicts transition from fully restrained (left) to fully unrestrained (right).

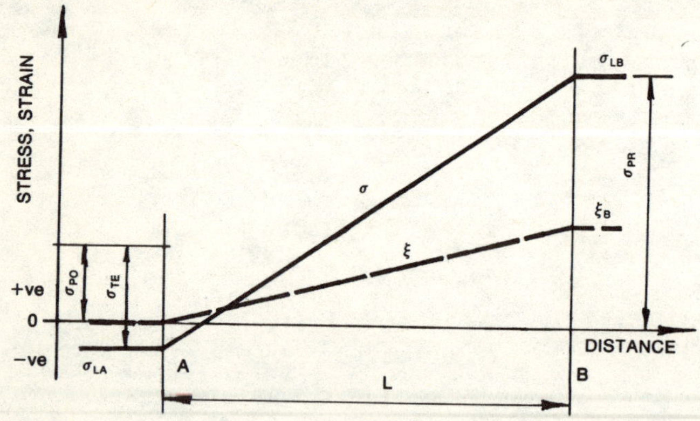

Figure 2. Transition of stress and strain between points A and B varies as linear function of length.

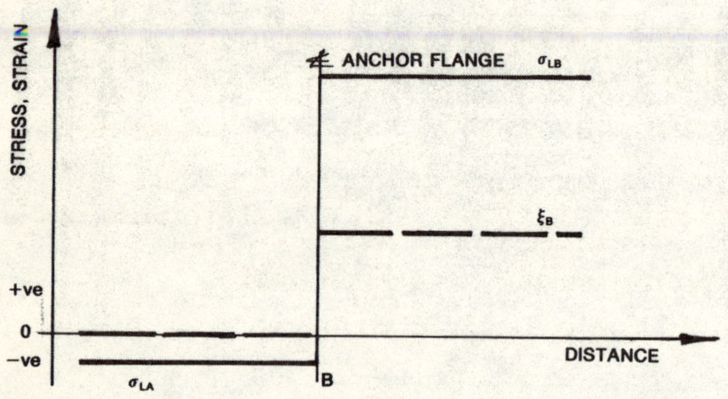

Figure 3. Distribution of stress where anchor is required to contain longitudinal deflections.

section A-B is a transition from fully restrained to unrestrained and beyond point B is unrestrained. (Fully restrained its taken to be a condition of zero longitudinal strain).

Analyzing a short section of fully restrained pipe, subject to internal pressure and a temperature change, there is a tensile stress due to a Poisson effect of

$$\sigma_{PO} = v\sigma_H \text{ where } \sigma_H = \frac{P\,Di}{2t}$$

and a compressive stress due to a temperature change of

$$\sigma_{TE} = E\,\alpha\,\Delta T$$

The net longitudinal stress therefore at Point A will be

$$\sigma_{LA} = v\,\sigma_H - E\,\alpha\,\Delta T \text{ psi} \qquad (1)$$

and the strain at point A, being fully restrained, will be zero

$$\varepsilon_A = O \qquad (2)$$

In a section of unrestrained pipe at B, the longitudinal stress caused by only internal pressure will be half the hoop stress.

$$\sigma_{LB} = \frac{\sigma_H}{2} \text{ psi} \qquad (3)$$

The associated strain will be made up of a

(i) Temperature effect

$$\varepsilon_{TE} = \alpha\,\Delta T$$

(ii) Pressure effect

$$\varepsilon_{PR} = \frac{\sigma_{LB}}{E} = \frac{\sigma_H}{2E}$$

(iii) Poisson effect

$$\varepsilon_{PO} = -\frac{v\,\sigma_H}{E}$$

The net longitudinal strain at Point B will therefore be

$$\varepsilon_B = \alpha\,\Delta T + \frac{\sigma_H}{E}\,(\tfrac{1}{2} - v) \text{ in./in.} \qquad (4)$$

The transition of stress and strain between point A and B is assumed to vary as a linear function of length as shown in Fig. 2.

In order to establish the length L over which the transition occurs, the longitudinal resistance of the soil needs to be known. For this part, it is assumed that any tendency to move will be counteracted by constant and opposite soil force. Wilbur[1] has recommended a design value for average soils of

$$F_s = 80\left(\frac{D_o}{12}\right)^2 \text{ lbf./ft.} \qquad (5)$$

Between A and B equilibrium of forces exists and therefore,

$$F_s\,L = A_m\,(\sigma_{LB} - \sigma_{LA})$$

From which

$$L = \frac{A_m\,(\sigma_{LB} - \sigma_{LA})}{F_s} \text{ ft.} \qquad (6)$$

Total movement at B will be the average strain between A and B over length L or

$$\delta = \frac{\varepsilon_B}{2}\,(12\,L) \text{ in.} \qquad (7)$$

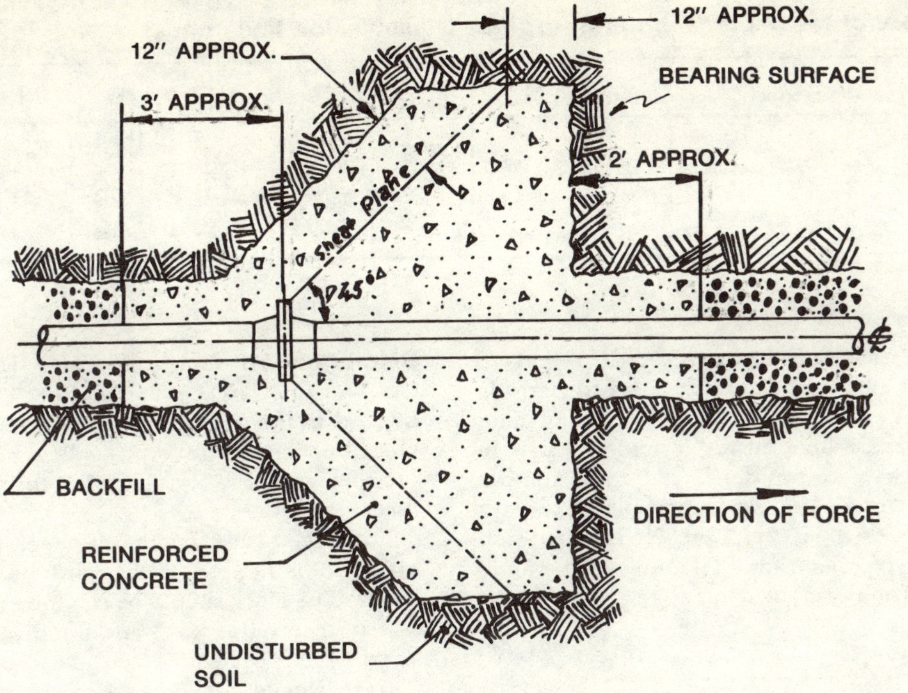

Figure 4. Sectional plan for anchor block design.

NOMENCLATURE

A_m	Pipe metal area $= \pi (D_o - t)t$, sq. in.
D_o	Pipe outside diameter, in.
D_i	Pipe inside diameter, in.
t	Pipe wall thickness, in.
E	Young's modulus of elasticity, psi
F	Anchor force, lb.
F_s	Soil resistance, lb./ft.
L	Transition length, ft.
P	Design pressure, psig
T_i	Installation temperature, °F
T_o	Operating temperature, °F
ΔT	Temperature difference $= (T_o - T_i)$, F°
α	Coefficient of thermal expansion, in./in.-F°
σ	Stress, psi
ε	Strain, in./in.
υ	Poissons ratio
δ	Deflection, in.

Subscripts (for Greek letters only)

L	Longitudinal
H	Hoop
A	Point A, fully restrained
B	Point B, unrestrained
PR	Pressure
TE	Temperature
PO	Poisson

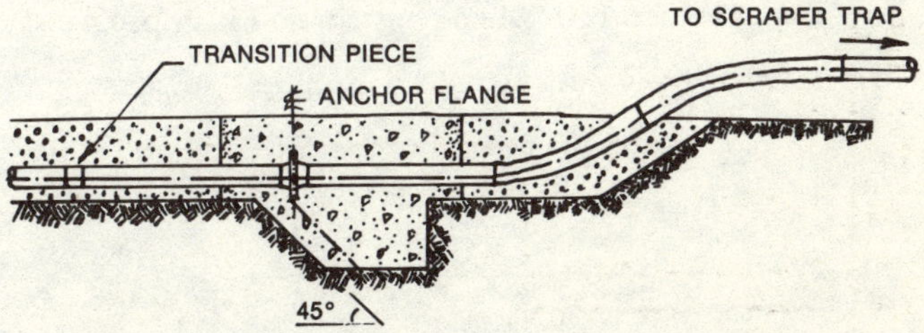

Figure 5. Sectional elevation for anchor block design.

Case II—with anchor. Where an anchor is required to contain longitudinal deflections, the stress distribution will be as shown in Fig. 3. The transition from being fully restrained to unrestrained occurring at the anchor. Resultant force on the anchor will simply be the difference in stress on each side times the pipe metal area (equilibrium of forces) or

$$F = (\sigma_{LB} - \sigma_{LA})\ A_m \text{ lb.} \tag{8}$$

Note that for the case of an increase in wall thickness at and beyond the anchor block, the result will be essentially the same—the decrease in stress being compensated by an increase in pipe metal area.

Table 1
Summary of results for high pressure gas transmission lines under design in Iran

No.	Pipe (All API-5LX-60)	P psig	T_O °F	F_S lb./ft.	σ_H psi	σ_{LA} psi	σ_{LB} psi	L ft.	δ in.	F lb.
1	16″ OD × 0.312″ w.t.	1150	162	142.2	28,334	−6957	14,167	2285	9.81	324,890
2	20″ OD × 0.344″ w.t.	1000	162	222.2	28,070	−7036	14,035	2014	8.62	447,550
3	16″ OD × 0.281″ w.t.	1000	162	142.2	27,470	−7216	13,735	2045	8.71	290,800
4	16″ OD × 0.250″ w.t.	880	162	142.2	27,280	−7273	13,640	1819	7.73	258,690
5	16″ OD × 0.688″ w.t.	2800	150	142.2	29,758	−4268	14,879	4457	17.66	633,770
6	10″ NB × 0.500″ w.t.	2800	150	64.2	28,700	−4585	14,350	4728	18.60	304,850
7	6″ NB × 0.432″ w.t.	2800	150	24.4	18,670	−7594	9,335	5828	20.41	142,200

It can be shown that this force is equal to that to produce a deflection of 2δ in Equation 7 over length *L*.

Table 1 is a summary of results for a number of high pressure gas transmission lines currently under design in Iran. Lines 1 through 4 are transfer lines from a number of production units to an NGL separation plant. Lines 5 through 7 are injection lines from the NGL separation plant to a number of wellheads. The following values were used for the various parameters:

Young's modulus, E, 29×10^6, psi

Poissons ratio, ν, 0.3

Thermal expansion coefficient, 6.5×10^{-6}, in./in. F°

Installation temperature, T_i, 80° F (minimum summer temperature when construction is in progress).

Negative stress values denote a compressive stress.

Discussion. As shown in Table 1, anchor block forces can be quite large and therefore, careful attention must be paid to the design. In order to minimize the size of the block, upper design limit values of lateral soil bearing pressure with due consideration to possible slip have been used.

For large blocks, friction between the block and the soil may also be figured in obtaining its resistance to imposed forces. The block should be reinforced concrete cast against undisturbed soil. Figs. 4 and 5 indicate in broad detail the anchor block design.

Care should also be taken to ensure that connected surface piping has sufficient flexibility to absorb a degree of lateral anchor movement. Scraper traps should be installed such that they can move with the piping rather than being rigidly attached to blocks. Numerous examples of traps together with their support blocks being displaced a few inches are not uncommon.

LITERATURE CITED

[1] Wilbur, W. E., *Pipe Line Industry*, February 1963.
[2] Timoshenko, S., Theory of Elasticity.

14. Oil & Products—Pump Stations

Estimate the discharge of a centrifugal pump at various speeds

Within the design limits of a pump, the discharge will be approximately proportionate to the speed.

$$Q_2 = Q_1 \frac{N_2}{N_1}$$

where Q_1 is the gallons per minute at a particular speed N_1, and Q_2 is the gallons per minute at N_2.

Example: If a pump delivers 350 gpm at 1,200 rpm what will it deliver at 1,800 rpm?

$$Q_2 = 350 \times \frac{1800}{1200} = 525 \text{ gpm.}$$

Similarly, for a given pump design head varies as the square of the impeller speed and power as the cube of the speed.

How to estimate the rate of liquid discharge from a pipe

A quick and surprisingly accurate method of determining liquid discharge from an open ended horizontal pipe makes use of the formula:

$$gpm = \frac{2.56\, X\, d^2}{\sqrt{Y}}$$

Where:

gpm = Gallons per minute
d = diameter of pipe, inches
X = inches
Y = inches

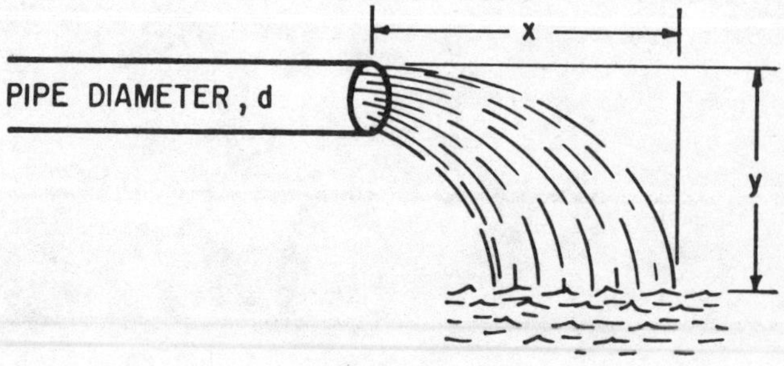

Example:

From the drawing:

X = 24 inches
Y = 25 inches
d = 2 inches

$$gpm = \frac{(2.56)\,(24)\,(2^2)}{\sqrt{25}} = \frac{245.76}{5} = 49.15$$

How to estimate the hp required to pump at a given rate at a desired discharge pressure

Ignoring effects of temperature, viscosity etc.

$$HP = \text{psi} \times \frac{\text{barrels per day}}{1000} \times .022$$

The factor .022 is arrived at by using a pump efficiency of 80 percent and a gear efficiency of 97 percent.

Example: What horsepower would be required to pump 15,000 barrels per day with an 800 psi discharge?

$$HP = 800 \times 15 \times .022 = 264 \text{ horsepower}$$

The factor .022 was derived from the formula:

$$HP = \frac{(\text{psi})\ (\text{BPD})}{58{,}800 \times \text{pump efficiency} \times \text{gear efficiency}}$$

How to select motors for field gathering pumps

The following simple formulas will provide a reasonably close approximation of the motor hp necessary to drive duplex double acting pumps normally used in the field.

Rule:

1. $hp = \dfrac{bpd \times psi}{48{,}000}$

2. $hp = \dfrac{bph \times psi}{2{,}000}$

3. $hp = \dfrac{gpm \times psi}{1{,}400}$

Motors selected with these formulas will fit delivery rates used in the formula if they are within capacity of the pump. If anticipated rate is below pump capacity, a savings may be made in first cost and power bills by selecting a motor corresponding to the actual rate rather than to full pump capacity.

It should be noted that net bpd values are to be used in Formula 1. That is, if 600 barrels are to be pumped each day, but it is desired to accomplish this in only 8 hours, the rate to be used in the formula is 1,800 bpd.

How to estimate the head for an average centrifugal pump

By slightly modifying the basic pump formula, $H = \left(\frac{DN}{1840\ K_u}\right)^2$ it is possible to **find a means of estimating the head a pump will develop for each impeller at various speeds.** The terms are defined as follows:

H = Head in feet
D = Diameter of the impeller in inches
N = Rpm of the impeller
K_u = Pump design coefficient. The values of this expression may vary from 0.95 to 1.15. For estimating purposes substitute 1,900 for 1,840 K_u. The formula becomes, $H = \left(\frac{DN}{1900}\right)^2 \times$ (the number of impellers)

Example: What head will be developed by a well designed centrifugal pump with four 11.75-inch impellers turning at 1200 rpm?

$$H = 4 \times \left(\frac{11.75 \times 1200}{1900}\right)^2 = 4 \times 7.4^2 = 220 \text{ feet of head.}$$

Find the capacity of reciprocating pump

Multiply the diameter of the plunger by itself, multiply this by the stroke, then by the number of strokes per minute, and finally by 0.11; the result is the capacity in barrels per day.

$C = B^2 \times S \times N \times 0.11$

Example: four-inch bore, six-inch stroke, 80 strokes per minute.

$4 \times 4 \times 6 \times 80 \times 0.11 = 845$ barrels per day.

Note: the rule is approximate only; it allows about 5 percent for "slip," which is actually somewhat variable. Also note that it is applicable only to single-acting pumps; for a double-acting piston type pump, N must be twice the rpm, and a reduction must be made for the area of the piston rod on the crank end of the pump.

Nomograph for determining horsepower requirement of pumps

In the selection of pumps it is necessary to calculate the horsepower requirement to handle a specified delivery rate of fluid, against the specified head. The formula for the horsepower required by a pump is as follows:

$$HP = \frac{G. \times H. \times Sp.\ Gr.}{3960 \times e}$$

where:

HP = pump horsepower
G = flow rate, gals./min.
H = total dynamic head (change in head), ft.
Sp. Gr. = specific gravity of fluid
e = pump efficiency

Figure 1 solves this equation.

Example

Given:

G = 200 U.S. gals./min.
H = 60 feet
Sp. Gr. = 1.1
e = .80

What is the required horsepower?

Connect	With	Mark or Read
e = .80	H = 60 feet	C
C	G = 200 gals./min.	HP = 0.78 horsepower

Source

Macdonald, J.O.S., *British Chemical Engineering,* 1, 6, 290 (1956).

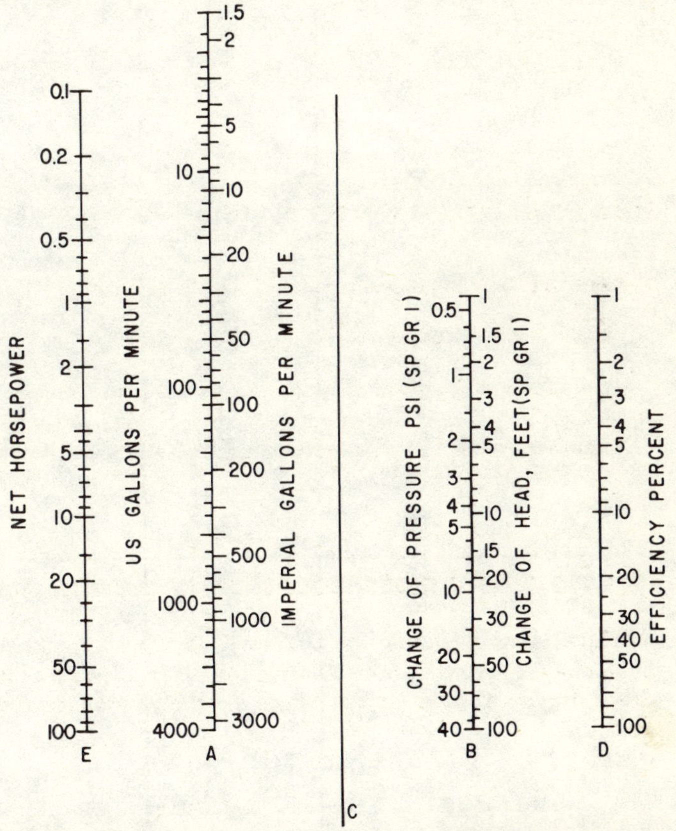

Figure 1. Pump horsepower.

Nomograph for determining reciprocating pump capacity

The theoretical capacity of reciprocating pumps can be calculated by the following equation:

$$C = 6/55\ D^2 \ .\ L.N$$

where:

C = theoretical pump capacity (delivery), gals./min.
D = diameter of cylinder, ft.
L = length of stroke, ft.
N = number of strokes/min.

The nomograph (Figure 1) is based on this equation. The diameter and length-of-stroke scales are calibrated in inch units instead of feet. The nomograph gives also the delivery in British imperial gallons.

Example

What is the theoretical capacity of a reciprocating pump having the following characteristics:

D = 3 inches   N = 60 strokes/min.
L = 3 inches

Connect	With	Mark or Read
D = 3 inches	L = 3 inches	Reference Scale
Ref. point	N = 60	C = 5 gal./min.

Source

Macdonald, J.O.S., *British Chemical Engineering,* 1, 6, 301 (1956).

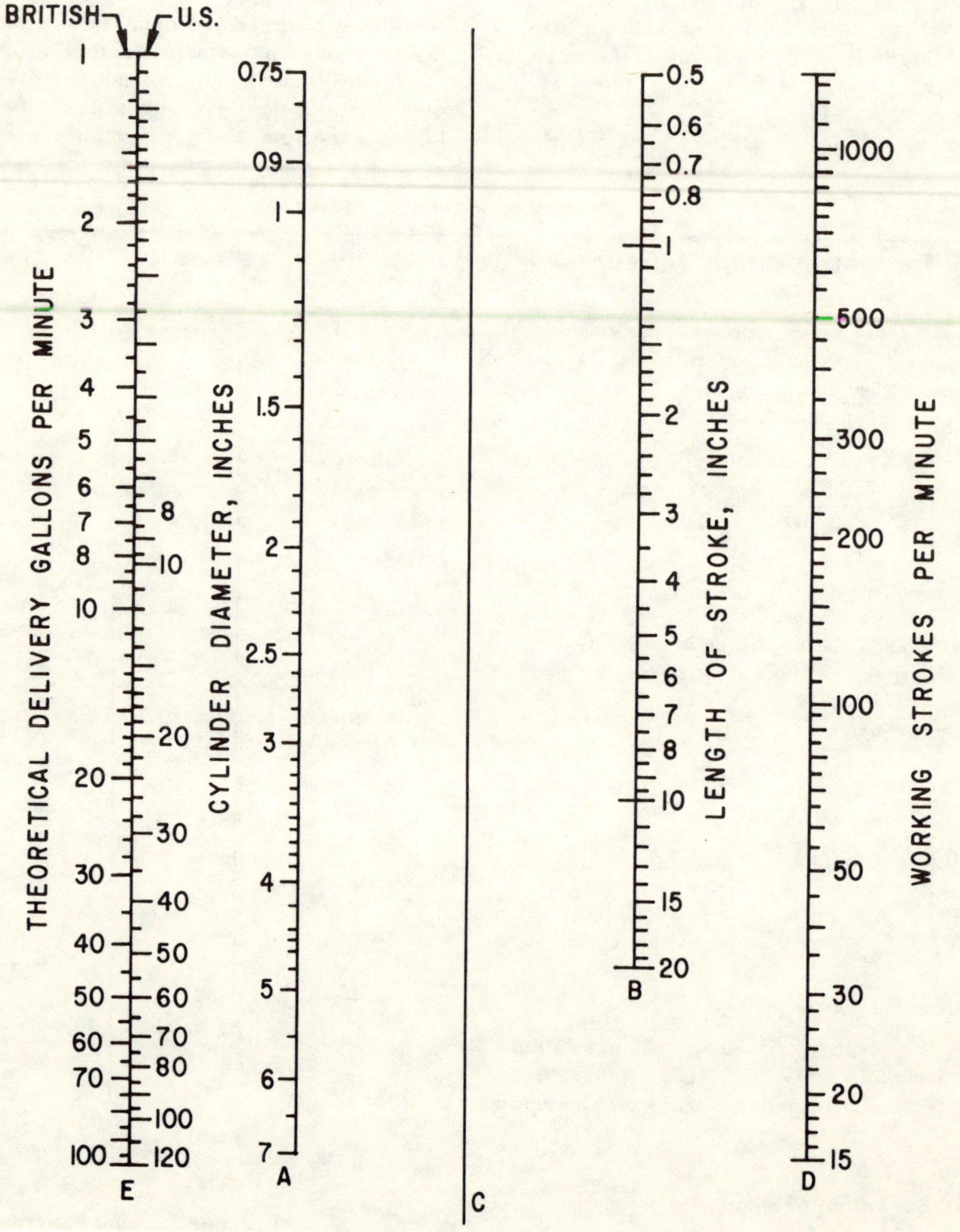

Figure 1. Theoretical capacity of reciprocating pumps.

Nomograph for determining specific speed of pumps

The specific speed is a criterion used in the selection of pumps for a given service. It can be used to (a) avoid cavitation troubles, (b) to select the most economic pump for a given process layout and (c) to assist in redesigning process conditions.

For single-stage, side-entering pumps, the NPSH is defined by the following equation:[1, 2]

$$S = N \frac{Q^{1/2}}{H^{3/4}} \tag{1}$$

where:

S = specific speed, no units
Q = pump capacity
N = pump speed revs./min.
H = Net positive suction head above the liquid vapor pressure (NPSH) measured at the pump suction flange, feet of fluid

H (in feet of liquid) is given by:

$$H = h_a \pm h_h - h_P - h_v \tag{2}$$

where:

h_a = ambient pressure at the liquid surface
h_h = static head between liquid surfaces
h_P = combined entrance, piping and fitting friction loss
h_v = proper pressure at pumping temperature

The specific speed of centrifugal pumps may vary between 1,100 to 15,000. Values above 12,000 are not considered practical and, where possible, the use of a specific speed below 8,500 is recommended. Specific speed limits for each type of pump are usually available from manufacturers of pumps.

The nomogram (Figure 1) solves Equation 1.

The specific speed is also defined in terms of a dimensionless number, using consistent units, by the following equation:

$$S = \frac{Q^{0.5} N}{(\Delta H \cdot g_c)^{0.75}} \tag{3}$$

g_c = gravitational conversion constant, 32.2 (lb. mass/lb. force) (ft./sec.^{2})

The range for this equation is 0.03 to 0.87. Results of Equation 1 can be converted to these units by dividing by 17,200.

Example

The following data are given for a single-suction reboiler pump:

Q = 400 gals./min. N = 3,500 revs./min.

The reboiler liquid level is 18 feet above the pump and friction from entrance, pipe and fitting loss add up to 5 feet of liquid.

For a reboiler, the liquid is at its boiling point, so $h_a = h_v$, using Equation 2, H = 13 feet.

Connect	With	Mark or Read
NPSH = 13	Q = 400 gals./min.	Reference Line
Ref. line pivot	N = 3,500	S = 10,200

Source

Jacobs, J. K., *Petroleum Refiner,* 36, 10, 145 (1957).

References

1. *Hydraulic Institute Standards,* 10th Edition, New York.
2. Brown, et. al., *Unit Operations,* 190, John Wiley and Sons, New York (1950).

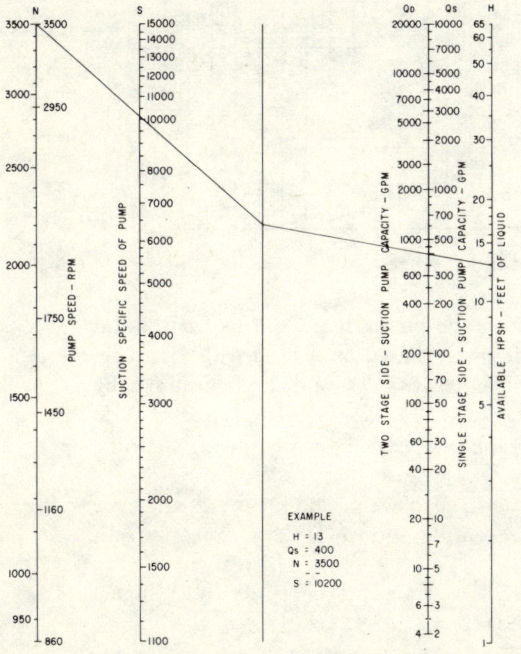

Figure 1. Pump specific speed from net positive suction head (Reproduced by permission, *Petroleum Refiner,* Vol 36, No. 10, copyright 1957).

Chart gives electric motor horsepower for pumping units.

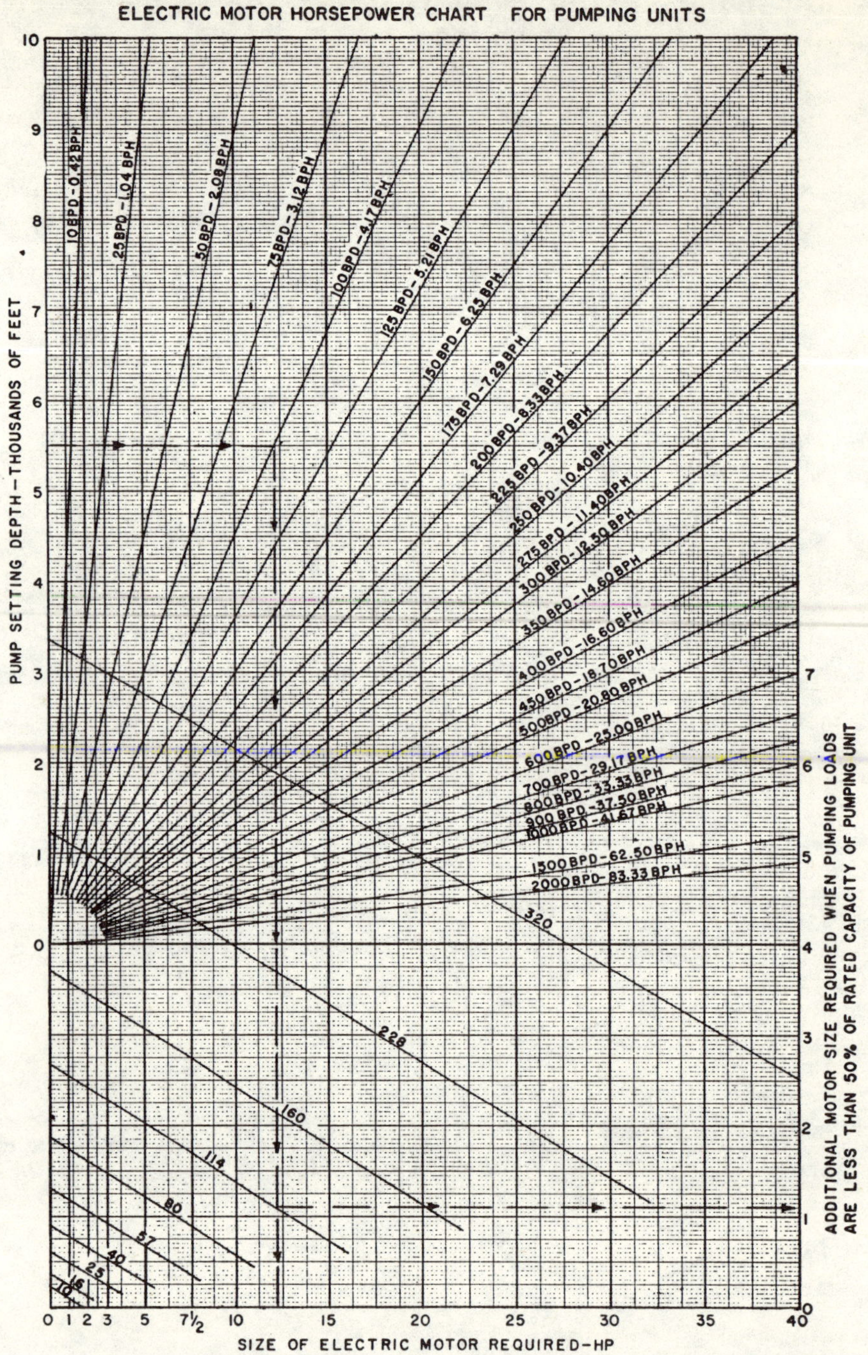

This chart provides a means of determining the power requirements for a beam pumping installation powered by an electric motor based on a fluid with a specific gravity of one, with fluid at the pump. The power requirement determined by the chart includes a mechanical efficiency factor of 0.45 and a cyclic factor of 0.75, which factors are frequently applied to motors used in sucker rod pumping service.

An arrangement is available for correcting the power requirement in the case of an underloaded pumping unit. The example shown in the chart is self-explanatory.

After the power requirements are determined from the chart the next higher size of commercially available motor is used.

Pumping stations:

Table gives capacitor multipliers for kilowatt loads for different desired power factor improvements

Original Power Factor, Percent	DESIRED POWER FACTOR—Percent				
	100	95	90	85	80
50	1.732	1.403	1.248	1.112	0.982
51	1.687	1.358	1.203	1.067	0.937
52	1.643	1.314	1.159	1.023	0.893
53	1.600	1.271	1.116	0.980	0.850
54	1.559	1.230	1.075	0.939	0.809
55	1.518	1.190	1.035	0.898	0.769
56	1.480	1.151	0.996	0.860	0.730
57	1.442	1.113	0.958	0.822	0.692
58	1.405	1.076	0.921	0.785	0.655
59	1.369	1.040	0.885	0.749	0.619
60	1.333	1.004	0.849	0.713	0.583
61	1.299	0.970	0.815	0.679	0.549
62	1.266	0.937	0.782	0.646	0.516
63	1.233	0.904	0.749	0.613	0.483
64	1.201	0.872	0.717	0.581	0.451
65	1.169	0.840	0.685	0.549	0.419
66	1.138	0.809	0.654	0.518	0 388
67	1.108	0.779	0.624	0.488	0.358
68	1.078	0.749	0.594	0.458	0.328
69	1.049	0.720	0.565	0.429	0.299
70	1.020	0.691	0.536	0.400	0.270
71	0.992	0.663	0.508	0.372	0.242
72	0.964	0.635	0.480	0.344	0.214
73	0.936	0.607	0.452	0.316	0.186
74	0.909	0.580	0.425	0.289	0.159
75	0.882	0.553	0.398	0.262	0.132
76	0.855	0.526	0.371	0.235	0.105
77	0.829	0.500	0.345	0.209	0.079
78	0.802	0.473	0.318	0.182	0.052
79	0.776	0.447	0.292	0.156	0.026
80	0.750	0.421	0.266	0.130	....
81	0.724	0.395	0.240	0.104	....
82	0.698	0.369	0.214	0.078	....
83	0.672	0.343	0.188	0.052	....
84	0.646	0.317	0.162	0.026	....
85	0.620	0.291	0.136	....	....
86	0.593	0.264	0.109	....	....
87	0.567	0.238	0.083	....	....
88	0.540	0.211	0.056	....	....
89	0.512	0.183	0.028	....	....
90	0.484	0.155			
91	0.456	0.127			
92	0.426	0.097			
93	0.395	0.066			
94	0.363	0.034			
95	0.329	....			
96	0.292	....			
97	0.251	....			
98	0.203	....			
99	0.143	....			

Example

Assume total plant load is 100 kw at 60 percent power factor. Capacitor kvar rating necessary to improve power factor to 80 percent is found by multiplying kw (100) by multiplier in table (0.583), which gives kvar (58.3). Nearest standard rating (60 kvar) should be recommended.

Courtesy General Electric Company.

Table gives velocity heads for various pipe diameters and different rates of discharge

Rate of Flow in Gallons per Minute	NOMINAL DIAMETER OF PIPE (Velocity Heads in Feet)														Rate of Flow in Gallons per Minute
	⅛″	¼″	⅜″	½″	¾″	1″	1¼″	1½″	2″	2½″	3″	4″	5″	6″	
1	.488	.148	.043	.017											1
2		.590	.175	.069	.022										2
3			.392	.155	.050	.020									3
4			.698	.276	.090	.035	.012	.006							4
5			1.089	.430	.141	.054	.018	.010							5
10				1.720	.564	.215	.071	.038	.016	.007	.003				10
15					1.265	.488	.160	.087	.036	.015	.007				15
20					2.250	.861	.286	.154	.065	.027	.013				20
25						1.345	.447	.241	.101	.041	.020				25
30						1.933	.643	.346	.146	.060	.029				30
35						2.635	.877	.472	.198	.082	.039				35
40						3.343	1.145	.618	.259	.106	.052	.016			40
45							1.548	.779	.329	.134	.065	.021			45
50							1.787	.963	.406	.166	.080	.026			50
70							3.503	1.888	.795	.326	.157	.050	.020		70
75								2.165	.912	.373	.180	.057	.023		75
100								3.851	1.621	.665	.320	.101	.041	.020	100
120								5.548	2.333	.956	.462	.146	.060	.029	120
125									2.528	1.035	.498	.158	.065	.031	125
150									3.649	1.493	.736	.229	.093	.045	150
175									4.954	2.021	.980	.309	.127	.061	175
200	**8″**								6.483	2.656	1.281	.406	.166	.080	200
225										3.359	1.618	.514	.209	.101	225
250	.040									4.131	1.992	.633	.259	.125	250
270	.047										2.333	.740	.302	.146	270
275	.048										2.429	.769	.313	.151	275
300	.057										2.883	.912	.373	.180	300
350	.078	**10″**										1.242	.509	.246	350
400	.102											1.621	.665	.320	400
450	.129	.053										2.053	.839	.406	450
470	.141	.057	**12″**									2.239	.917	.443	470
475	.144	.059										2.276	.934	.453	475
500	.159	.065	.031									2.535	1.038	.502	500
550	.193	.079	.038										1.256	.606	550
600	.229	.093	.045										1.493	.721	600
				14″											
650	.269	.109	.053	.026									1.752	.846	650
700	.312	.127	.062	.031									2.035	.980	700
750	.358	.146	.071	.035									2.337	1.123	750
800	.407	.166	.080	.040										1.281	800
850	.462	.187	.090	.045										1.548	850
					15″										
900	.516	.209	.102	.050										1.618	900
950	.574	.234	.113	.056	.043									1.813	950
1000	.637	.259	.125	.062	.048									1.999	1000
1050	.703	.296	.138	.069	.053									2.202	1050
1100	.771	.313	.152	.075	.058									2.421	1100
1150	.841	.343	.166	.082	.064	**16″**								2.648	1150
1200	.917	.373	.181	.090	.069									2.880	1200
1250	.995	.406	.196	.097	.075	.062								3.135	1250
1500	1.433	.582	.282	.145	.108	.088									1500
2000	2.547	1.035	.502	.249	.193	.158									2000
							20″								
2500		1.618	.784	.391	.301	.246		**24″**							2500
3000		2.329	1.128	.562	.434	.355	.147		**30″**						3000
3500		3.224	1.540	.766	.590	.502	.199	.096							3500
4000			2.007	1.000	.771	.633	.261	.125	.051						4000
4200			2.209	1.102	.851	.698	.288	.137	.056						4200
4500			2.539	1.265	.975	.799	.330	.158	.065						4500
5000			3.135	1.561	1.204	.985	.407	.195	.079	**36″**					5000
5500				1.884	1.460	1.196	.493	.227	096						5500
6000				2.239	1.733	1.421	.586	.281	.115	.056					6000
6500				2.627	2.035	1.669	.690	.329	.134	.065					6500
7000				3.056	2.360	1.936	.797	.384	.156	.076					7000
7200					2.709	2.056	.846	.404	.165	.080					7200
7500					3.047	2.224	.917	.440	.180	.087					7500
8000							1.045	.498	.204	.099					8000
8500							1.177	.564	.230	.112					8500
9000							1.319	.633	.259	.125	**42″**				9000
9500							1.469	.702	.289	.140					9500
10000							1.630	.779	.319	.154	.083				10000
10500								.858	.352	.169	.092				10500
11000								.943	.387	.186	.101				11000
11500								1.033	.423	.204	.110				11500
12000								1.123	.460	.222	.120				12000
12500								1.218	.500	.240	.130				12500
13000								1.316	.541	.260	.141				13000
13500								1.421	.582	.281	.152				13500

F. G. Switzer "Pump-Fax," Goulds Pumps, Inc.

15. Corrosion

Chart aids in calculating ground bed resistance and rectifier power costs

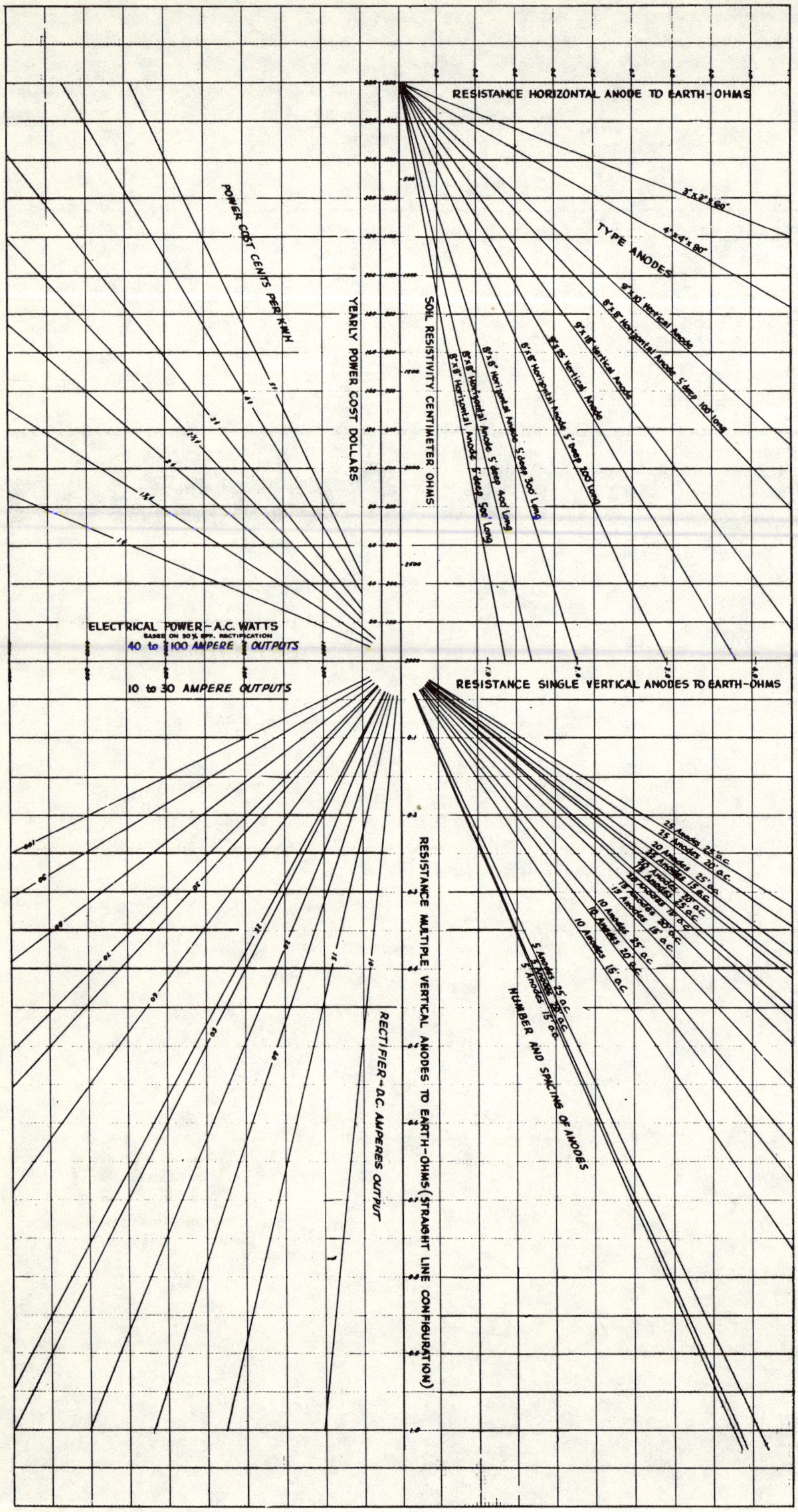

For conventional type ground beds this chart permits a rapid determination of the resistance to earth of various ground beds and the yearly power costs based on 50 percent efficient rectification. These curves are specifically designed for standard 3-inch by 60-inch and 4-inch by 80-inch graphite anodes centered in a vertical or horizontal excavation and backfilled to the dimensions indicated with well tamped coal coke breeze.

Example:

Soil resistivity—1,000 ohm-cms.

Power rate—3 cents per kwh.

Rectifier rating—20-ampere.

To determine the optimum size ground bed, enter the chart at the "Soil Resistivity-ohms-cms." line and extend a vertical line from the 1,000-ohm-cms. point to intersect all the anode types. Proceed horizontally to the left for the resistance to earth of horizontal type anodes and horizontally to the right for vertical type anodes. Various groups of vertical anodes with 15, 20 and 25-foot spacings are represented by curves.

Resistance values for horizontal anodes obtained from the ordinate of the upper left hand section should be transferred to the abscissa of the upper right hand section to determine power consumption and annual costs from the lower right hand and lower left hand section. In obtaining power costs for the two lower sections, the ordinate shows power consumption in AC watts based on 50 percent efficient rectification, and the abscissa of points on the lower left hand section indicated annual power costs in dollars.

How can output of magnesium anodes be predicted?

For a single 17-pound magnesium anode, working against a fully protected pipe, divide the soil resistivity in ohm-cm units into 180,000; the answer is the current output in milliamperes.

Example: Soil resistivity 3200 ohm-cm; 180,000/3200 = 56.25; current output will be about 56 milliamperes.

The rule works for *single* anodes against fully protected pipe; anode groups will have lower current output per anode, and if the pipe is not fully protected, the current output will be greater. For 32-pound anodes use 195,000; for 50-pound anodes use 204,000.

How to determine the efficiency of a cathodic protection rectifier

Multiply DC voltage output times DC amperes times 100 and divide by the AC watts input. The answer is percent efficiency.

Example: A rectifier has an output of 50 amperes DC at 15 volts. The wattmeter gives a direct reading of an AC input of 1155 watts.

$$\text{Efficiency} = \frac{(\text{DC volts})\ (\text{DC amperes})\ (100)}{(\text{AC watts input})}$$

$$= \frac{(15)\ (50)\ (100)}{1155} = 65 \text{ percent.}$$

On those rectifier units that do not have a wattmeter, the AC input may be obtained from the watt hour meter mounted on the pole. To do this, count the number of revolutions made by the watt hour meter in so many seconds, multiply by the meter constant stamped on the meter times 3600 and divide by the number of seconds timed. This gives AC input in watts. Then go ahead and use the formula as above.

Example: A rectifier unit does not have a watt meter. The wheel of the watt hour meter revolves 11 times in one minute. Meter constant is shown as 1.8. Output of the rectifier is the same as in the first case, 50 amperes at 15 volts DC.

$$\text{AC watts} = \frac{K \times N \times 3600}{T}$$

where K is the meter constant and N is the number of meter revolutions in T seconds.

$$\text{AC watts} = \frac{(1.8)\ (11)\ (3600)}{60} = \mathbf{1188 \text{ watts AC}}$$

$$\text{Efficiency} = \frac{(15)\ (50)\ (100)}{1188} = \mathbf{63 \text{ percent.}}$$

How to calculate the voltage drop in ground bed cable quickly

Use the following formula for calculating the voltage drop in a copper ground bed cable:

$$V_T = \frac{R_{1000}}{1000}\left[I_T D_{GB} + D_A\left[(N-1)(I_T) - I_A(1+2+3+4 \ldots N-1)\right]\right]$$

where V_T = Total voltage drop in volts
R_{1000} = Cable resistance per 1000 feet
I_T = Total current in amperes
$I_A = \frac{I_T}{N}$ Current per anode
D_{GB} = Distance from rectifier to first anode in feet
D_A = Distance between anodes in feet
N = Number of anodes in ground bed.

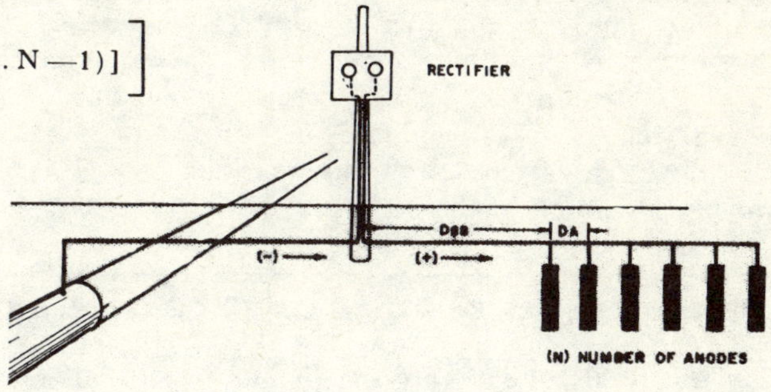

Example: What is the voltage drop in a copper wire ground bed cable in a cathodic protection system having a rectifier output of 225 amperes? The distance from rectifier to ground bed is 100 feet and 16 anodes are spaced on 20 feet centers. The cable used is No. 4 copper having a resistance of 0.2487 ohms per 1000 feet.

Then $R_{1000} = 0.2487$ ohms
$I_T = 225$ amperes
$I_A = \frac{225}{16} = 14$ amperes
$D_{GB} = 100$ feet
$D_A = 20$ feet
$N = 16$

$$V_T = \frac{.2487}{1000}\left[(225)\ (100) + 20\ [(15)\ (225) - (14)\ (120)]\right]$$

$$V_T = \frac{.2487}{1000}\ (22{,}500 + 20\ [1695])$$

$$V_T = \left(\frac{.2487}{1000}\right)(56{,}400)$$

$$V_T = (.2487)\ (56.4)$$

$$V_T = 14.00 \text{ volts drop}$$

What is the most economical size for a rectifier cable?

Choosing the economical size for a rectifier cable involves balancing the first cost of the cable against the power losses. A rule which does not include all of the factors, but which usually gives a very close approximation, is as follows:

$$a = 2280\ I\ \sqrt{P},$$

where "a" is the area of the cable in circular mils (refer to a wire table for size), I is the current to be carried, in amperes, and P is the cost of power in cents per kilowatt-hour. This formula is based on the present prices of copper cables, and assumes the use of a good grade of direct burial cable.

Example: Suppose I = 60 amps

P = 4 cents

a = (2280) (60) ($\sqrt{4}$)

a = 273,600 circular mils

Cable is sized 250,000 and 300,000 circular mils. Use the latter.

How to estimate the number of magnesium anodes required and their spacing for a bare line or for a corrosion "hot spots"

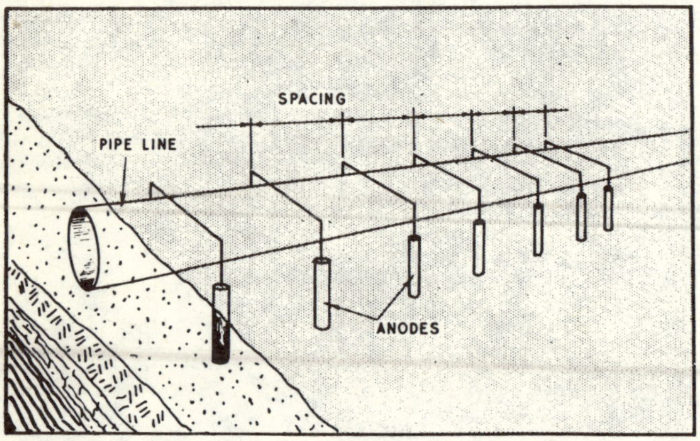

Soil and pipe line conditions will vary widely but where these given assumptions hold, an estimate can be made for test installations using this method. First find the area of pipe to be protected, multiply by the milliamperes per square foot necessary to protect the line and divide by design output of an anode in this type soil. This will give the number of anodes required; to find spacing, divide number of anodes into the length of line.

Example: Find the number of 17-pound magnesium anodes needed and spacing used to protect 10,000 feet of bare 4-inch pipe line in a corrosive soil having a resistance of 1000 ohms per cubic centimeter. Assume a current requirement of one milliampere per square foot, and that the anode output curve shows an output of 100 milliamperes per anode in this type soil.

$$\text{Area of pipe} = \frac{4 \times 3.14}{12} \times 10{,}000 = 10{,}500 \text{ square feet}$$

Total current requirement = 10,500 × 1.0 = 10,500 milliamperes

$$\text{Anodes required} = \frac{10{,}500 \text{ milliamperes}}{100 \text{ ma/anode}} = 105 \text{ anodes distributed}$$

$$\text{Spacing} = \frac{10{,}000}{105} = 95 \text{ feet}$$

Similarly find the number of anodes required to protect ten 100-foot long "hot-spots" on a 4-inch line assuming a 100 milliampere output from a 17-pound magnesium anode. Estimate as follows:

$$\text{Area of each ``hot-spot''} = \frac{4 \times 3.14}{12} = 105 \text{ sq. ft.}$$

$$\text{Anodes required} = \frac{105 \text{ sq. ft.} \times 1.0 \text{ ma per sq. ft.}}{100 \text{ ma per anode}} = 1.05$$

one 17-pound magnesium anode will produce approximately enough current in 1000-ohm soil to protect one of these "hot-spots."

Ten anodes required for "hot-spots" on this section of line.

(Note: For these rules the protective criterion is —0.85 volts pipe-to-soil potential using a copper-copper sulfate reference electrode; or a negative change of potential of as much as —0.25 volts. Soil and pipe line conditions vary widely, so it is well to consult an experienced corrosion engineer in the design of a complete cathodic protection system.)

How can resistivity of fresh water be determined from chemical analysis

Divide the total solids expressed in parts per million into 500,000; the result is the resistivity in ohm-centimeter units.

Example: Total solid content is 400 ppm.; 500,000 ÷ 400 = 1250 ohm centimeters resistivity.

This rule of thumb will give very accurate results for any naturally occurring fresh waters; severely contaminated waters may differ markedly, however, depending on the kind of contamination.

What will be the resistance to earth of a single graphite anode?

For the most common size, 3 inches by 60 inches, installed in ten feet of coke breeze backfill in an 8 inch hole, multiply the soil resistivity in ohm-cm by 0.002; this will give the resistance to earth in ohms.

Example: 1800 ohm-cm soil; 1800 × 0.002 = 3.6 ohms.

This assumes that the resistivity is uniform to considerable depth; if it increases with depth, the resistance will be slightly higher.

How to estimate the monthly power bill for a cathodic protection rectifier

Multiply the rectifier output voltage and current together, and multiply this by 1½; the result is the power consumption for one month, in kilowatt-hours. If the unit is on a meter of its own, figure the bill by the utilities rate schedule; if it represents an additional load, use the rate per kilowatt-hour for the highest bracket reached.

Example: The terminal voltage of a rectifier is 10.2 volts, and the current output is 21 amperes; $10.2 \times 21 \times 1.5 = 321.2$; the power consumption will be about 320 kwh per month; at 3 cents this would amount to $9.60, at 1.1 cents it would be $3.42, etc.

Note that the actual current and voltage values are used; if the rating of the unit is used, the result will be the maximum power consumption within the range of the equipment. The formula is approximate; it is based on an efficiency of about 48 percent.

What will be the resistance to earth of a group of graphite anodes, in terms of the resistance of a single anode

When the spacing is 20 feet (the most common value), the resistance of the group will be equal to the resistance of a single anode, divided by the square root of the number of anodes.

Example: Resistance of 4 anodes = ½ resistance of single; Resistance of 16 anodes = ¼ resistance of single, etc.

This rule is reasonably accurate, for soils which are uniform to a considerable depth; if the resistivity increases with depth, the "crowding" effect will be greater, whereas if it decreases, the individual anodes will not affect each other so much.

How can the current output of magnesium rod used for the cathodic protection of heat exchanger shells be predicted?

Divide the total dissolved solids content of the water by 13; the answer is the current output per foot of 1.315" (1" pipe size) magnesium rod to be expected.

This is an approximation to the current flow which will be found after stabilized conditions have been reached, if protection is obtained. If insufficient magnesium is installed, the current flow will probably be greater, and if too much is installed, the current will be less.

What spacing for test leads to measure current on a pipe line?

Multiply the pipe weight, in pounds per foot, by 4; the result is a spacing which will give a resistance of 0.001 ohm, or one ampere per millivolt IR drop.

Example:

Pipe weight is 40.48 lbs. per foot; $40.5 \times 4 = 162$. Install test leads at 162 foot spacing.

The rule works for welded lines of grade B seamless; variations in steel may introduce errors as great as 15 percent. It is always better to calibrate a line section if accurate results are to be had, but if the above rule is used, the spacing will not be too small for satisfactory measurement, or too great for convenience in installation.

How many magnesium anodes are needed for supplementary protection to a short-circuited bare casing

For 17-pound anodes, use A ρ/50,000 to give the number needed for the casing alone; A is the area of the casing, in square feet, and ρ is the soil resistivity in ohm-centimeters. The life of the anodes may be estimated from $\rho/180$. If the life in years, as given by this expression, is too short, refigure for 32-pound anodes, using A ρ/55,000 for the number and $\rho/100$ for the life. If this is still too short, figure for 50-pound anodes; number is A ρ/60,000 and life is $\rho/65$.

Such an installation, grouped on both sides of the line at each end of the casing, will enable the main line protection unit to carry beyond the casing; the values obtained above are approximate only, and should be checked by potential measurements after installation.

Example: Suppose the pipe line has shorted out to 45 feet of 24-inch bare casing. Soil resistivity is 1100 ohm-centimeters. How many 17-pound anodes would be required for supplementary protection?

$$\text{Number of Anodes} = \frac{A\rho}{50{,}000}$$

$$A = 2 \times 3.14 \times 45 = 283$$

$$\text{Number of Anodes} = \frac{(284)\ (1100)}{50{,}000}$$

$$= 6.21$$

Place three anodes at each end of casing. Their life expectancy will be

$$\frac{1100}{180} = 6 \text{ years}$$

Group installation of sacrificial anodes

When anodes are installed in groups, a spacing correction factor must be taken into account to determine the current output. The chart below shows the correction factor for multiple anode installations.

Example:
Single anode 8 inches by 60 inches in 1,500 ohm soil.
Resistance = 4.849 ohms.
Potential = 0.85 volts.
Current output =

$$\frac{E}{R} = \frac{17.5 - 0.85}{4.849} = 0.185 \text{ amps.}$$

Current output of 5 anodes on 15 foot spacing under above conditions. Combined resistance of group $= \frac{4.849}{5} \times 1.22$ (correction factor) = 1.17 ohms. Total current output =

$$\frac{1.75 - 0.85}{1.17} = 0.769 \text{ amps.}$$

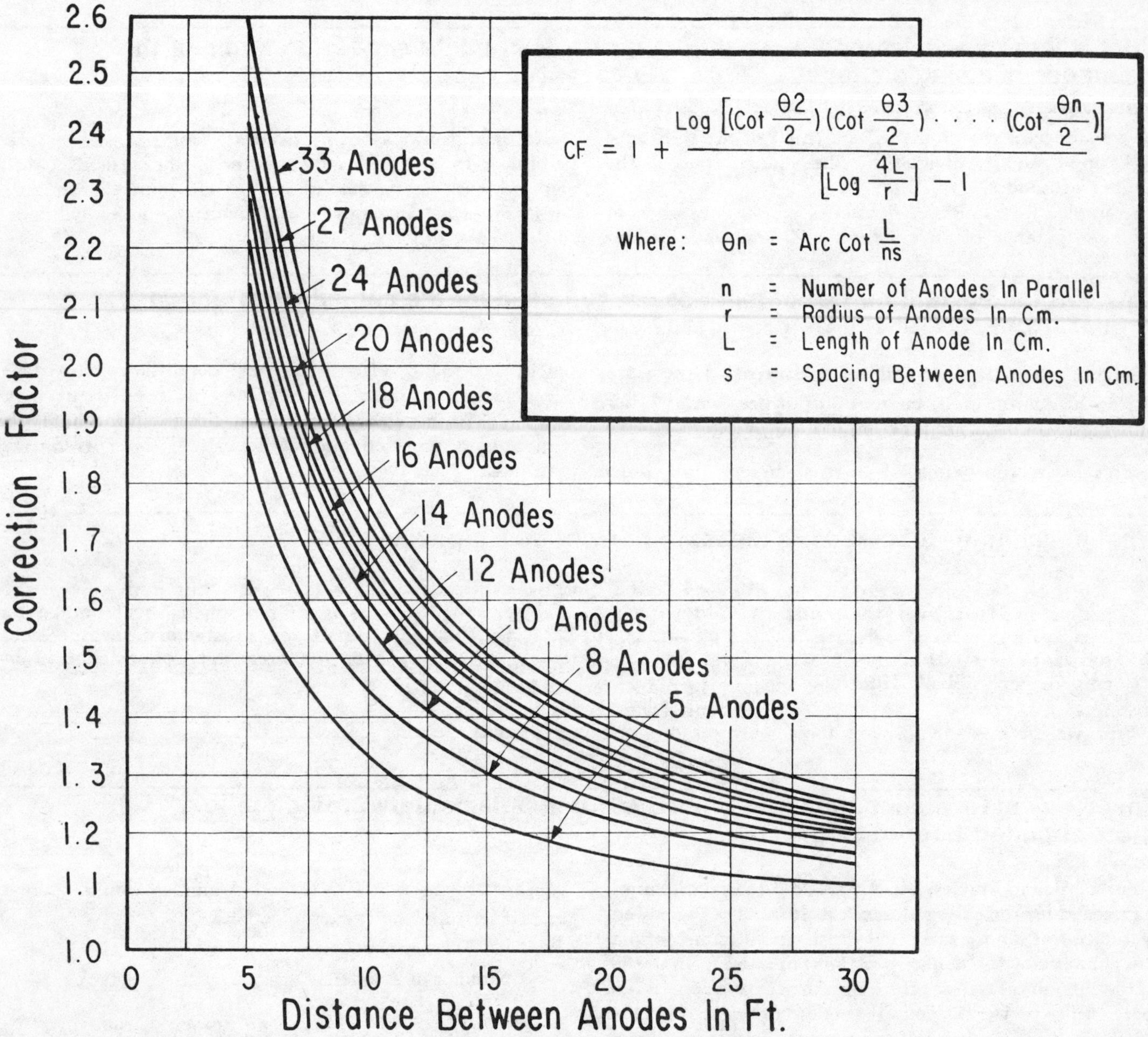

How can the life of magnesium anodes be predicted?

Multiply the anode weight in pounds by 57, and divide by the current output in milliamperes.

Example:
A 17-pound anode has a current output of 83 milliamperes

What is its projected life?
17 × 57/83 = 11.7; the projected life is 11.7 years.

The rule is based on 50 percent current efficiency, and on stabilized conditions. If more anodes are to be added, or if time has not been allowed for polarization (one to four weeks), then the current output will probably fall off, and the actual life will be 10 to 40 percent longer than indicated by the rule.

How to find the voltage rating of a rectifier if it is to deliver a given amount of current thru a given ground bed (graphite or carbon)

Multiply the desired current, in amperes, by the total resistance of the ground bed, in ohms (either calculated or measured); add two volts, to overcome the galvanic difference between graphite and protected steel.

Example: A current of 20 amperes is desired from a ground bed whose resistance is 0.9 ohms.

$(20 \times 0.9) + 2 = 20$ volts; specify the next larger size, which will probably be (depending upon the manufacturer) 24 volts.

Note: where the resistance is measured rather than calculated, and is the total loop resistance between pipe and ground bed, this method is quite precise; for calculated values, caution should be observed.

Determining current requirements for coated lines

Survey along the line after a polarization run of three or four hours, and locate the two points whose pipe-to-soil potentials are 1.0 volt and 0.8 volt (see Fig. 1). The section of line lying between these two points will then be at an average potential of about 0.9 volt, which is approximately the average of a line protected with distributed magnesium anodes. If, then, the line current is measured at each of these two points, the difference between these two line currents will be the total amount of current picked up on the line in the section. This quantity, divided by the length of the section, will then give the current requirements of the line, in amperes per mile. Naturally this figure can only be used for an entire line if the section chosen is truly representative; judgment must be used at this point.

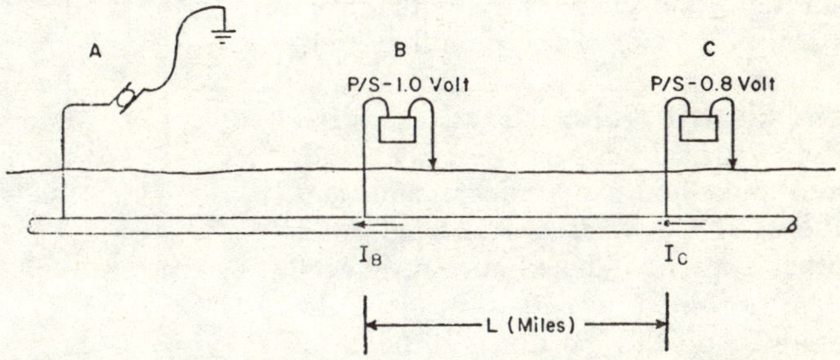

Fig. 1. Current is drained from the line at A, by a welding machine or other source of direct current. Points B and C, having P/S potentials of 1.0 volt and 0.8 volt respectively, are located; the average P/S between them is 0.9 volt. The current required to bring L miles of line to 0.9 volt is then $I_B - I_C$, and the average current required for the line is $\frac{I_B - I_C}{L}$, in amperes per mile.

Determining current requirements for coated lines when pipe-to-soil potential values are estimated

If it is not possible or convenient to find the points which have the exact values of 1.0 and 0.8, then readings can be taken which approximate these values, and both potential and line current can then be adjusted to find the desired quantity; this is illustrated in Fig. 1; readings were taken at points "A" and "B," and, by assuming a static potential of 0.6 volt and linear attenuation, it is determined that points "C" and "D" have the values of 1.0 and 0.8; since line current varies according to the same laws as potential, it is possible to determine the current values at the two latter points, and so to arrive at the desired quantity.

It is not to be expected that the estimate thus made will be highly accurate; it is intended to be merely preliminary.

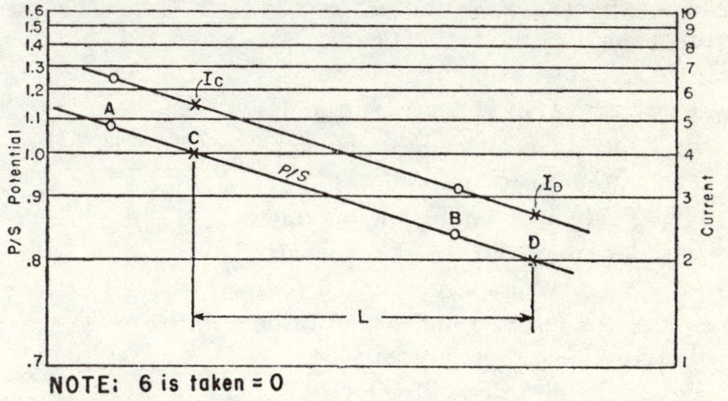

Fig. 1. Measurements of P/S potential and of Line Current are taken at A and B; it is then assumed (in the absence of other information) that the static potential of the line was originally at 0.6 volt. On this basis, the P/S and current curves are drawn as straight lines on semi-logarithmic paper, and the points whose P/S are 1.0 and 0.8 are located, at C and D. The difference between the current values for these two points then gives the current demand of the line for the distance L.

HVDC effects on pipe lines

Latest methods to calculate quantitative characteristics of the earth's gradient and High Voltage Direct Current effects on pipe lines

Thomas B. Hays, George W. Silkworth, Robert D. Schilling
Cornell, Howell, Hayes & Merryfield, Inc.

A PIPE LINE collects, conducts and discharges electric current of high voltage direct current (HVDC) origin as a result of the electric field established by the DC current transmitted through the earth and by the electrical properties of the pipe line itself.

The Electric Field. It is assumed that the electrical properties of the earth and pipe line are uniform and that the HVDC effects are isolated from other stray or induced current effects. This assumption makes possible the analysis of this portion of the total protection problem and assists in visualizing the unique effects of HVDC earth currents.

Ultimately, the total electrical current conducted by a pipe line will be the sum of all the various components. This includes protective current, telluric current, and other stray or interference currents.

Conduction of a direct current through the earth is an electrolytic phenomena, characterized by an electric field that surrounds each of the two earth electrodes and is directly proportional to the HVDC earth current (Fig. 1). When the electrodes are spaced several hundred miles apart, the field appears to be concentric around each ground electrode.

The fields surrounding the opposite electrodes are of the opposite polarity. If the polarity of the ground current reverses, the surrounding electric fields will also reverse. The potential at any point in the electric field with respect to remote earth (V), due to an electrode current (I) at a distance (r) from an electrode in a homogeneous earth of the resistivity (ρ) is found by the relationship:

$$V = \frac{\rho I}{2 \pi r}$$

This relationship shows a simplified representation of what actually occurs. The earth's resistivity is not uni-

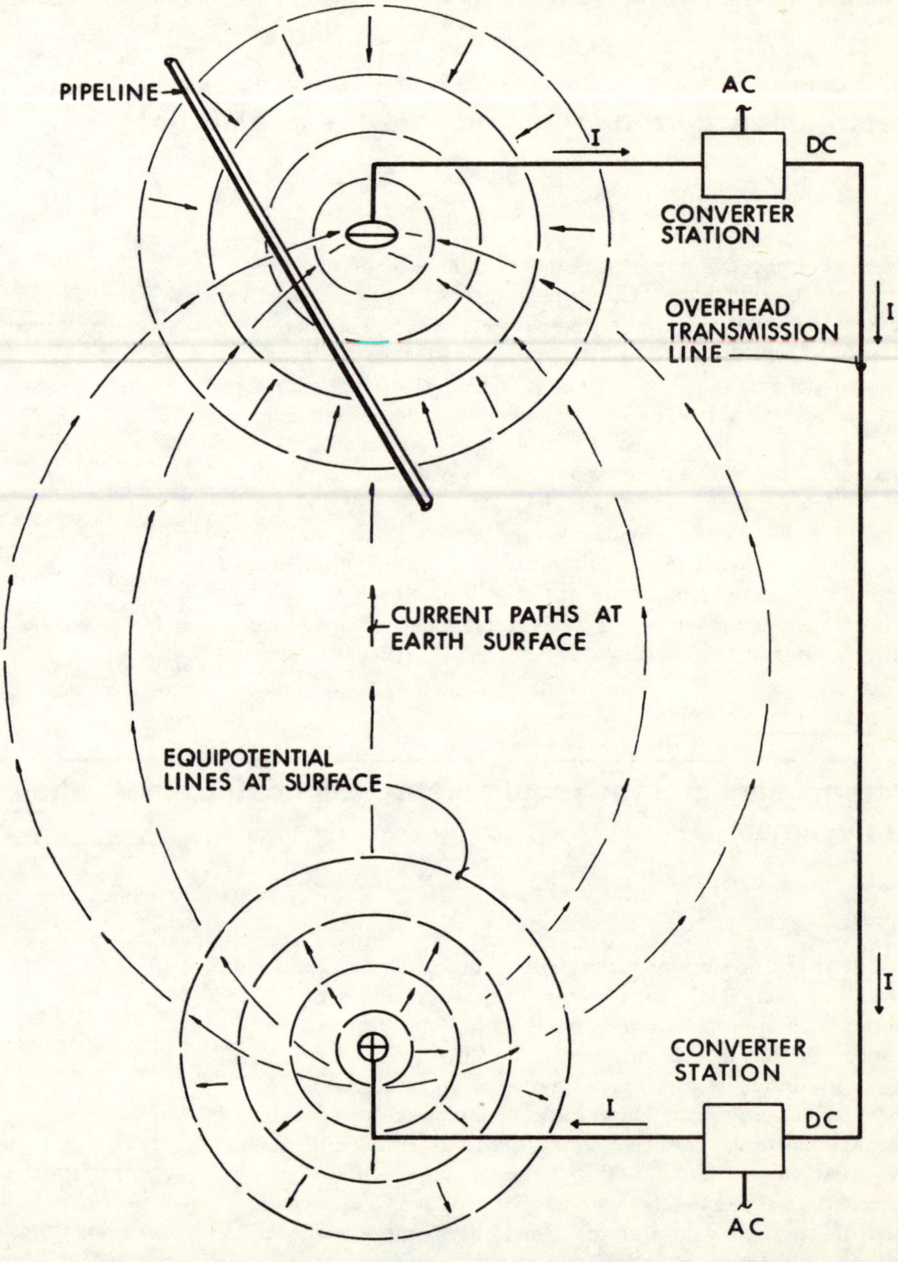

Fig. 1—Conduction of a direct current through the earth is an electrolytic phenomena, characterized by an electric field that surrounds each of the two earth electrodes and is directly proportional to the HVDC earth current.

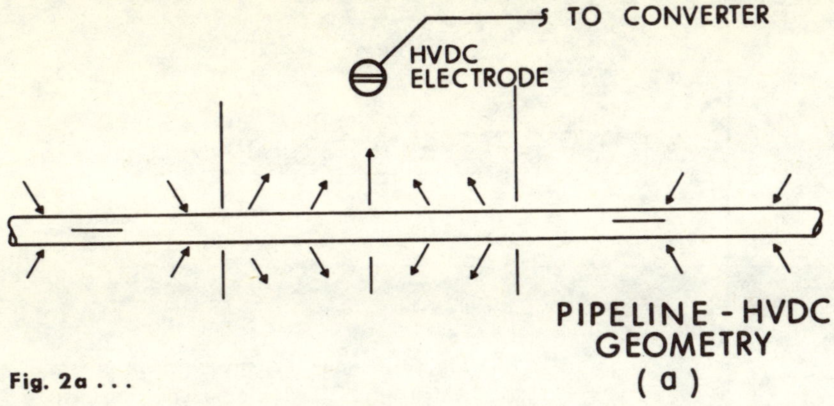

Fig. 2a . . .

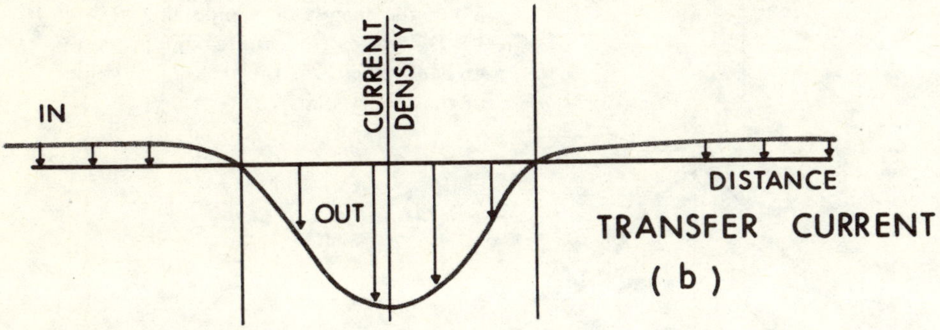

Fig. 2b . . .

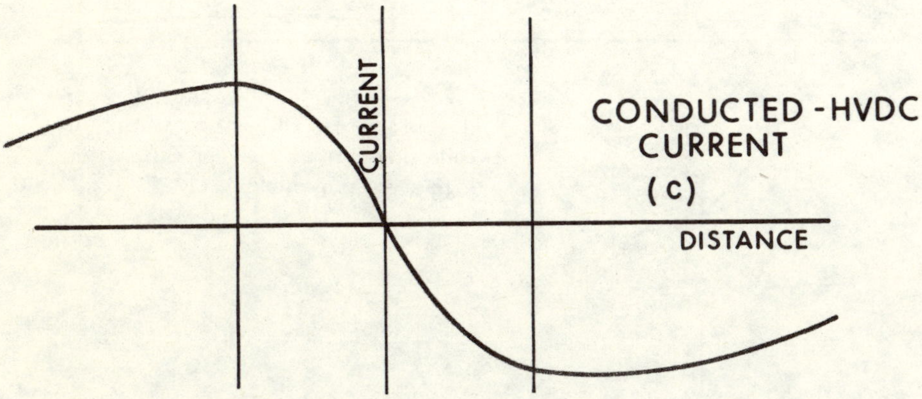

Fig. 2a—Geometric plan between the HVDC electrode and the pipe line. Arrows indicate the direction of the HDVC component current. These would be reversed if the polarity of the electrode were reversed. **Fig. 2b**—Curve shows characteristics of the transfer of HVDC into and out of the pipe line along its length. The transfer current is equivalent to current density. **Fig. 2c**—Conducted current in the pipe line which results in the summation of the transfer current along the pipe.

form at all depths or horizontal distances, which causes irregularities to the ideal voltage distribution existing around an electrode in a homogeneous earth.

To refine the earth current analysis and to account for different resistivities, computer techniques have been developed to estimate the earth's gradients existing for conditions of multilayers of earth of different resistivities. Pipe line effects have also been predicted by this method. Reasonable agreement between these models and field data has been reported. Techniques are available to calculate the quantitative characteristics of both the earth's gradient and HVDC effects on pipe line.

HVDC Effects on a Pipe Line. Whether the HVDC component of the total current existing on the line causes a corrosion problem depends on the level of protection along the line and the effect of the added component to the level of protection. Fig. 2a illustrates the HVDC component of the total conducted current along the pipe line for a given separation distance under the condition where the HVDC electrode is collecting current from the earth.

In determining protection requirements for the pipe line, the characteristic of the current transfer relationship is important (Fig. 2b). In those regions of the pipe line where the current transfer is from the pipe line to the soil—along the center in this case—corrosion could occur if the HVDC component of the total transfer current to the soil is great enough to exceed any local level of protection against corrosion.

The pipe line HVDC arrangement shown in Fig. 2a will serve as a basis to develop and illustrate the performance requirements for protecting a pipe line from the detrimental effects of HVDC earth currents. The hypothetical problem, as shown, is one that may be encountered in the field. In this example, the following assumptions are made:

- The pipe line is long, straight, electrically continuous, and uniform in wall thickness, diameter, material and coating.
- It is approximately 30 miles away from the nearest HVDC earth electrode.
- The earth's resistivity is uniform over the general area.
- The maximum value of the HVDC earth current in either polarity causes the pipe line effects.

While these conditions may not exist in any actual pipe line HVDC system situation, they will serve to identify and evaluate the requirements for protection against HVDC effects in an actual situation.

Fig. 3 shows the range or envelope of effects as the HVDC varies from its maximum value in one polarity to the maximum value of the other polarity. The coordinates on this graph are DISTANCE along the *X* axis and VOLTAGE on the *Y* axis.

The voltage corresponds to the effect which would usually be measured on an actual pipe line: pipe-to-soil voltage measured with a copper-copper sulfate half-cell. The value of the voltage effect at any point on the line is the product of the current transfer, and the electrical resistance between the pipe material and earth over the same area.

Fig. 3 also shows reasonable hypothetical scale values illustrating the various combinations of protection requirements which may exist in an actual installation. The voltage values give the change in pipe-to-soil reading that would be caused by the HVDC transfer current. If this HVDC is considered as the only source of corrosion,

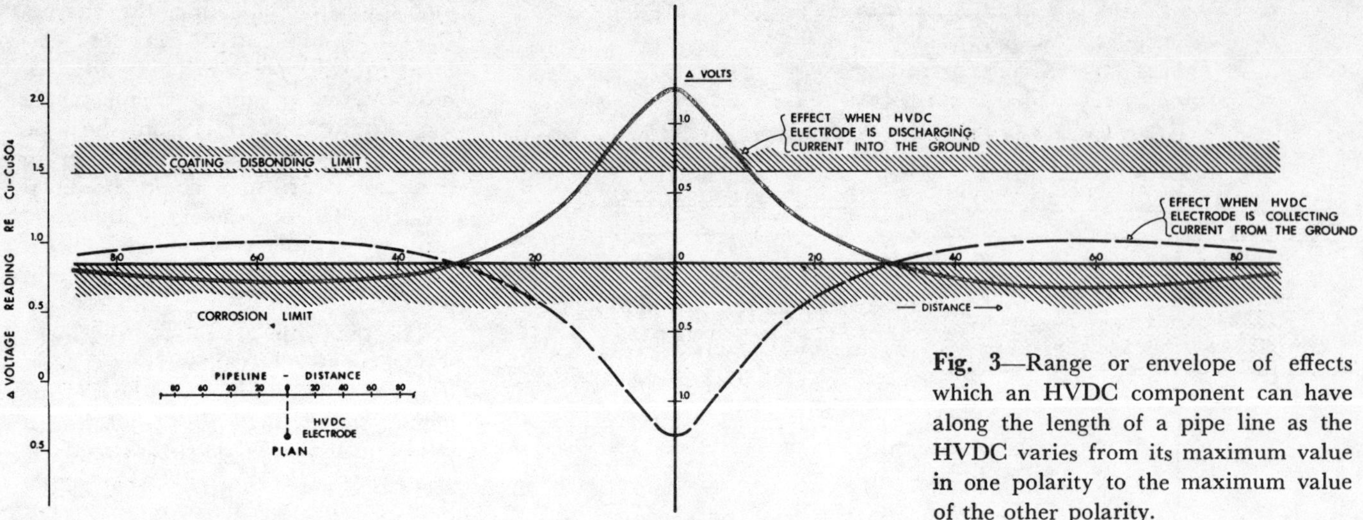

Fig. 3—Range or envelope of effects which an HVDC component can have along the length of a pipe line as the HVDC varies from its maximum value in one polarity to the maximum value of the other polarity.

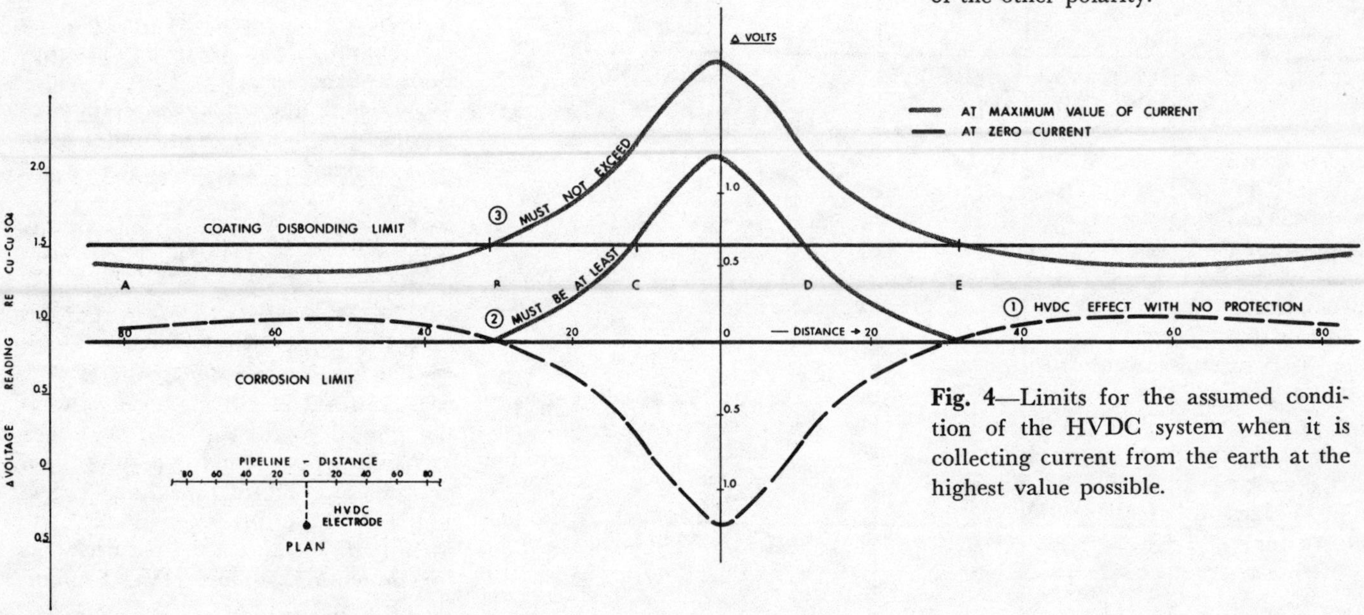

Fig. 4—Limits for the assumed condition of the HVDC system when it is collecting current from the earth at the highest value possible.

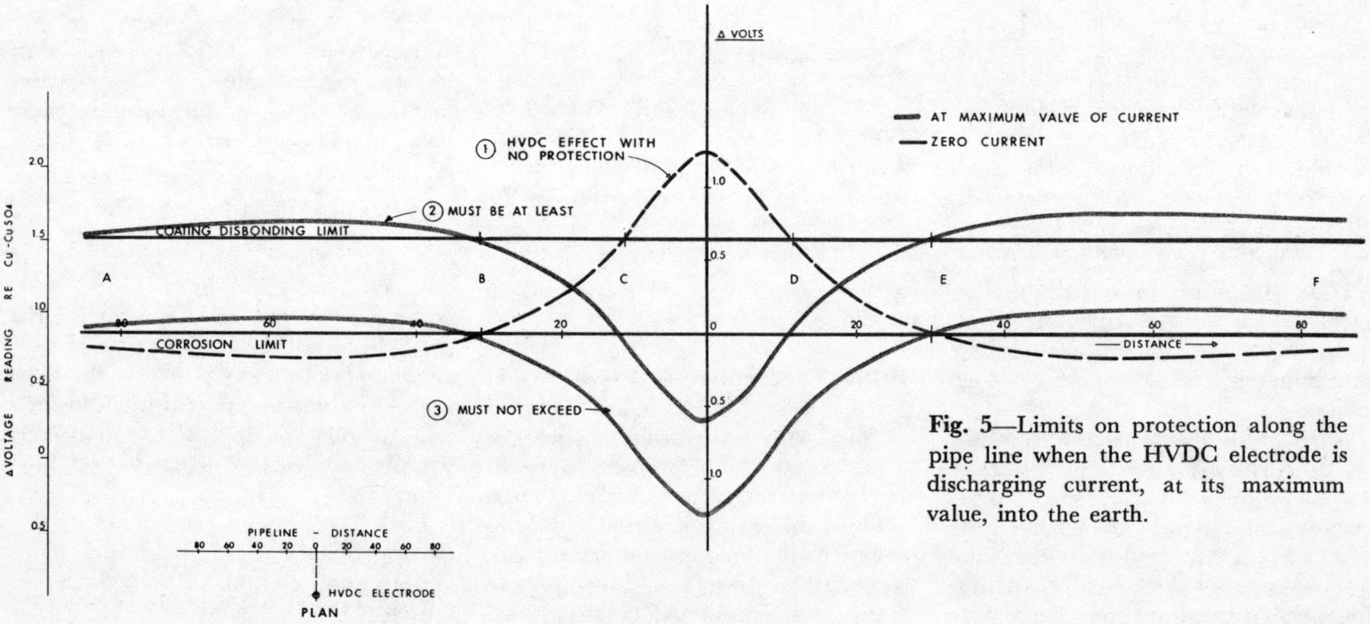

Fig. 5—Limits on protection along the pipe line when the HVDC electrode is discharging current, at its maximum value, into the earth.

then a pipe-to-soil voltage reading with a copper-copper sulfate half-cell would be approximately 0.85 volts at no transfer current. This scale is shown on the left side of the figure. Corrosion occurs for any reading below this axis, since the current is leaving the pipe line and entering the soil to cause this change in the pipe-to-soil reading.

At locations on the pipe line where HVDC causes conditions which increase the volt-meter reading, current is collecting on the pipe line. The horizontal axis, therefore, represents

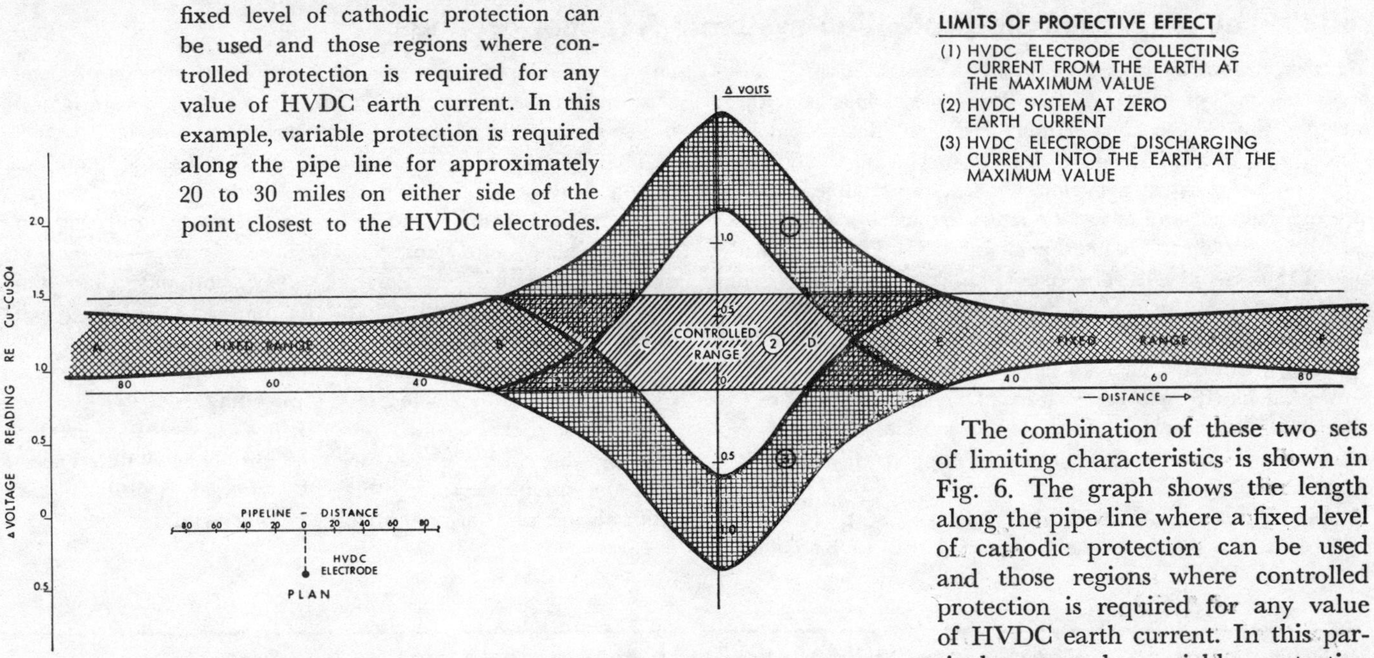

Fig. 6—Combination of Figs. 4 and 5. The length along the pipe line where a fixed level of cathodic protection can be used and those regions where controlled protection is required for any value of HVDC earth current. In this example, variable protection is required along the pipe line for approximately 20 to 30 miles on either side of the point closest to the HVDC electrodes.

a corrosion limit. Above this axis, a coating disbonding limit can exist where the density of the current transferred from the earth to the pipe line causes gas evolution, which may blister the coating.

In this example, a pipe-to-soil reading with a copper-copper sulfate half-cell of 1.5 volts is shown as the coating disbonding limit. The difference between these limits gives an assumed safe operating band width of 0.65 volts. Values, other than these may be used for specific applications. These limits, however, will serve to illustrate the principles.

Protective System Limits. The requirements for protection along the line from either corrosion or disbonding is developed from examination of Fig. 3 and the results are plotted in the subsequent figures. The upper limit of protection is the maximum amount of current that can be collected by the pipe line without exceeding the disbonding limit. The lower limit is the minimum amount of current that can be collected by the pipe line to keep it above the corrosion limit.

At this value of HVDC (Fig. 4), the uncorrected pipe line effect is shown by Curve 1. No protection is needed along either end of the pipe line in Regions *A-B* or *E-F*. Corrosion protection is needed between *B* and *E* The minimum protection effect along the line is shown by Curve 2. The resultant of the HVDC and this protection effect lies on the corrosion limit. The maximum protective effect that can be applied along the line, without exceeding the coating disbonding limit, is shown by Curve 3. Under this maximum value and polarity of HVDC current, the region between these two protective limits, therefore, defines the range of acceptable protective system performance at any location on the pipe line.

When the HVDC effect decreases, due to a reduction in the ground current, the required level of protection on the line also decreases. Along the pipe line length from *C to D,* the protection **must** be reduced to avoid exceeding the disbonding limit. At zero HVDC effect, the protective system effect must fall between the corrosion and disbonding limits all along the line.

Examination of the limits for these two values of HVDC system operation shows that the protective effect must be varied in response to the HVDC effect over the length of the pipe line from *C* to *D*: automatic regulation of the protective current is needed over this length of the pipe line. With only this polarity of HVDC current, a fixed level of protection can be applied along the pipe line from *A* to *C* and from *D* to *F*.

When the earth current can reverse polarity, as with the Pacific Northwest (PNW)-Pacific Southwest (PSW) intertie, then the same process that was used to develop the limits shown in Fig. 4 can be repeated for the opposite HVDC polarity. The limits on protection along the pipe line when the HVDC electrode is discharging current, at its maximum value, into the earth are shown in Fig. 5.

The combination of these two sets of limiting characteristics is shown in Fig. 6. The graph shows the length along the pipe line where a fixed level of cathodic protection can be used and those regions where controlled protection is required for any value of HVDC earth current. In this particular example, variable protection is required along the pipe line for approximately 20 to 30 miles on either side of the point closest to the HVDC electrode. Cathodic protection is required when the HVDC electrode is collecting current from the earth. That is, the HVDC component of current, conducted by the pipe line, is discharging from the line near the earth electrode. The required protective current along the pipe line to counteract the HVDC effect under this condition is shown as Region 1 in Fig. 6.

When the HVDC polarity is reversed and the magnitude of the ground current is increased to the maximum value, a different problem is encountered. The electric field which is established around the electrode and pipe line causes current to enter the pipe line through the coating. In this example, the disbonding limit is exceeded over the length of the pipe line from *C* to *D* at the maximum value of HVDC earth current. The required correction for this detrimental effect is similar to the effect for the previous HVDC polarity: namely, the uncorrected electric field must be modified by an added field to give an acceptable total. Theoretically, in this situation, a portion of the HVDC earth current can be collected in a protective ground bed and conducted to the pipe line to satisfactorily modify the original field into an acceptable one.

Trouble shooting cathodic protection systems: Rectifier-ground bed

When the monitor, or a pipe-to-soil potential survey, indicates inadequacy of protection, the first place to look is at the protective unit. The current output of the rectifier should be checked; if it is normal, the trouble is on the line itself; if it is high, and accompanied by low voltage, the trouble is certainly on the line, and is caused either by increased current demand or by a short-circuit to parasitic metal. If the current output is low, with voltage normal or high, the trouble is in the ground bed or connecting cables. A pipe-to-soil potential over the ground bed will show a peak over every anode which is working; a disconnected anode will not show at all. This test is particularly useful if comparison can be made to a similar test made at the time of the original installation. When inactive anodes are found, only digging will uncover the cause.

If the rectifier and its anode bed appear to be performing satisfactorily, the source of the low potentials must be sought on the line itself. Any area in which work has been done recently should be investigated; for example, if a new lateral has been connected, the insulation should certainly be checked. If investigation of such suspected sources discloses nothing, then a more detailed search must be made. First, the pipe-to-soil potentials should be studied, to see if the failure seems to be localized. A more thorough, but slower, approach, is that of making a detailed line current survey; find out where the drained current is coming onto the line. This will be much more easily interpreted if a similar survey, made when the line was in satisfactory condition, is on record; a comparison will often very quickly locate the offending parasite. On the other hand, if the current collected in each section is greater than on the earlier survey, with no pronounced differences, then the trouble is simply that of increased overall current requirement, probably due to coating deterioration.

Trouble shooting cathodic protection systems: Magnesium anode system

The basic technique is the same as that outlined above; more measurements are required, because of the multiplicity of drain points. First, the current output of stations nearest the point of low potential should be checked; if these are satisfactory, a similar check should be extended in both directions until it is clear that the trouble must be on the line. When a given anode group shows a marked drop in current output, the cause may be drying out, shrinkage of backfill, or severed or broken lead wires. If the current is zero, the pipe-to-soil potential of the lead wire will show whether it is still connected to the pipe or the anode, and thus indicate the direction to the failure. If the current is low, there may be a loss of one or more anodes, by a severed wire; a pipe-to-soil potential survey over the anodes will show which are active, just as in the case of rectifier anodes.

Trouble indicated on the line, rather than at the anode stations, is tracked down in the same manner as that used for line protected by rectifiers; first investigating locations where something has been done which might be responsible, next checking potentials, and finally making a line current survey.

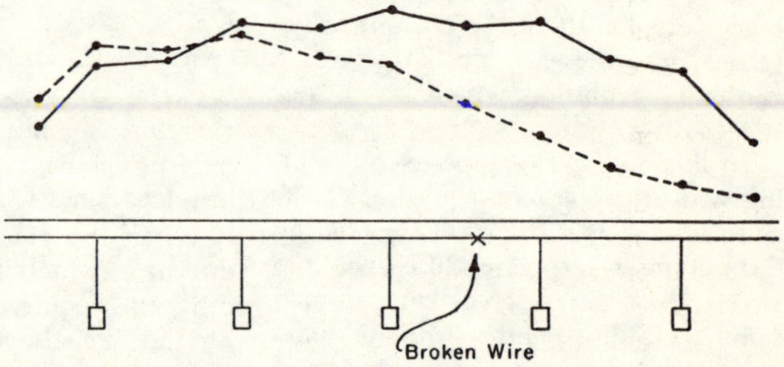

Fig. 1. Locating Idle Anodes by Surface Potentials. The solid line shows the potentials found along a line of anodes when all are delivering current; the dotted line exhibits the change when there is a break in the anode lead at the point indicated. Single disconnected anodes may also be located by this method. A driven ground rod, a pipe lead, or even a rectifier terminal may be used for the reference ground; all readings should be referred to the same reference.

Corrosion: How to control corrosion at compressor stations

Eight factors design engineers should consider

R. E. Hodge, Superintendent of Corrosion Control, Natural Gas Pipeline Co. of America

Corrosion failures at compressor stations result from carelessness on the part of the user, or poor choice of material/configuration by the designer.

Design engineers should consider the following:

- Coat everything underground, except ground rods, with a coating properly selected for the conditions
- Use proper coating application and inspection
- Use ground rods anodic to steel and insulated (coated) ground wires
- Avoid shielding as much as possible and allow space for corrosion control additions
- Use a minimum of insulating fittings
- Protect against water system corrosion
- Select materials carefully, with corrosion in mind as well as the more obvious considerations
- Not to retire facilities in place—take them out.

Coatings. Coatings are one of the most important considerations for control of underground corrosion. Generally, all underground metallic structures, except ground rods, should be coated. This would include gas piping, control lines, tubing, water lines, conduit, air lines, braces, etc.

There is enough buried conduit around a compressor station to soak up large amounts of cathodic protection if not coated. Non-metallic carriers can be used in some cases instead of coating.

Proper selection and use of coatings are important. The best of coatings are no good if used in the wrong place or improperly applied. The emphasis here is on the selection of coating for the temperature and environment expected.

For gas discharge lines, temperature becomes a dominant consideration. Be sure to get a coating that will withstand gas discharge temperature. Be careful, sag temperatures listed in coating literature are not maximum operating temperatures. Very few coatings on the market can withstand operating temperatures above 120° without damage. Any bare metal underground becomes a drain on the cathodic protection system and a potential shielding problem.

There are other areas for coating consideration besides underground. For instance, tank interiors and atmospheric coatings. These must also be matched to the situation.

Grounding. Grounds should be of materials anodic to steel. These might include magnesium or zinc anodes, or galvanized ground rods.

Copper ground rods should not be used. Copper is cathodic to steel. Some compressor stations have been installed with massive bare copper grounding systems. These are large galvanic cells with the steel gas piping as the anode (corroding element). Bare ground wires and rods also take cathodic protection current. This bad situation can be avoided by using galvanized rods, for example, and using copper ground wire with insulation. The junction of copper ground wire and galvanized ground rod must be coated. Grounding systems should be checked periodically for proper functioning.

Shielding. Shielding poses a major problem at compressor stations. It could be described as the blocking of cathodic protection due to geometric configuration.

Congested areas, where buried metal concentration is high, are difficult to cathodically protect. Piping, or other metal structures, will tend to shield cathodic protection currents from reaching all of the buried structures to be protected.

Cathodic protection ground beds of various styles have been used at compressor stations. Remote conventional beds and deep beds are popular for many applications. But NGPL has found that distributed anodes seem to be the best way to get adequate cathodic protection to all the shielded areas in a station. This type bed can be described as a ground bed with the anodes distributed in a nongeometrical pattern, interspersed through the piping.

Designers should allow space between lines where possible; avoid congestion. This would cause less shielding and allow space for distributed anode installations where required. It would also allow space for access to work equipment for anode installations.

Insulation. Use insulating fittings (flanges, couplings, unions, etc.) only where absolutely necessary.

Hold insulating fittings to minimum. These can often cause more problems than they solve. You might want to insulate water wells from the adjacent piping for control of cathodic protection current to the wells. This may not even be necessary if distributed ground beds are used.

Some companies insulate compressor stations from the lines entering and leaving the station. Other companies do not. This is a matter of individual company preference.

Water systems. Each water system (jacket water, piston water, boiler water, etc.) should have corrosion monitoring provisions designed into the system, such as coupons or corrosion rate probes. If there is a gas treating plant in conjunction with the compressor station, corrosion monitoring provisions should be designed into that system also.

Water treatment for corrosion control would have to be considered, depending on individual circumstances. Some treatment facilities may be designed into the systems.

Provisions should be considered for cathodic protection of the internal surfaces of storage tanks and water softeners. One problem often encountered on smaller tanks is adequate openings. Enough openings of sufficient size should be allowed on any tank for possible future work.

Water should be kept out of some systems. For instance, use air dryers on control lines.

Materials selection. Proper selection of materials is important throughout design.

Two key considerations from a corrosion standpoint are: (1) avoid the use of dissimilar metals; (2) match the material to its environment. Examples of these would include the use of a brass valve in a buried steel water line; or use of mechanically strong bolts in the wrong electrolyte, resulting in subsequent failure from stress corrosion cracking.

Select materials carefully, with corrosion in mind as well as the more obvious considerations. If design engineers do this, many subsequent failures can be eliminated.

Retirements. Many compressor station designs involve retiring something that is already there.

If something must be retired, especially underground, take it out. Many times a company pays unnecessarily to continue protection of facilities retired in place.

Even if the retired facilities can be completely disconnected from the protective systems, shielding is still a problem. Shielding is a problem by foundations also, so if they are retired they should be removed.

Thus, attacking the corrosion problems requires a truly interdisciplinary effort. In fact, corrosion offers one of the richest fields for interdisciplinary activity.

(Based on a paper, "Construction Design to Facilitate Corrosion Control-Compressor Stations," presented by the author at AGA Distribution Conference, May 8-10, 1972, Atlanta, Ga.)

16. Economics—Pipe Line & Station Operation

Estimate the cost of a pipe line in the U.S. (based on 1964 data)

A series of cost factors made available by the Federal Power Commission based on gas industry construction during 1964 yield the averages shown below for various diameter pipe lines.

To arrive at the cost of a particular project, multiply the diameter cost per inch mile times the diameter of the pipe to be installed times the number of miles to be installed. The resultant product represents the total cost of the pipe, coatings, average right of way easements and damages, miscellaneous materials, cost of labor and normal cost of construction. These figures represent averages based on projects reported during 1964 and any estimates derived would vary with distances from manufacturers, terrain, locality, labor costs, season of the year and other variables.

Diameter of pipe	Average cost/inch diameter/mile
10-inch	$2059
16-inch	$2849
20-inch	$2632
24-inch	$3325
30-inch	$4085
36-inch	$4260

Example. To find the cost of 100 miles of 36-inch diameter pipe to be installed, multiply 4,260 times 36 times 100 equals $15,336,000.

How to compare the cost of operating an engine on diesel and natural gas

RULE: For a rough estimate of the cost of diesel fuel per hour of operation, multiply: (.05) (engine horsepower) (cost per gallon of diesel fuel). To find the cost of natural gas per hour's operation, multiply: (.008) (engine horsepower) (cost of natural gas per Mcf).

Example: What would it cost per hour to operate a 500 horsepower engine on diesel fuel? on natural gas? Diesel fuel costs 10 cents per gallon and natural gas costs 20 cents per mcf.

Solution:

Cost/hour for diesel fuel operation = (.05) (500) (.10) = $2.50

(A more accurate answer can be obtained by substituting .0533 for .05. This would make the cost of diesel fuel operation $2.67 per hour.)

Cost/hour for natural gas = (.008) (500) (.20) = $0.80

Example: Compare the cost of operating the engine on straight diesel fuel with dual fuel consisting of 90 percent natural gas and 10 percent diesel pilot fuel, if the conditions are the same as above.

Solution: Cost per hour = (.10) (2.50) + (.90) (.80) = $0.97.

Generally speaking, 1000 cu. ft. of natural gas has a Btu content of 1,000,000 and a gallon of diesel fuel has a Btu content of 150,000. Thus 1 Mcf of natural gas is equivalent to 6⅔ gallons of diesel fuel. To arrive at the figures above, the engine was assumed to require an average of 8000 Btu per brake horsepower hour. This is an arbitrary figure which is about the average for engines in good mechanical condition operating under high percentage of load. Some pipe line companies are equipped with engines that consume about 7000 Btu/bhp; others run as high as 11,000 Btu/bhp. Condition of the engine and the loading account for most of the difference.

If an engine used an average of 7000 Btu/bhp in developing its full horsepower of say, 1000, naturally it would use 7,000,000 Btu per hour. This would require 7 Mcf of natural gas or about 46.7 gallons of diesel fuel.

How to estimate energy costs for different pipe line thruputs

The rule of thumb given below will apply to liquid pipe lines where the flow is in the turbulent region only. It will serve to estimate energy requirements for pumping through pipe lines between increments of plus 15 percent and decrements of minus 15 percent.

Rule: Multiply the normal throughput in barrels times the cost of energy per barrel. Then multiply the increment or decrement (provided it is within a range of plus 15 percent or minus 15 percent) times the cost of energy per barrel at normal throughput times 3. Then add or subtract the latter amount from the cost of normal throughput.

Or restated in simpler form:

Cost of energy per barrel at normal throughput = Y

Cost of energy per barrel of increment or decrement within a range of ± 15 percent = 3Y

Example: Assume energy requirements for pumping 2000 barrels per hour through a 12-inch pipe line are $.01 per barrel. What are energy requirements for pumping 2200 barrels per hour?

Energy Cost for 2000 B/H: 2000 × .01 = $20 per hour

Energy Cost for Additional 200 B/H:

200 × .01 (3) = 6 per hour

Energy Cost for 2200 B/H = $26 per hour

Comparing fuel costs for diesel and electric prime movers

Here's a formula which will give you a quick comparison of fuel costs of a pump station—for diesel engine or electric motor.

Just find out which of the expressions below, C_e or C_d is the smaller.

$$C_e = \frac{0.7457}{n} \times E$$

$$C_d = 0.065\,D + 0.04$$

where,

C_e = power consumption cost of electric motors, cents/BHP-hour.
C_d = fuel and lube oil cost of diesel engines, cents/BHP-hour.
E = electric power rate, cents/KW-hour.
D = diesel fuel cost, cents/gallon.
n = efficiency of electric motor.

Example:

Let's see which is more economical: running a 1,000 BHP unit on diesel or electric driver.

The data are:

n = 95%
E = 1 ¢/KW-hr.
D = 10 ¢/gallon

So, we have:

$$C_e = \frac{0.7457}{0.95} \times 1 = 0.785$$

$$C_d = 0.065 \times 10 + 0.04 = 0.690$$

Then, for a 1,000 BHP unit the estimated operating cost per hour would be:

motor: $7.85/hour
engine: $6.90/hour

Keep in mind that this rule deals only with costs of power and fuel consumption. It does not take into consideration such expenses as operating personnel, repairs and renewals, depreciation, overhead, and so on.

Nomograph for calculating scale-up of equipment or plant costs

In correlating the cost of equipment or complete plants as a function of size, it has been shown that very often the cost varies as a fractional power of the ratio of capacities (or equipment sizes), as follows:

$$\left(\frac{C_2}{C_1}\right)^n = \frac{I_2}{I_1}$$

where C_1 and C_2 are the capacities (or sizes) of the equipment or plants in question;

I_1 and I_2 are the cost or fixed investment of the known unit and the new unit, respectively; n is the exponent that applies to the particular case at hand.

The nomogram enables one to calculate the capacity ratio raised to the fractional power n and the equipment or plant cost if the size for one capacity level is known.

Example

A rotary drum filter costs $20,000 when its filtering area is 50 sq. ft. Assuming the cost varies as the 0.6 power of the capacity, how much will a 100 sq. ft. filter cost?

The capacity ratio is C_2/C_1 = 100/50 = 2. Enter 2 on the capacity ratio scale (A) and align with 0.6 to the intersection point on the middle scale with $20,000 on the (D) scale and read I_2 = $30,000 on the (E) scale.

Source

Kuong, J. F., *Brit. Chem. Eng.* 7, No. 7, 546 (1962).

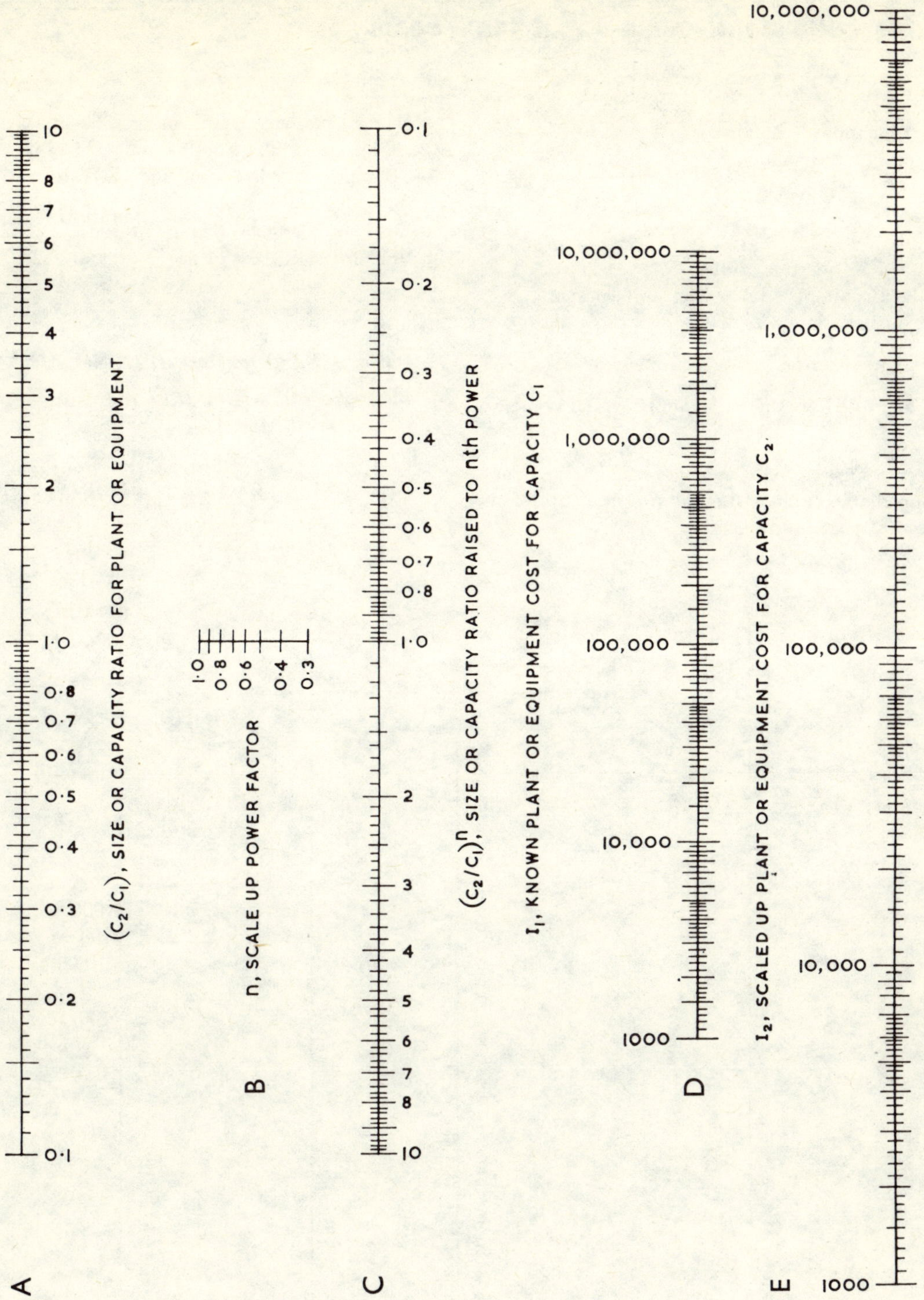

Figure 1. Scale-up of equipment and plant costs.

Nomograph for calculating scale-up of tank costs

The nomogram enables one to estimate the cost of a steel container of given volume when one knows the cost and volume of another or standard container.

The chart is based on the equation:[1]

$$C = cV_s^{0.3}V^{0.7}$$

where:

C = cost of container in question, dollars
c = cost of standard container, dollars per unit volume
V_s = volume of standard container
V = volume of container in question

V_s and V are expressed in the same units: gallons cubic feet, barrels.

Based on Dickson's suggestions,[1] the chart can be used for drums of the same diameter and internal pressure, and for spheres of the same internal pressure.

Example

If a 30,000-barrel tank costs $36,000, what is the cost of a 15,000-barrel tank? For the standard tank, c = 36,000/30,000 or $1.2 per barrel. Follow the key; connect 30,000 on the V_s scale with 15,000 on the V-scale and note the intersection with the α-axis. Connect this point with 1.2 on the c-scale and read the desired cost of approximately $22,000 on the C-scale.

Source

Davis, D. S., *Petrol Refiner,* 37, No. 4200 (1958).

Reference

1. Dickson, R. A., *Chem. Eng.,* 54, (8) 122 (1947).

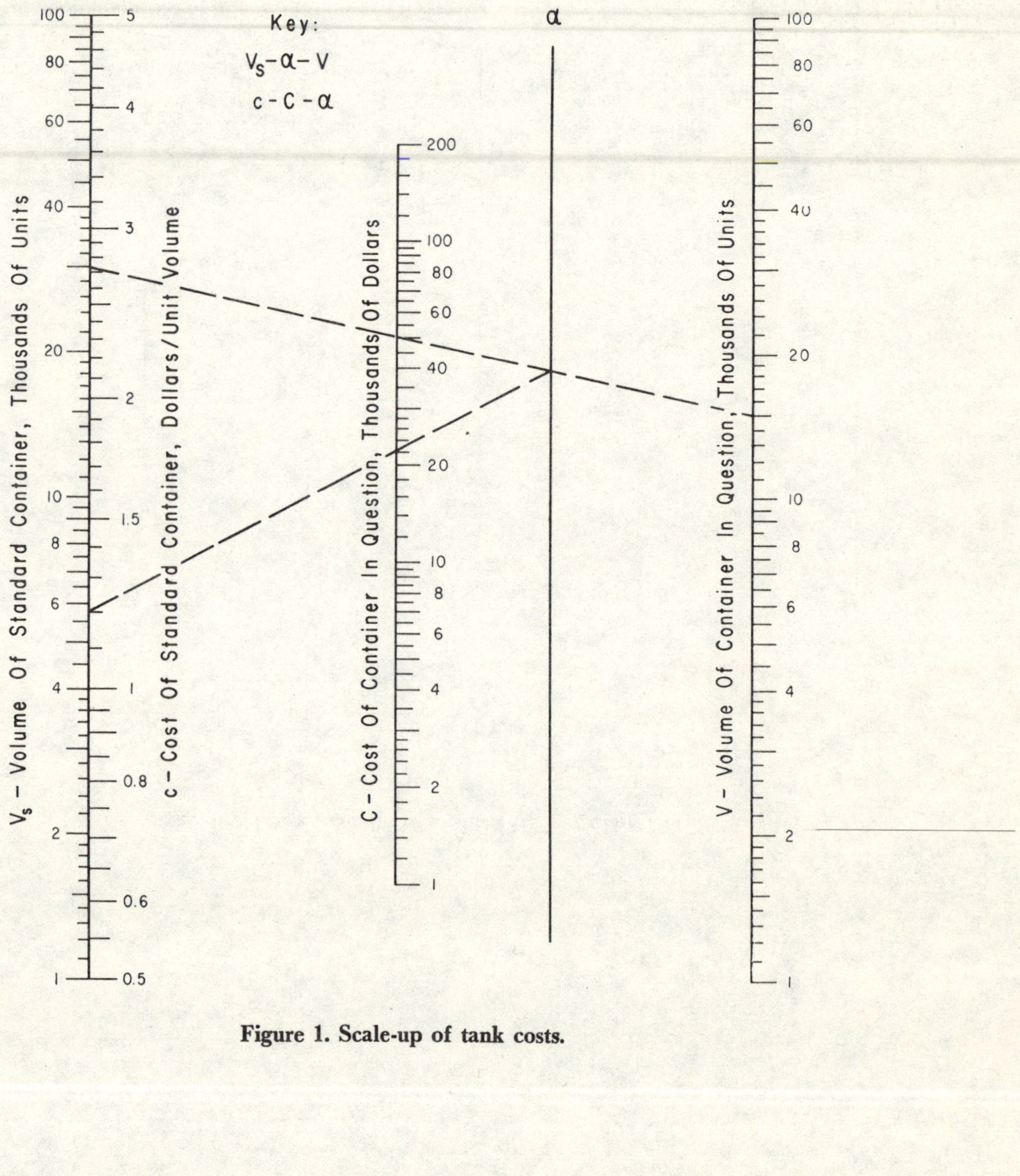

Figure 1. Scale-up of tank costs.

Nomograph for determining sum-of-years depreciation

In estimating annual depreciation of fixed capital assets, the sum-of-the-years-digits depreciation method is commonly used. This method allows for the effect of accelerated depreciation in the earlier years of a project. It is based on the number of years of service remaining each year and on the sum of the arithmetic series of the years from 1 to N. The depreciation factor for each year is calculated by dividing the number of remaining service years by the sum of the digits corresponding to the total service years. The depreciation charge can then be obtained by multiplying the depreciation factor each year by the total fixed asset cost.

The nomogram was constructed from the following equation:

$$D = \frac{N_D}{\sum_{I}^{N} Ni} \cdot V = \frac{N_D}{1+2+3+\ldots N} \cdot V$$

where:

D = annual depreciation, dollars
N = number of years of service of the depreciable item
N_D = number of years of service remaining each year ($N-0$, $N-1$, ... $N-(N-1)$ for the first second and Nth year of depreciation)
N_i = digit corresponding to the ith year

Example

The value of a piece of equipment is $5,200. Its salvage value at the end of 10 years is $200. What is the annual depreciation for the first, second and third years?

Solution

The depreciable amount is $5200 - 200 = \$5000$. The first, second and third years correspond to 10, 9, and 8 years of remaining service life, respectively.

Connect	With	Mark or Read
55 for 10 yrs.	$N_D = 10$ (1st yr.)	R
R	$V = 5000$	$D = \$910$

Repeat the procedure for $N_D = 9$ and 8 and obtain $D = 820$ and 720 for the second and third years of depreciation, respectively.

If the total depreciable value V is larger than the range of the scale, divide by a factor that brings it within the range of the scale, then read the D value, but remembering that the result must be multiplied by the same factor.

Source

Kuong, J. F., *Chem. Eng.,* 67, No. 23, 237 (1960).

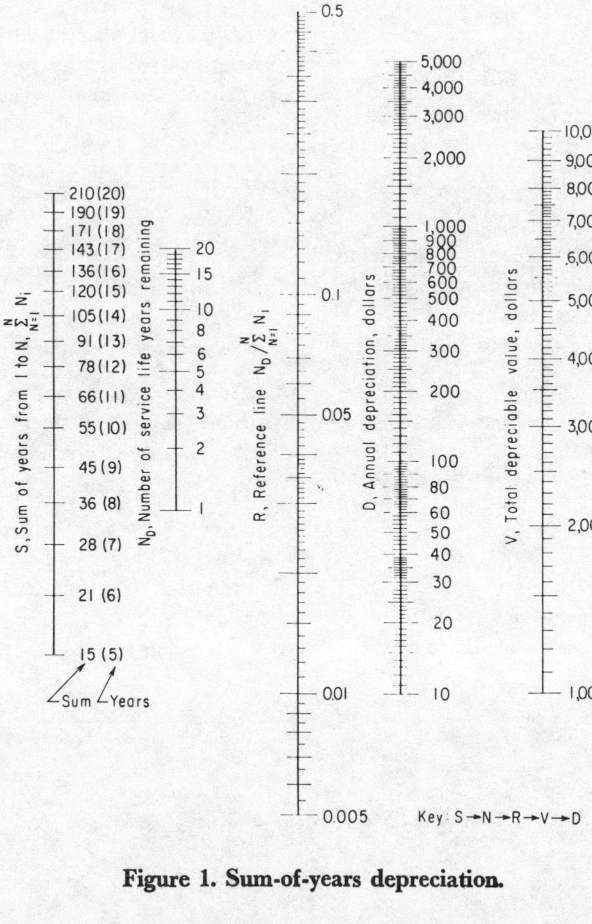

Figure 1. Sum-of-years depreciation.

Nomograph for estimating interest rate of return on investment ("profitability index")

Several methods are used for evaluating the profitability of a given project. A method for estimating profitability which has received considerable emphasis in the past few years is the "interest rate of return," Fig. 1, also known as "profitability index" or "discount cash flow" method. This profitability method takes into account the time value of money. Mathematically, interest rate of return may be defined as that rate at which the company's outstanding investment is repaid by proceeds from the project.[1]

The nomograph was developed for the case in which the yearly profits (P) (or the difference between income and costs) are equal through the life of a project. The nomograph is based on the following assumptions:

1. Depreciation method is sum of the years-digits.
2. Project economic life is equal to depreciable life (T).
3. Construction time is zero, i.e., total investment occurs instantaneously at time zero.
4. There is no salvage value at the end of the project life.
5. There is no start-up expense, (S = 0).

Where these conditions apply, the nomograph gives the interest rate of return R, as a function of P and T, fixed investment (F), and working capital (W).

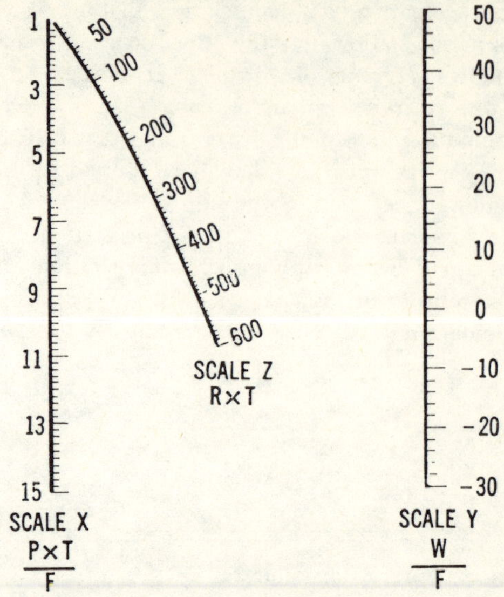

Figure 1. Interest rate of return.

Example 1

Basic Profitability. For the preliminary evaluation of a project the following estimates are available:

Plant capacity and sales	10,000,000 lbs./yr.
Average price	\$0.20/lb.
Unit cost (at capacity)	\$0.10/lb.
Overhead 15% of price	\$0.03/lb.
Project life 10 years (T)	
Depreciation is sum of the years digits	
Plant investment (F)	\$1,000,000
Working Capital (W)	\$ 200,000

It may be assumed that no start-up expense is incurred.

Estimate the interest rate of return if the project is undertaken. The reference or base case is not to invest.

First compute P the yearly before-tax and depreciation profit.

Gross Income: 10,000,000 × 0.20	= \$2,000,000
Less Overhead @ 15% 10,000,000 × 0.03	= 300,000
Less operating cost 10,000,000 × 0.10	= 1,000,000
P, Profit before tax and depreciation	= 700,000

Compute:

$$\frac{P \times T}{F} = \frac{700{,}000 \times 10}{1{,}000{,}000} = 7$$

Compute:

$$\frac{W}{F} = \frac{200{,}000}{1{,}000{,}000} = 0.2$$

Connect	With	Read
$\frac{P \times T}{F} = 7$	$\frac{W}{F} = 0.2$	$R \times T = 330$

Since T = 10, the interest rate of return for the project is 33%/yr.

This is equivalent to saying that the outlay of capital of \$1,000,000 in plant facilities plus 200,000 in working capital when invested in the project would generate enough cash to pay back the total investment plus interest at the rate of 33%/yr.

Example 2

Comparison of alternatives. Consider the following alternatives for replacing a pump.

Alternative A. Replace existing pump with one of the same type at a cost of \$3,000, working capital of \$50 and maintenance expense of \$300/yr.

Alternative B. Install a new-type pump costing \$3500 with working capital of \$100 and maintenance costs of \$100/yr. Economic and depreciable life for both pumps is 10 years.

The saving per year for using pump B versus pump A is \$300 − 100 = \$100/yr. to justify an increment of investment of \$500 (\$3500 − 3000). What is the Profitability Index or interest rate of return on the differential investments.

Calculate: $\frac{P \times T}{F} = \frac{200 \times 10}{500} = 4$

$$\frac{100(W)}{F} = \frac{50}{500} = 0.10$$

Connect	With	Read
$\frac{P \times T}{F} = 4$	0.1 on $\frac{W}{F}$ = 0.1 Scale	$R \times T = 210$

Since T = 10, R = 21%/Yr. = profitability index for project.

If 21% is considered an adequate rate of return, after consideration of risk, then pump B would be the preferred one.

Source

Greist, W. H., Bates, A. G., and Weaver, J. B., *Ind Eng. Chem.,* 53, 52A (1961).

References

1. Dean, J., *Harvard Bus. Rev., 32,* 120-130 Jan., Feb. (1954).
2. Martin, J. C., in *Chemical Process Economics in Practice,* Hur, J. J., Ed., Reinhold Publishing Corp., New York, 1956.
3. Weaver, J. B., and Reilly, R. J., *Chem. Eng. Progr., 52,* 405 (1956).

Nomograph for determining break-even point

The nomograph gives the break even point for a plant as per cent of design capacity when the unit fixed cost, Cf, the sales price, S, and the unit variable cost, Cv are known. It is based on the following equation:

$$BE = 100.Cf / (S\text{-}Cv)$$

The nomograph can also be used to determine the sensitivity of the break even point to any combination of fixed variable costs or sales price or to determine the maximum fixed cost that can be absorbed for a given break even point.

Example

Determine the break even point when:

Design Output	10,000 Lb./Yr.
Unit Fixed Cost	$0.045/Lb.
Unit Variable Cost	$0.075/Lb.
Sales Price	$0.200/Lb.

Calculate $\Delta = 0.200 - 0.075 = 0.125$ or 12.5¢/Lb.

Align Cf = 4.5¢ with $\Delta = 12.5$ and obtain

BE = 36% of plant capacity.

Source

Kuong, J. F., *Petro/Chem. Engineer, 35,* No. 1, 44 (1963).

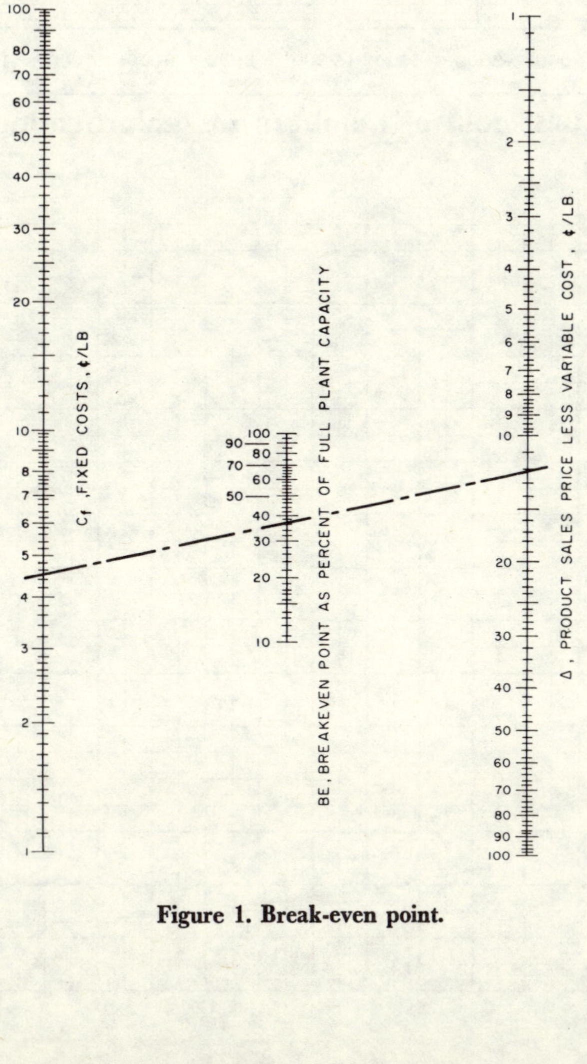

Figure 1. Break-even point.

Chart gives unit cost per brake horsepower of reciprocating compressors with various types of prime movers

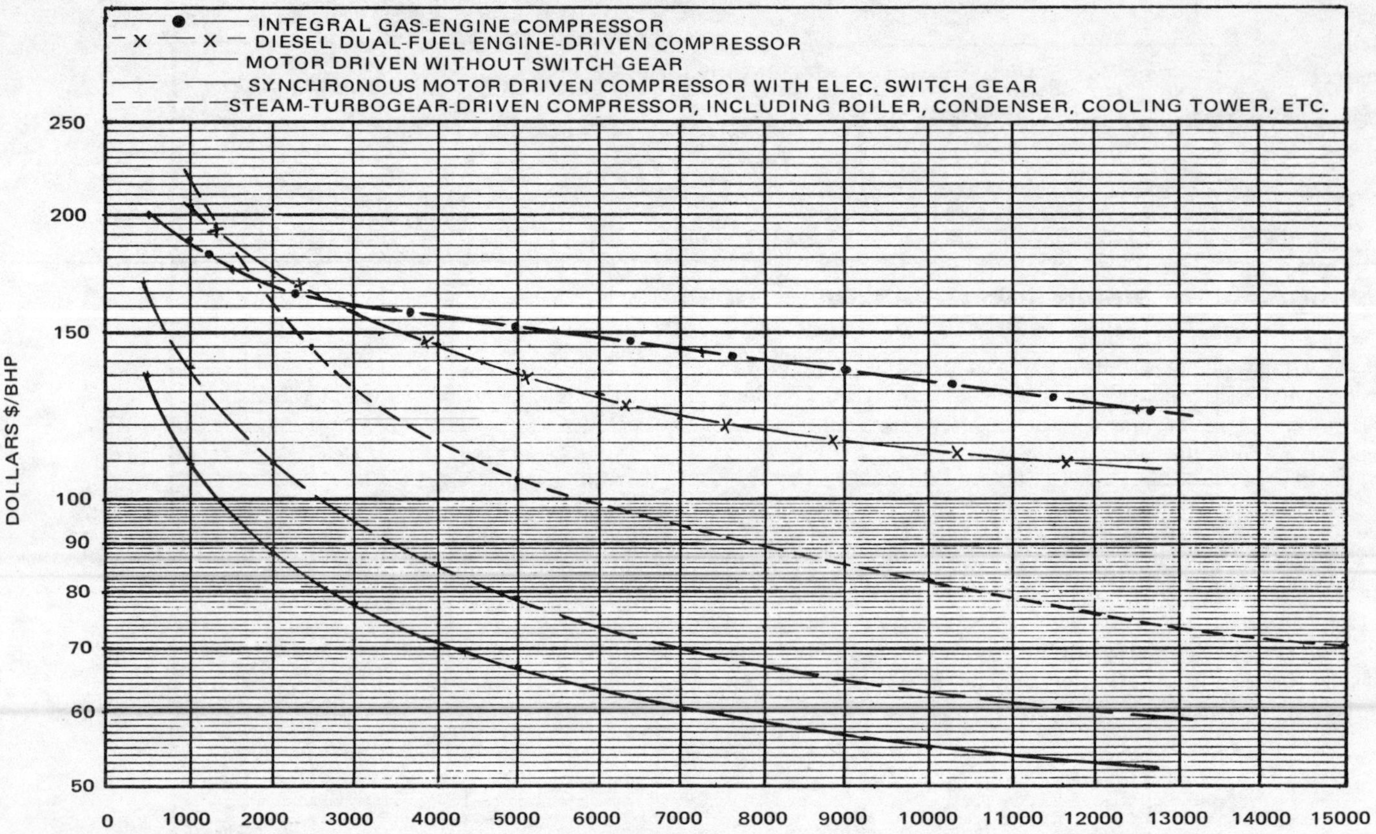

Chart shows influence on unit cost of numbers of reciprocating compressor units installed in one station

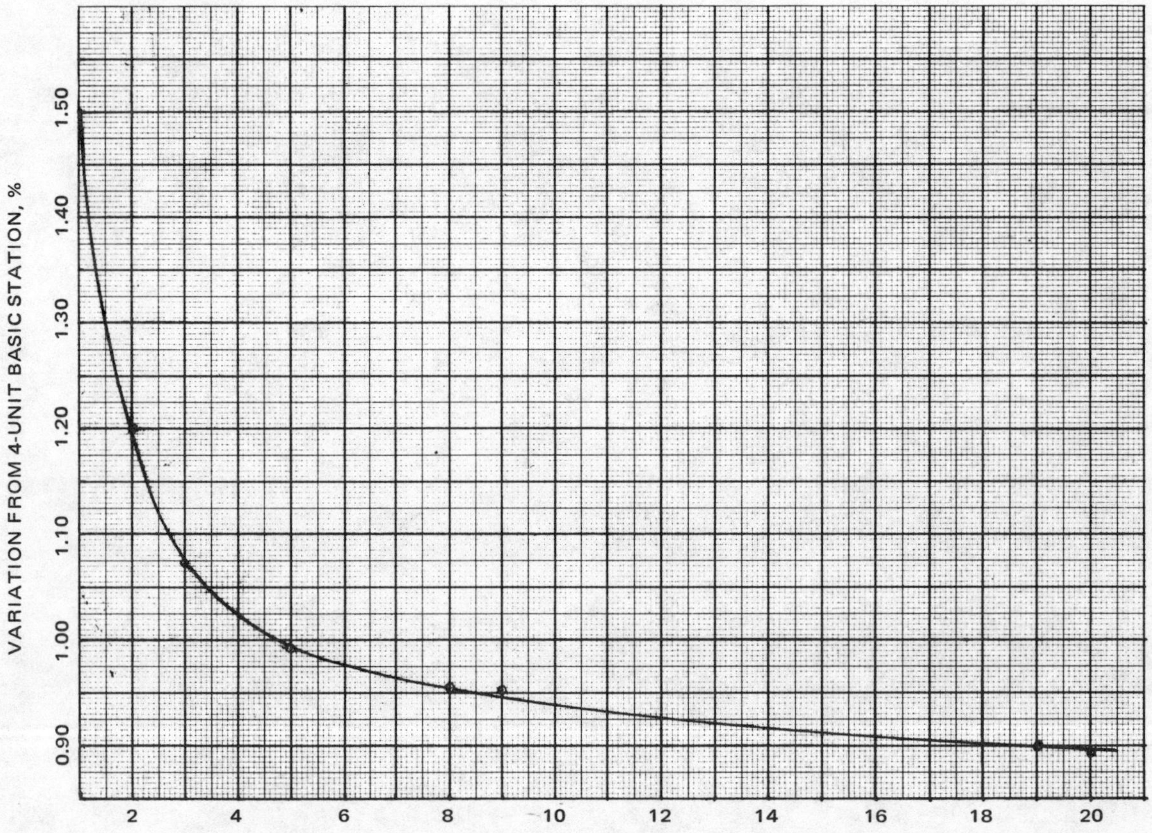

Chart gives unit cost per brake horsepower of centrifugal compressors with various types of prime movers

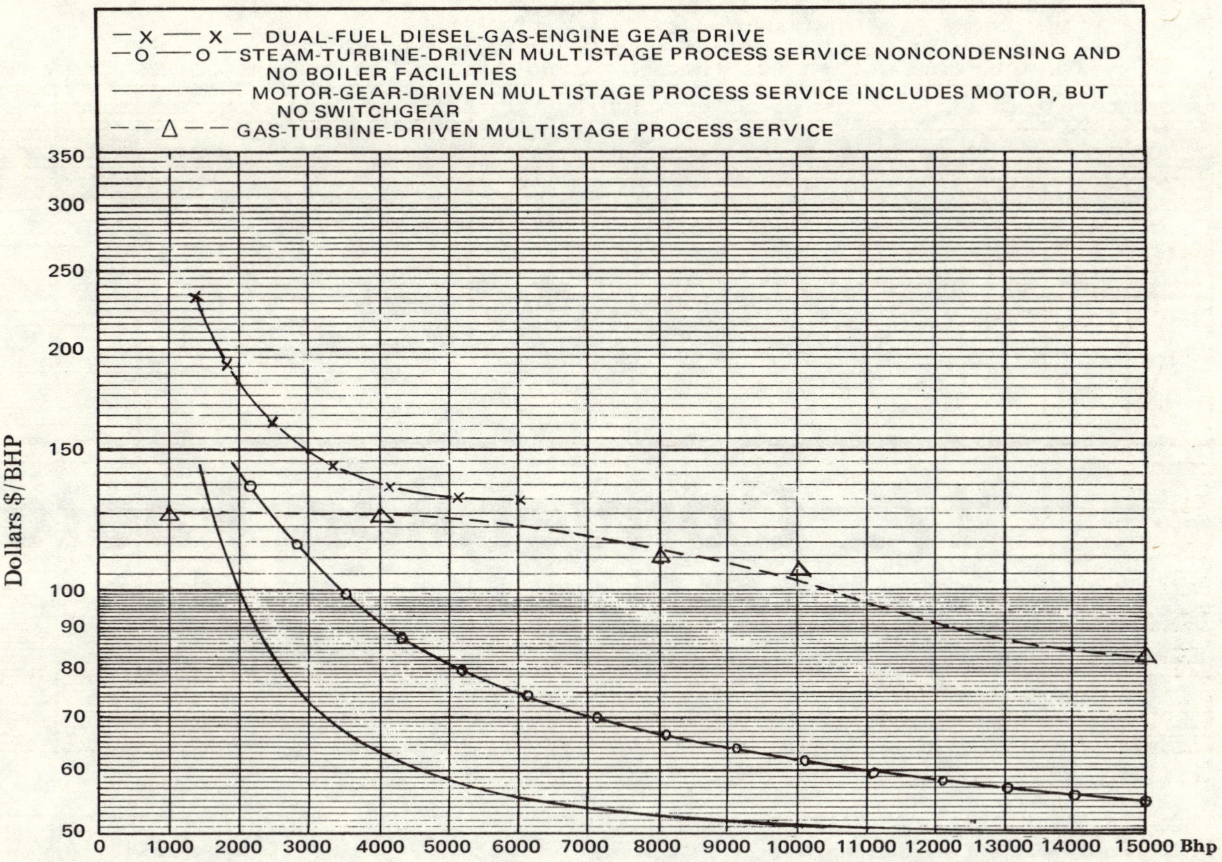

17. Conversion Factors

General liquid density nomograph

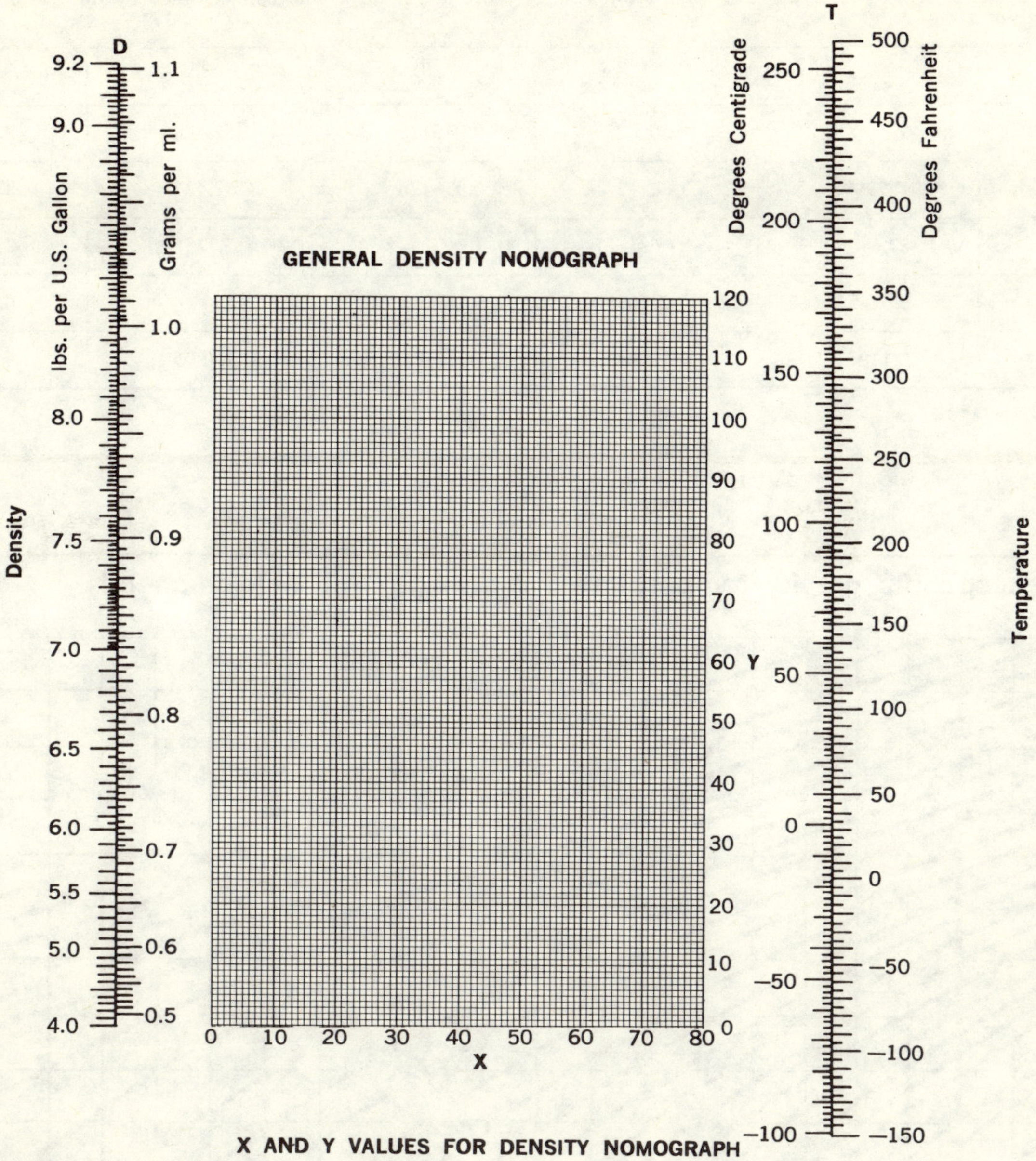

X AND Y VALUES FOR DENSITY NOMOGRAPH

Compound	X	Y	Compound	X	Y	Compound	X	Y
Acetic Acid	40.6	93.5	Ethyl chloride	42.7	62.4	Methyl sulfide	31.9	57.4
Acetone	26.1	47.8	Ethylene	17.0	3.5	n-Nonane	16.2	36.5
Acetonitrile	21.8	44.9	Ethyl ether	22.6	35.8	n-Octadecane	16.2	46.5
Acetylene	20.8	10.1	Ethyl formate	37.6	68.4	n-Octane	12.7	32.5
Ammonia	22.4	24.6	Ethyl propionate	32.1	63.9	n-Pentadecane	15.8	44.2
Isoamyl alcohol	20.5	52.0	Ethyl propyl ether	20.0	37.0	n-Pentane	12.6	22.6
Ammine	33.5	92.5	Ethyl auindo	25.7	55.3	n-Nonadecane	14.9	47.0
Benzene	32.7	63.0	Fluorobenzene	41.9	87.6	Isopentane	13.5	22.5
n-Butyric acid	31.3	78.7	n-Heptadecane	15.6	45.7	Phenol	36.7	103.8
Isobutane	13.7	16.5	n-Heptane	12.6	29.8	Phosphine	28.0	22.1
Isobutyric acid	31.5	75.9	n-Hexadecane	15.8	45.0	Propane	14.2	12.2
Carbon dioxide	78.6	45.4	n-Hexane	13.5	27.0	Propionic acid	35.0	83.5
Chlorobenzene	41.7	105.0	Methanethiol	37.3	59.5	Piperidene	27.5	60.0
Cyclohexane	19.6	44.0	Methyl acetate	40.1	70.3	Propionitrile	20.1	44.6
n-Decane	16.0	38.2	Methyl alcohol	25.8	49.1	Propyl acetate	33.0	65.5
n-Dedecane	14.3	41.4	Methyl n-butyrate	31.5	65.5	Propyl alcohol	23.8	50.8
Diethylamine	17.8	33.5	Methyl isobutyrate	33.0	64.1	Propyl formate	33.8	66.7
n-Elconane	14.8	47.5	Methyl chloride	52.3	62.9	n-Tetradecane	15.8	43.3
Ethane	10.8	4.4	Methyl ether	27.2	30.1	n-Tridecane	15.3	42.4
Ethanethiol	32.0	55.5	Methyl ethyl ether	25.0	34.4	Triethylamine	17.9	37.0
Ethyl acetate	35.0	95.0	Methyl formate	46.4	74.6	n-Undecane	14.4	39.2
Ethyl alcohol	24.2	48.6	Methyl propionate	36.5	68.3			

Reprinted by permission of Tubular Exchanger Manufacturers Association (TEMA).

Density Conversions

To convert the numerical value of a property expressed in one of the units in the left-hand column of the table to the numerical value expressed in one of the units in the top row of the table, multiply the former value by the factor in the block common to both units.

Units→ ↓	g/cu cm	g/ml	lb/cu in.	lb/cu ft	lb/gal (U.S.)
g/cu cm	1	1.000028	0.036128	62.428	8.3455
g/ml	0.99997	1	0.036126	62.427	8.3452
lb/cu in.	27.680	27.681	1	1,728	231
lb/cu ft	0.016018	0.016019	5.7870×10^{-4}	1	0.13368
lb/gal (U.S.)	0.11983	0.11983	4.3290×10^{-3}	7.4805	1

Chart gives specific gravity/temperature relationship for petroleum oils

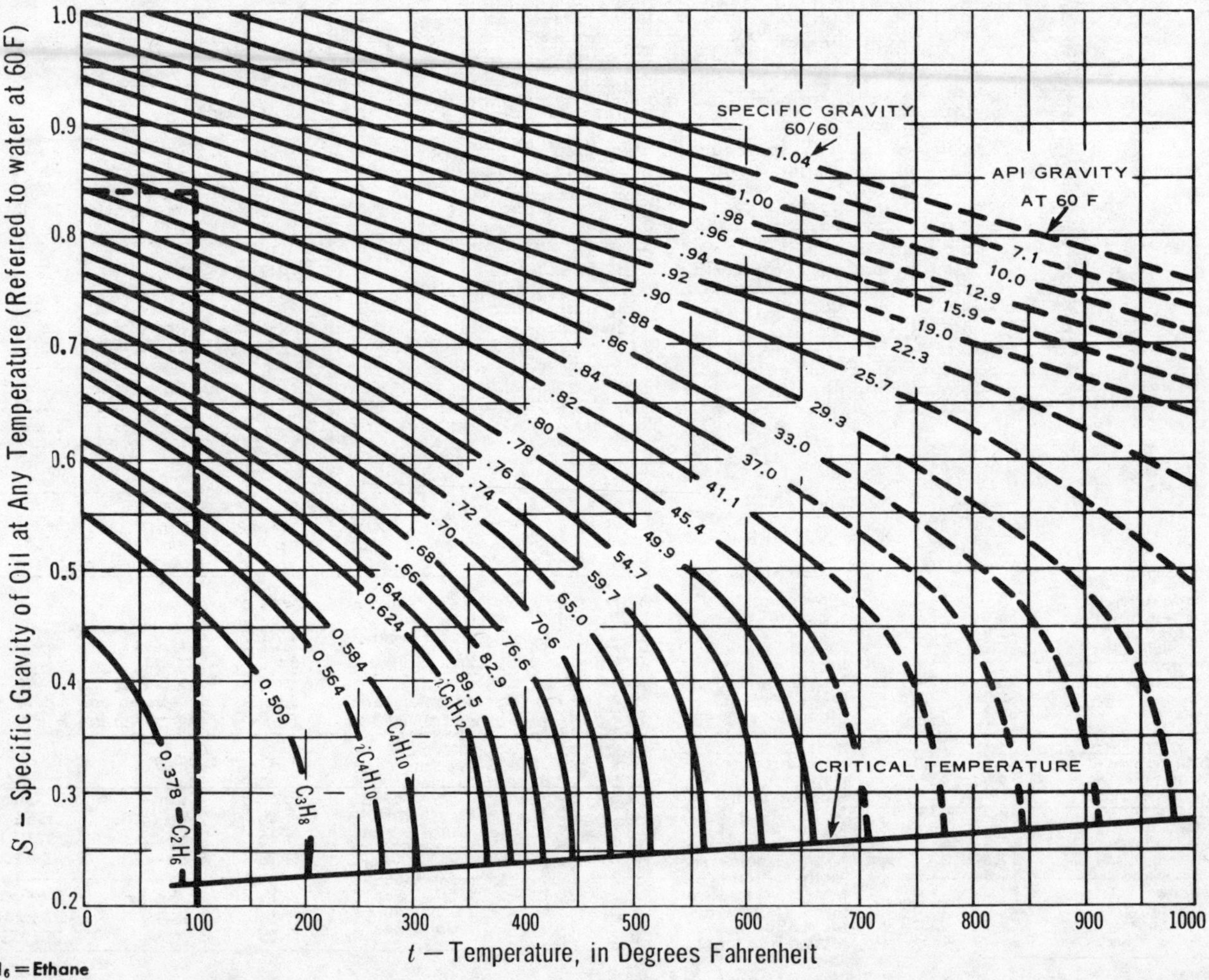

C_2H_6 = Ethane
C_3H_8 = Propane
C_4H_{10} = Butane
iC_4H_{10} = Isobutane
iC_5H_{12} = Isopentane

Example: The specific gravity of an oil at 60 F is 0.85. The specific gravity at 100 F = 0.83.

Courtesy Crane Co. (Technical Paper #410).

To find the weight density of a petroleum oil at its flowing temperature when the specific gravity at 60 F/60 F is known, multiply the specific gravity of the oil at flowing temperature (see chart above) by 62.4, the density of water at 60 F.

Weight Density and Specific Gravity of Various Liquids

Liquid	Temp. t Deg. Fahr.	Weight Density ρ Lbs. per Cu. Ft.	Specific Gravity S	Liquid	Temp. t Deg. Fahr.	Weight Density ρ Lbs. per Cu. Ft.	Specific Gravity S
Acetone	60	49.4	0.792	Mercury	20	849.74	...
Ammonia, Saturated	10	40.9	...	Mercury	40	848.03	...
Benzene	32	56.1	...	Mercury	60	846.32	13.570
Brine, 10% Ca Cl	32	68.05	...	Mercury	80	844.62	...
Brine, 10% Na Cl	32	67.24	...	Mercury	100	842.93	...
Bunkers C Fuel Max.	60	63.25	1.014	Milk	...	†	...
Carbon Disulphide	32	80.6	...	Olive Oil	59	57.3	0.919
Distillate	60	52.99	0.850	Pentane	59	38.9	0.624
Fuel 3 Max.	60	56.02	0.898	SAE 10 Lube‡	60	54.64	0.876
Fuel 5 Min.	60	60.23	0.966	SAE 30 Lube‡	60	56.02	0.898
Fuel 5 Max.	60	61.92	0.993	SAE 70 Lube‡	60	57.12	0.916
Fuel 6 Min.	60	61.92	0.993	Salt Creek Crude	60	52.56	0.843
Gasoline	60	46.81	0.751	32.6° API Crude	60	53.77	0.862
Gasoline, Natural	60	42.42	0.680	35.6° API Crude	60	52.81	0.847
Kerosene	60	50.85	0.815	40° API Crude	60	51.45	0.825
M. C. Residuum	60	58.32	0.935	48° API Crude	60	49.16	0.788

*Liquid at 60 F referred to water at 60 F.

†Milk has a weight density of 64.2 to 64.6.

‡100 Viscosity Index.

Courtesy Crane Co. (Technical Paper #410).

Metric conversions—metric to English, English to metric

To convert from	To	Multiply by	To convert from	To	Multiply by
Length			**Length**		
mm	inches	.03937	inches	mm	25.40
cm	inches	.3937	inches	cm	2.540
meters	inches	39.37	inches	meters	.0254
meters	feet	3.281	feet	meters	.3048
meters	yards	1.0936	feet	km	.0003048
km	feet	3280.8	yards	meters	.9144
km	yards	1093.6	yards	km	.0009144
km	miles	.6214	miles	km	1.609
Area			**Area**		
sq mm	sq inches	.00155	sq inches	sq mm	645.2
sq cm	sq inches	.155	sq inches	sq cm	6.452
sq meters	sq feet	10.764	sq feet	sq meters	.09290
sq meters	sq yards	1.196	sq yards	sq meters	.8361
sq km	sq miles	.3861	sq miles	sq km	2.590
hectares	acres	2.471	acres	hectares	.4047
Volume			**Volume**		
cu cm	cu inches	.06102	cu inches	cu cm	16.387
cu cm	fl ounces	.03381	cu inches	liters	.01639
cu meters	cu feet	35.314	cu feet	cu meters	.02832
cu meters	cu yards	1.308	cu feet	liters	28.317
cu meters	US gal	264.2	cu yards	cu meters	.7646
liters	cu inches	61.023	fl ounces	cu cm	29.57
liters	cu feet	.03531	US gal	cu meters	.003785
liters	US gal	.2642	US gal	liters	3.785
Weight			**Weight**		
grams	grains	15.432	grains	grams	.0648
grams	ounces†	.0353	ounces†	grams	28.350
kg	ounces†	35.27	ounces†	kg	.02835
kg	pounds†	2.2046	pounds†	kg	.4536
kg	US tons	.001102	pounds†	tonnes	.000454
kg	long tons	.000984	US tons	kg	907.2
tonnes	pounds†	2204.6	US tons	tonnes	.9072
tonnes	US tons	1.1023	long tons	kg	1016
tonnes	long tons	.9842	long tons	tonnes	1.0160

To convert from	To	Multiply by	To convert from	To	Multiply by
Unit weight			**Unit weight**		
gr/sq cm	lb/sq in	.01422	lb/ft	kg/m	1.4881
gr/cu cm	lb/cu in	.0361	lb/sq in	gr/sq cm	70.31
kg/sq cm	lb/sq in	14.22	lb/sq in	kg/sq cm	.07031
kg/cu m	lb/cu ft	.0624	lb/cu in	gr/cu cm	27.68
kg/m	lb/ft	.6720	lb/cu ft	kg/cu m	16.018
Unit volume			**Unit volume**		
liters/min	US gpm	.2642	US gpm	liters/min	3.785
liters/min	cfm	.03531	US gpm	liters/hr	227.1
liters/hr	US gpm	.0044	US gpm	cu m/hr	.2271
cu m/min	cfm	35.314	cfm	liters/min	28.317
cu m/hr	cfm	.5886	cfm	cu m/min	.02832
cu m/hr	US gpm	4.4028	cfm	cu m/hr	1.6992
Power			**Power**		
watts	ft-lb/sec	.7376	ft-lb/sec	watts	1.356
watts	hp	.00134	hp	watts	745.7
kw	hp	1.3410	hp	kw	.7457
cheval-vap	hp	.9863	hp	cheval-vap	1.0139
Heat			**Heat**		
gr-cal	Btu	.003969	Btu	gr-cal	252
kg-cal	Btu	3.9693	Btu	kg-cal	.252
kg-cal/kg	Btu/lb	1.800	Btu/lb	kg-cal/kg	.5556
gr-cal/sq cm	Btu/sq ft	3.687	Btu/sq ft	gr-cal/sq cm	.2713
kg-cal/cu m	But/cu ft	.1124	Btu/cu ft	kg-cal/cu m	8.899
Work or energy			**Work or energy**		
joule	ft-lb	.7376	ft-lb	joule	1.356
meter-kg	ft-lb	7.2330	ft-lb	meter-kg	.1383
gr-cal	ft-lb	3.087	ft-lb	gr-cal	.3239
kg-cal	ft-lb	3087	ft-lb	kg-cal	.000324
hp-hr	ft-lb	1,980,000	ft-lb	hp-hr	5.05×10^{-7}
kwhr	ft-lb	2,655,000	ft-lb	kwhr	3.766×10^{-7}
Btu	ft-lb	778.0	ft-lb	Btu	.001285

†avoirdupois pounds and ounces.

Courtesy Cameron Hydraulic Data Book, *Ingersoll-Rand Company.*

Temperature conversion—centigrade to Fahrenheit or Fahrenheit to centigrade

NOTE: The numbers in boldface refer to the temperature in degrees, either Centigrade or Fahrenheit, which it is desired to convert into the other scale. If converting from Fahrenheit to Centigrade degrees, the equivalent temperature will be found in the left column; while if converting from degrees Centigrade to degrees Fahrenheit, the answer will be found in the column on the right.

Centigrade		*Fahrenheit*	*Centigrade*		*Fahrenheit*	*Centigrade*		*Fahrenheit*	*Centigrade*		*Fahrenheit*
−73.3	**−100**	−148.0	2.8	**37**	98.6	33.3	**92**	197.6	293	**560**	1040
−67.8	**−90**	−130.0	3.3	**38**	100.4	33.9	**93**	199.4	299	**570**	1058
−62.2	**−80**	−112.0	3.9	**39**	102.2	34.4	**94**	201.2	304	**580**	1076
−59.4	**−75**	−103.0	4.4	**40**	104.0	35.0	**95**	203.0	310	**590**	1094
−56.7	**−70**	−94.0	5.0	**41**	105.8	35.6	**96**	204.8	316	**600**	1112
−53.9	**−65**	−85.0	5.6	**42**	107.6	36.1	**97**	206.6	321	**610**	1130
−51.1	**−60**	−76.0	6.1	**43**	109.4	36.7	**98**	208.4	327	**620**	1148
−48.3	**−55**	−67.0	6.7	**44**	111.2	37.2	**99**	210.2	332	**630**	1166
−45.6	**−50**	−58.0	7.2	**45**	113.0	37.8	**100**	212.0	338	**640**	1184
−42.8	**−45**	−49.0	7.8	**46**	114.8	43	**110**	230	343	**650**	1202
−40.0	**−40**	−40.0	8.3	**47**	116.6	49	**120**	248	349	**660**	1220
−37.2	**−35**	−31.0	8.9	**48**	118.4	54	**130**	266	354	**670**	1238
−34.4	**−30**	−22.0	9.4	**49**	120.2	60	**140**	284	360	**680**	1256
−31.7	**−25**	−13.0	10.0	**50**	122.0	66	**150**	302	366	**690**	1274
−28.9	**−20**	−4.0	10.6	**51**	123.8	71	**160**	320	371	**700**	1292
−26.1	**−15**	5.0	11.1	**52**	125.6	77	**170**	338	377	**710**	1310
−23.3	**−10**	14.0	11.7	**53**	127.4	82	**180**	356	382	**720**	1328
−20.6	**−5**	23.0	12.2	**54**	129.2	88	**190**	374	388	**730**	1346
−17.8	**0**	32.0	12.8	**55**	131.0	93	**200**	392	393	**740**	1364
−17.2	**1**	33.8	13.3	**56**	132.8	99	**210**	410	399	**750**	1382
−16.7	**2**	35.6	13.9	**57**	134.6	100	**212**	414	404	**760**	1400
−16.1	**3**	37.4	14.4	**58**	136.4	104	**220**	428	410	**770**	1418
−15.6	**4**	39.2	15.0	**59**	138.2	110	**230**	446	416	**780**	1436
−15.0	**5**	41.0	15.6	**60**	140.0	116	**240**	464	421	**790**	1454
−14.4	**6**	42.8	16.1	**61**	141.8	121	**250**	482	427	**800**	1472
−13.9	**7**	44.6	16.7	**62**	143.6	127	**260**	500	432	**810**	1490
−13.3	**8**	46.4	17.2	**63**	145.4	132	**270**	518	438	**820**	1508
−12.8	**9**	48.2	17.8	**64**	147.2	138	**280**	536	443	**830**	1526
−12.2	**10**	50.0	18.3	**65**	149.0	143	**290**	554	449	**840**	1544
−11.7	**11**	51.8	18.9	**66**	150.8	149	**300**	572	454	**850**	1562
−11.1	**12**	53.6	19.4	**67**	152.6	154	**310**	590	460	**860**	1580
−10.6	**13**	55.4	20.0	**68**	154.4	160	**320**	608	466	**870**	1598
−10.0	**14**	57.2	20.6	**69**	156.2	166	**330**	626	471	**880**	1616
−9.4	**15**	59.0	21.1	**70**	158.0	171	**340**	644	477	**890**	1634
−8.9	**16**	60.8	21.7	**71**	159.8	177	**350**	662	482	**900**	1652
−8.3	**17**	62.6	22.2	**72**	161.6	182	**360**	680	488	**910**	1670
−7.8	**18**	64.4	22.8	**73**	163.4	188	**370**	698	493	**920**	1688
−7.2	**19**	66.2	23.3	**74**	165.2	193	**380**	716	499	**930**	1706
−6.7	**20**	68.0	23.9	**75**	167.0	199	**390**	734	504	**940**	1724
−6.1	**21**	69.8	24.4	**76**	168.8	204	**400**	752	510	**950**	1742
−5.6	**22**	71.6	25.0	**77**	170.6	210	**410**	770	516	**960**	1760
−5.0	**23**	73.4	25.6	**78**	172.4	216	**420**	788	521	**970**	1778
−4.4	**24**	75.2	26.1	**79**	174.2	221	**430**	806	527	**980**	1796
−3.9	**25**	77.0	26.7	**80**	176.0	227	**440**	824	532	**990**	1814
−3.3	**26**	78.8	27.2	**81**	177.8	232	**450**	842	538	**1000**	1832
−2.8	**27**	80.6	27.8	**82**	179.6	238	**460**	860	566	**1050**	1922
−2.2	**28**	82.4	28.3	**83**	181.4	243	**470**	878	593	**1100**	2012
−1.7	**29**	84.2	28.9	**84**	183.2	249	**480**	896	621	**1150**	2102
−1.1	**30**	86.0	29.4	**85**	185.0	254	**490**	914	649	**1200**	2192
−0.6	**31**	87.8	30.0	**86**	186.8	260	**500**	932	677	**1250**	2282
0.0	**32**	89.6	30.6	**87**	188.6	266	**510**	950	704	**1300**	2372
0.6	**33**	91.4	31.1	**88**	190.4	271	**520**	968	732	**1350**	2462
1.1	**34**	93.2	31.7	**89**	192.2	277	**530**	986	760	**1400**	2552
1.7	**35**	95.0	32.2	**90**	194.0	282	**540**	1004	788	**1450**	2642
2.2	**36**	96.8	32.8	**91**	195.8	288	**550**	1022	816	**1500**	2732

The formulas at the right may also be used for converting Centigrade or Fahrenheit degrees into the other scale.

Degrees Cent., $C^\circ = \frac{5}{9}(F^\circ + 40) - 40$ Degrees Fahr., $F^\circ = \frac{9}{5}(C^\circ + 40) - 40$

Courtesy Natural Gas Processors Suppliers Association.

Units of measurement from one system to another

TO CONVERT	INTO	MULTIPLY BY
	A	
Abcoulomb	Statcoulombs	2.998×10^{10}
Acre	Sq. chain (Gunters)	10
Acre	Rods	160
Acre	Square links (Gunters)	1×10^{5}
Acre	Hectare or sq. hectometer	.4047
acres	sq feet	43,560.0
acres	sq meters	4,047.
acres	sq miles	1.562×10^{-3}
acres	sq yards	4,840.
acre-feet	cu feet	43,560.0
acre-feet	gallons	3.259×10^{5}
amperes/sq cm	amps/sq in.	6.452
amperes/sq cm	amps/sq meter	10^{4}
amperes/sq in.	amps/sq cm	0.1550
amperes/sq in.	amps/sq meter	1,550.0
amperes/sq meter	amps/sq cm	10^{-4}
amperes/sq meter	amps/sq in.	6.452×10^{-4}
ampere-hours	coulombs	3,600.0
ampere-hours	faradays	0.03731
ampere-turns	gilberts	1.257
ampere-turns/cm	amp-turns/in.	2.540
ampere-turns/cm	amp-turns/meter	100.0
ampere-turns/cm	gilberts/cm	1.257
ampere-turns/in.	amp-turns/cm	0.3937
ampere-turns/in.	amp-turns/meter	39.37
ampere-turns/in.	gilberts/cm	0.4950
ampere-turns/meter	amp/turns/cm	0.01
ampere-turns/meter	amp-turns/in.	0.0254
ampere-turns/meter	gilberts/cm	0.01257
Angstrom unit	Inch	3937×10^{-9}
Angstrom unit	Meter	1×10^{-10}
Angstrom unit	Micron or (Mu)	1×10^{-4}
Are	Acre (US)	.02471
Ares	sq. yards	119.60
ares	acres	0.02471
ares	sq meters	100.0
Astronomical Unit	Kilometers	1.495×10^{8}
Atmospheres	Ton/sq. inch	.007348
atmospheres	cms of mercury	76.0
atmospheres	ft of water (at 4°C)	33.90
atmospheres	in. of mercury (at 0°C)	29.92
atmospheres	kgs/sq cm	1.0333
atmospheres	kgs/sq meter	10,332.
atmospheres	pounds/sq in.	14.70
atmospheres	tons/sq ft	1.058
	B	
Barrels (U.S., dry)	cu. inches	7056.
Barrels (U.S., dry)	quarts (dry)	105.0
Barrels (U.S., liquid)	gallons	31.5
barrels (oil)	gallons (oil)	42.0
bars	atmospheres	0.9869
bars	dynes/sq cm	10^{6}
bars	kgs/sq meter	1.020×10^{4}
bars	pounds/sq ft	2,089.
bars	pounds/sq in.	14.50
Baryl	Dyne/sq. cm.	1.000
Bolt (US Cloth)	Meters	36.576
BTU	Liter–Atmosphere	10.409
Btu	ergs	1.0550×10^{10}
Btu	foot-lbs	778.3
Btu	gram-calories	252.0
Btu	horsepower-hrs	3.931×10^{-4}
Btu	joules	1,054.8
Btu	kilogram-calories	0.2520
Btu	kilogram-meters	107.5
Btu	kilowatt-hrs	2.928×10^{-4}
Btu/hr	foot-pounds/sec	0.2162
Btu/hr	gram-cal/sec	0.0700
Btu/hr	horsepower-hrs	3.929×10^{-4}
Btu/hr	watts	0.2931
Btu/min	foot-lbs/sec	12.96
Btu/min	horsepower	0.02356
Btu/min	kilowatts	0.01757
Btu/min	watts	17.57
Btu/sq ft/min	watts/sq in.	0.1221
Bucket (Br. dry)	Cubic Cm.	1.818×10^{4}
bushels	cu ft	1.2445
bushels	cu in.	2,150.4
bushels	cu meters	0.03524
bushels	liters	35.24
bushels	pecks	4.0
bushels	pints (dry)	64.0
bushels	quarts (dry)	32.0
	C	
Calories, gram (mean)	B.T.U. (mean)	3.9685×10^{-3}
Candle/sq. cm	Lamberts	3.142
Candle/sq. inch	Lamberts	.4870
centares (centiares)	sq meters	1.0
Centigrade	Fahrenheit	(C°x9/5)+32
centigrams	grams	0.01
Centiliter	Ounce fluid (US)	.3382
Centiliter	Cubic inch	.6103
Centiliter	drams	2.705
centiliters	liters	0.01
centimeters	feet	3.281×10^{-2}
centimeters	inches	0.3937
centimeters	kilometers	10^{-5}
centimeters	meters	0.01
centimeters	miles	6.214×10^{-6}
centimeters	millimeters	10.0
centimeters	mils	393.7
centimeters	yards	1.094×10^{-2}
centimeter-dynes	cm-grams	1.020×10^{-3}
centimeter-dynes	meter-kgs	1.020×10^{-8}
centimeter-dynes	pound-feet	7.376×10^{-8}
centimeter-grams	cm-dynes	980.7
centimeter-grams	meter-kgs	10^{-5}
centimeter-grams	pound-feet	7.233×10^{-5}
centimeters of mercury	atmospheres	0.01316
centimeters of mercury	feet of water	0.4461
centimeters of mercury	kgs/sq meter	136.0
centimeters of mercury	pounds/sq ft	27.85
centimeters of mercury	pounds/sq in.	0.1934
centimeters/sec	feet/min	1.1969
centimeters/sec	feet/sec	0.03281
centimeters/sec	kilometers/hr	0.036
centimeters/sec	knots	0.1943
centimeters/sec	meters/min	0.6
centimeters/sec	miles/hr	0.02237
centimeters/sec	miles/min	3.728×10^{-4}
centimeters/sec/sec	feet/sec/sec	0.03281
centimeters/sec/sec	kms/hr/sec	0.036
centimeters/sec/sec	meters/sec/sec	0.01
centimeters/sec/sec	miles/hr/sec	0.02237
Chain	Inches	792.00
Chain	meters	20.12
Chains (surveyors' or Gunter's)	yards	22.00
circular mils	sq cms	5.067×10^{-6}
circular mils	sq mils	0.7854
Circumference	Radians	6.283
circular mils	sq inches	7.854×10^{-7}
Cords	cord feet	8
Cord feet	cu. feet	16
Coulomb	Statcoulombs	2.998×10^{9}
coulombs	faradays	1.036×10^{-5}

Units of measurement from one system to another

TO CONVERT	INTO	MULTIPLY BY
coulombs/sq cm	coulombs/sq in.	64.52
coulombs/sq cm	coulombs/sq meter	10^4
coulombs/sq in.	coulombs/sq cm	0.1550
coulombs/sq in.	coulombs/sq meter	1,550.
coulombs/sq meter	coulombs/sq cm	10^{-4}
coulombs/sq meter	coulombs/sq in.	6.452×10^{-4}
cubic centimeters	cu feet	3.531×10^{-5}
cubic centimeters	cu inches	0.06102
cubic centimeters	cu meters	10^{-6}
cubic centimeters	cu yards	1.308×10^{-6}
cubic centimeters	gallons (U. S. liq.)	2.642×10^{-4}
cubic centimeters	liters	0.001
cubic centimeters	pints (U.S. liq.)	2.113×10^{-3}
cubic centimeters	quarts (U.S. liq.)	1.057×10^{-3}
cubic feet	bushels (dry)	0.8036
cubic feet	cu cms	28,320.0
cubic feet	cu inches	1,728.0
cubic feet	cu meters	0.02832
cubic feet	cu yards	0.03704
cubic feet	gallons (U.S. liq.)	7.48052
cubic feet	liters	28.32
cubic feet	pints (U.S. liq.)	59.84
cubic feet	quarts (U.S. liq.)	29.92
cubic feet/min	cu cms/sec	472.0
cubic feet/min	gallons/sec	0.1247
cubic feet/min	liters/sec	0.4720
cubic feet/min	pounds of water/min	62.43
cubic feet/sec	million gals/day	0.646317
cubic feet/sec	gallons/min	448.831
cubic inches	cu cms	16.39
cubic inches	cu feet	5.787×10^{-4}
cubic inches	cu meters	1.639×10^{-5}
cubic inches	cu yards	2.143×10^{-5}
cubic inches	gallons	4.329×10^{-3}
cubic inches	liters	0.01639
cubic inches	mil-feet	1.061×10^5
cubic inches	pints (U.S. liq.)	0.03463
cubic inches	quarts (U.S. liq.)	0.01732
cubic meters	bushels (dry)	28.38
cubic meters	cu cms	10^6
cubic meters	cu feet	35.31
cubic meters	cu inches	61,023.0
cubic meters	cu yards	1.308
cubic meters	gallons (U.S. liq.)	264.2
cubic meters	liters	1,000.0
cubic meters	pints (U.S. liq.)	2,113.0
cubic meters	quarts (U.S. liq.)	1,057.
cubic yards	cu cms	7.646×10^5
cubic yards	cu feet	27.0
cubic yards	cu inches	46,656.0
cubic yards	cu meters	0.7646
cubic yards	gallons (U.S. liq.)	202.0
cubic yards	liters	764.6
cubic yards	pints (U.S. liq.)	1,615.9
cubic yards	quarts (U.S. liq.)	807.9
cubic yards/min	cubic ft/sec	0.45
cubic yards/min	gallons/sec	3.367
cubic yards/min	liters/sec	12.74

D

TO CONVERT	INTO	MULTIPLY BY
Dalton	Gram	1.650×10^{-24}
days	seconds	86,400.0
decigrams	grams	0.1
deciliters	liters	0.1
decimeters	meters	0.1
degrees (angle)	quadrants	0.01111
degrees (angle)	radians	0.01745
degrees (angle)	seconds	3,600.0
degrees/sec	radians/sec	0.01745
degrees/sec	revolutions/min	0.1667
degrees/sec	revolutions/sec	2.778×10^{-3}
dekagrams	grams	10.0
dekaliters	liters	10.0
dekameters	meters	10.0
Drams (apothecaries' or troy)	ounces (avoidupois)	0.1371429
Drams (apothecaries' or troy)	ounces (troy)	0.125
Drams (U.S., fluid or apoth.)	cubic cm.	3.6967
drams	grams	1.7718
drams	grains	27.3437
drams	ounces	0.0625
Dyne/cm	Erg/sq. millimeter	.01
Dyne/sq. cm.	Atmospheres	9.869×10^{-7}
Dyne/sq. cm.	Inch of Mercury at 0°C	2.953×10^{-5}
Dyne/sq. cm.	Inch of Water at 4°C	4.015×10^{-4}
dynes	grams	1.020×10^{-3}
dynes	joules/cm	10^{-7}
dynes	joules/meter (newtons)	10^{-5}
dynes	kilograms	1.020×10^{-6}
dynes	poundals	7.233×10^{-5}
dynes	pounds	2.248×10^{-6}
dynes/sq cm	bars	10^{-6}

E

TO CONVERT	INTO	MULTIPLY BY
Ell	Cm.	114.30
Ell	Inches	45
Em, Pica	Inch	.167
Em, Pica	Cm.	.4233
Erg/sec	Dyne – cm/sec	1.000
ergs	Btu	9.480×10^{-11}
ergs	dyne-centimeters	1.0
ergs	foot-pounds	7.367×10^{-8}
ergs	gram-calories	0.2389×10^{-7}
ergs	gram-cms	1.020×10^{-3}
ergs	horsepower-hrs	3.7250×10^{-14}
ergs	joules	10^{-7}
ergs	kg-calories	2.389×10^{-11}
ergs	kg-meters	1.020×10^{-8}
ergs	kilowatt-hrs	0.2778×10^{-13}
ergs	watt-hours	0.2778×10^{-10}
ergs/sec	Btu/min	$5,688 \times 10^{-9}$
ergs/sec	ft-lbs/min	4.427×10^{-6}
ergs/sec	ft-lbs/sec	7.3756×10^{-8}
ergs/sec	horsepower	1.341×10^{-10}
ergs/sec	kg-calories/min	1.433×10^{-9}
ergs/sec	kilowatts	10^{-10}

F

TO CONVERT	INTO	MULTIPLY BY
farads	microfarads	10^6
Faraday/sec	Ampere (absolute)	9.6500×10^4
faradays	ampere-hours	26.80
faradays	coulombs	9.649×10^4
Fathom	Meter	1.828804
fathoms	feet	6.0
feet	centimeters	30.48
feet	kilometers	3.048×10^{-4}
feet	meters	0.3048
feet	miles (naut.)	1.645×10^{-4}
feet	miles (stat.)	1.894×10^{-4}
feet	millimeters	304.8
feet	mils	1.2×10^4
feet of water	atmospheres	0.02950
feet of water	in. of mercury	0.8826
feet of water	kgs/sq cm	0.03048

Units of measurement from one system to another

TO CONVERT	INTO	MULTIPLY BY
feet of water	kgs/sq meter	304.8
feet of water	pounds/sq ft	62.43
feet of water	pounds/sq in.	0.4335
feet/min	cms/sec	0.5080
feet/min	feet/sec	0.01667
feet/min	kms/hr	0.01829
feet/min	meters/min	0.3048
feet/min	miles/hr	0.01136
feet/sec	cms/sec	30.48
feet/sec	kms/hr	1.097
feet/sec	knots	0.5921
feet/sec	meters/min	18.29
feet/sec	miles/hr	0.6818
feet/sec	miles/min	0.01136
feet/sec/sec	cms/sec/sec	30.48
feet/sec/sec	kms/hr/sec	1.097
feet/sec/sec	meters/sec/sec	0.3048
feet/sec/sec	miles/hr/sec	0.6818
feet/100 feet	per cent grade	1.0
Foot – candle	Lumen/sq. meter	10.764
foot-pounds	Btu	1.286×10^{-3}
foot-pounds	ergs	1.356×10^{7}
foot-pounds	gram-calories	0.3238
foot-pounds	hp-hrs	5.050×10^{-7}
foot-pounds	joules	1.356
foot-pounds	kg-calories	3.24×10^{-4}
foot-pounds	kg-meters	0.1383
foot-pounds	kilowatt-hrs	3.766×10^{-7}
foot-pounds/min	Btu/min	1.286×10^{-3}
foot-pounds/min	foot-pounds/sec	0.01667
foot-pounds/min	horsepower	3.030×10^{-5}
foot-pounds/min	kg-calories/min	3.24×10^{-4}
foot-pounds/min	kilowatts	2.260×10^{-5}
foot-pounds/sec	Btu/hr	4.6263
foot-pounds/sec	Btu/min	0.07717
foot-pounds/sec	horsepower	1.818×10^{-3}
foot-pounds/sec	kg-calories/min	0.01945
foot-pounds/sec	kilowatts	1.356×10^{-3}
Furlongs	miles (U.S.)	0.125
furlongs	rods	40.0
furlongs	feet	660.0
	G	
gallons	cu cms	3,785.0
gallons	cu feet	0.1337
gallons	cu inches	231.0
gallons	cu meters	3.785×10^{-3}
gallons	cu yards	4.951×10^{-3}
gallons	liters	3.785
gallons (liq. Br. Imp.)	gallons (U.S. liq.)	1.20095
gallons (U.S.)	gallons (Imp.)	0.83267
gallons of water	pounds of water	8.3453
gallons/min	cu ft/sec	2.228×10^{-3}
gallons/min	liters/sec	0.06308
gallons/min	cu ft/hr	8.0208
gausses	lines/sq in.	6.452
gausses	webers/sq cm	10^{-8}
gausses	webers/sq in.	6.452×10^{-8}
gausses	webers/sq meter	10^{-4}
gilberts	ampere-turns	0.7958
gilberts/cm	amp-turns/cm	0.7958
gilberts/cm	amp-turns/in	2.021
gilberts/cm	amp-turns/meter	79.58
Gills (British)	cubic cm.	142.07
gills	liters	0.1183
gills	pints (liq.)	0.25
Grade	Radian	.01571
Grains	drams (avoirdupois)	0.03657143

TO CONVERT	INTO	MULTIPLY BY
grains (troy)	grains (avdp)	1.0
grains (troy)	grams	0.06480
grains (troy)	ounces (avdp)	2.0833×10^{-3}
grains (troy)	pennyweight (troy)	0.04167
grains/U.S. gal	parts/million	17.118
grains/U.S. gal	pounds/million gal	142.86
grains/Imp. gal	parts/million	14.286
grams	dynes	980.7
grams	grains	15.43
grams	joules/cm	9.807×10^{-5}
grams	joules/meter (newtons)	9.807×10^{-3}
grams	kilograms	0.001
grams	milligrams	1,000.
grams	ounces (avdp)	0.03527
grams	ounces (troy)	0.03215
grams	poundals	0.07093
grams	pounds	2.205×10^{-3}
grams/cm	pounds/inch	5.600×10^{-3}
grams/cu cm	pounds/cu ft	62.43
grams/cu cm	pounds/cu in	0.03613
grams/cu cm	pounds/mil-foot	3.405×10^{-7}
grams/liter	grains/gal	58.417
grams/liter	pounds/1,000 gal	8.345
grams/liter	pounds/cu ft	0.062427
grams/liter	parts/million	1,000.0
grams/sq cm	pounds/sq ft	2.0481
gram-calories	Btu	3.9683×10^{-3}
gram-calories	ergs	4.1868×10^{7}
gram-calories	foot-pounds	3.0880
gram-calories	horsepower-hrs	1.5596×10^{-6}
gram-calories	kilowatt-hrs	1.1630×10^{-6}
gram-calories	watt-hrs	1.1630×10^{-3}
gram-calories/sec	Btu/hr	14.286
gram-centimeters	Btu	9.297×10^{-8}
gram-centimeters	ergs	980.7
gram-centimeters	joules	9.807×10^{-5}
gram-centimeters	kg-cal	2.343×10^{-8}
gram-centimeters	kg-meters	10^{-5}
	H	
Hand	Cm.	10.16
hectares	acres	2.471
hectares	sq feet	1.076×10^{5}
hectograms	grams	100.0
hectoliters	liters	100.0
hectometers	meters	100.0
hectowatts	watts	100.0
henries	millihenries	1,000.0
Hogsheads (British)	cubic ft.	10.114
Hogsheads (U.S.)	cubic ft.	8.42184
Hogsheads (U.S.)	gallons (U.S.)	63
horsepower	Btu/min	42.44
horsepower	foot-lbs/min	33,000.
horsepower	foot-lbs/sec	550.0
horsepower (metric) (542.5 ft lb/sec)	horsepower (550 ft lb/sec)	0.9863
horsepower (550 ft lb/sec)	horsepower (metric) (542.5 ft lb/sec)	1.014
horsepower	kg-calories/min	10.68
horsepower	kilowatts	0.7457
horsepower	watts	745.7
horsepower (boiler)	Btu/hr	33.479
horsepower (boiler)	kilowatts	9.803
horsepower-hrs	Btu	2,547.
horsepower-hrs	ergs	2.6845×10^{13}
horsepower-hrs	foot-lbs	1.98×10^{6}
horsepower-hrs	gram-calories	641,190.
horsepower-hrs	joules	2.684×10^{6}

Units of measurement from one system to another

TO CONVERT	INTO	MULTIPLY BY
horsepower-hrs	kg-calories	641.1
horsepower-hrs	kg-meters	2.737×10^{5}
horsepower-hrs	kilowatt-hrs	0.7457
hours	days	4.167×10^{-2}
hours	weeks	5.952×10^{-3}
Hundredweights (long)	pounds	112
Hundredweights (long)	tons (long)	0.05
Hundredweights (short)	ounces (avoirdupois)	1600
Hundredweights (short)	pounds	100
Hundredweights (short)	tons (metric)	0.0453592
Hundredweights (short)	tons (long)	0.0446429
	I	
inches	centimeters	2.540
inches	meters	2.540×10^{-2}
inches	miles	1.578×10^{-5}
inches	millimeters	25.40
inches	mils	1,000.0
inches	yards	2.778×10^{-2}
inches of mercury	atmospheres	0.03342
inches of mercury	feet of water	1.133
inches of mercury	kgs/sq cm	0.03453
inches of mercury	kgs/sq meter	345.3
inches of mercury	pounds/sq ft	70.73
inches of mercury	pounds/sq in.	0.4912
inches of water (at 4°C)	atmospheres	2.458×10^{-3}
inches of water (at 4°C)	inches of mercury	0.07355
inches of water (at 4°C)	kgs/sq cm	2.540×10^{-3}
inches of water (at 4°C)	ounces/sq in.	0.5781
inches of water (at 4°C)	pounds/sq ft	5.204
inches of water (at 4°C)	pounds/sq in.	0.03613
International Ampere	Ampere (absolute)	.9998
International Volt	Volts (absolute)	1.0003
International volt	Joules (absolute)	$1\text{-}593 \times 10^{-19}$
International volt	Joules	9.654×10^{4}
	J	
joules	Btu	9.480×10^{-4}
joules	ergs	10^{7}
joules	foot-pounds	0.7376
joules	kg-calories	2.389×10^{-4}
joules	kg-meters	0.1020
joules	watt-hrs	2.778×10^{-4}
joules/cm	grams	1.020×10^{4}
joules/cm	dynes	10^{7}
joules/cm	joules/meter (newtons)	100.0
joules/cm	poundals	723.3
joules/cm	pounds	22.48
	K	
kilograms	dynes	980,665.
kilograms	grams	1,000.0
kilograms	joules/cm	0.09807
kilograms	joules/meter (newtons)	9.807
kilograms	poundals	70.93
kilograms	pounds	2.205
kilograms	tons (long)	9.842×10^{-4}
kilograms	tons (short)	1.102×10^{-3}
kilograms/cu meter	grams/cu cm	0.001
kilograms/cu meter	pounds/cu ft	0.06243
kilograms/cu meter	pounds/cu in.	3.613×10^{-5}
kilograms/cu meter	pounds/mil-foot	3.405×10^{-10}
kilograms/meter	pounds/ft	0.6720
Kilogram/sq. cm.	Dynes	980,665
kilograms/sq cm	atmospheres	0.9678
kilograms/sq cm	feet of water	32.81

TO CONVERT	INTO	MULTIPLY BY
kilograms/sq cm	inches of mercury	28.96
kilograms/sq cm	pounds/sq ft	2,048.
kilograms/sq cm	pounds/sq in.	14.22
kilograms/sq meter	atmospheres	9.678×10^{-5}
kilograms/sq meter	bars	98.07×10^{-6}
kilograms/sq meter	feet of water	3.281×10^{-3}
kilograms/sq meter	inches of mercury	2.896×10^{-3}
kilograms/sq meter	pounds/sq ft	0.2048
kilograms/sq meter	pounds/sq in.	1.422×10^{-3}
kilograms/sq mm	kgs/sq meter	10^{6}
kilogram-calories	Btu	3.968
kilogram-calories	foot-pounds	3,088.
kilogram-calories	hp-hrs	1.560×10^{-3}
kilogram-calories	joules	4,186.
kilogram-calories	kg-meters	426.9
kilogram-calories	kilojoules	4.186
kilogram-calories	kilowatt-hrs	1.163×10^{-3}
kilogram meters	Btu	9.294×10^{-3}
kilogram meters	ergs	9.804×10^{7}
kilogram meters	foot-pounds	7.233
kilogram meters	joules	9.804
kilogram meters	kg-calories	2.342×10^{-3}
kilogram meters	kilowatt-hrs	2.723×10^{-6}
kilolines	maxwells	1,000.0
kiloliters	liters	1,000.0
kilometers	centimeters	10^{5}
kilometers	feet	3,281.
kilometers	inches	3.937×10^{4}
kilometers	meters	1,000.0
kilometers	miles	0.6214
kilometers	millimeters	10^{6}
kilometers	yards	1,094.
kilometers/hr	cms/sec	27.78
kilometers/hr	feet/min	54.68
kilometers/hr	feet/sec	0.9113
kilometers/hr	knots	0.5396
kilometers/hr	meters/min	16.67
kilometers/hr	miles/hr	0.6214
kilometers/hr/sec	cms/sec/sec	27.78
kilometers/hr/sec	ft/sec/sec	0.9113
kilometers/hr/sec	meters/sec/sec	0.2778
kilometers/hr/sec	miles/hr/sec	0.6214
kilowatts	Btu/min	56.92
kilowatts	foot-lbs/min	4.426×10^{4}
kilowatts	foot-lbs/sec	737.6
kilowatts	horsepower	1.341
kilowatts	kg-calories/min	14.34
kilowatts	watts	1,000.0
kilowatt-hrs	Btu	3,413.
kilowatt-hrs	ergs	3.600×10^{13}
kilowatt-hrs	foot-lbs	2.655×10^{6}
kilowatt-hrs	gram-calories	859,850.
kilowatt-hrs	horsepower-hrs	1.341
kilowatt-hrs	joules	3.6×10^{6}
kilowatt-hrs	kg-calories	860.5
kilowatt-hrs	kg-meters	3.671×10^{5}
kilowatt-hrs	pounds of water evaporated from and at 212° F.	3.53
kilowatt-hrs	pounds of water raised from 62° to 212° F.	22.75
knots	feet/hr	6,080.
knots	kilometers/hr	1.8532
knots	nautical miles/hr	1.0
knots	statute miles/hr	1.151

Units of measurement from one system to another

TO CONVERT	INTO	MULTIPLY BY
knots	yards/hr	2,027.
knots	feet/sec	1.689
L		
league	miles (approx.)	3.0
Light year	Miles	5.9×10^{12}
Light year	Kilometers	9.46091×10^{12}
lines/sq cm	gausses	1.0
lines/sq in.	gausses	0.1550
lines/sq in.	webers/sq cm	1.550×10^{-9}
lines/sq in.	webers/sq in.	10^{-8}
lines/sq in.	webers/sq meter	1.550×10^{-5}
links (engineer's)	inches	12.0
links (surveyor's)	inches	7.92
liters	bushels (U.S. dry)	0.02838
liters	cu cm	1,000.0
liters	cu feet	0.03531
liters	cu inches	61.02
liters	cu meters	0.001
liters	cu yards	1.308×10^{-3}
liters	gallons (U.S. liq.)	0.2642
liters	pints (U.S. liq.)	2.113
liters	quarts (U.S. liq.)	1.057
liters/min	cu ft/sec	5.886×10^{-4}
liters/min	gals/sec	4.403×10^{-3}
lumens/sq ft	foot-candles	1.0
Lumen	Spherical candle power	.07958
Lumen	Watt	.001496
Lumen/sq. ft.	Lumen/sq. meter	10.76
lux	foot-candles	0.0929
M		
maxwells	kilolines	0.001
maxwells	webers	10^{-8}
megalines	maxwells	10^{6}
megohms	microhms	10^{12}
megohms	ohms	10^{6}
meters	centimeters	100.0
meters	feet	3.281
meters	inches	39.37
meters	kilometers	0.001
meters	miles (naut.)	5.396×10^{-4}
meters	miles (stat.)	6.214×10^{-4}
meters	millimeters	1,000.0
meters	yards	1.094
meters	varas	1.179
meters/min	cms/sec	1.667
meters/min	feet/min	3.281
meters/min	feet/sec	0.05468
meters/min	kms/hr	0.06
meters/min	knots	0.03238
meters/min	miles/hr	0.03728
meters/sec	feet/min	196.8
meters/sec	feet/sec	3.281
meters/sec	kilometers/hr	3.6
meters/sec	kilometers/min	0.06
meters/sec	miles/hr	2.237
meters/sec	miles/min	0.03728
meters/sec/sec	cms/sec/sec	100.0
meters/sec/sec	ft/sec/sec	3.281
meters/sec/sec	kms/hr/sec	3.6
meters/sec/sec	miles/hr/sec	2.237
meter-kilograms	cm-dynes	9.807×10^{7}
meter-kilograms	cm-grams	10^{5}
meter-kilograms	pound-feet	7.233
microfarad	farads	10^{-6}
micrograms	grams	10^{-6}
microhms	megohms	10^{-12}
microhms	ohms	10^{-6}
microliters	liters	10^{-6}
Microns	meters	1×10^{-6}
miles (naut.)	feet	6,080.27
miles (naut.)	kilometers	1.853
miles (naut.)	meters	1,853.
miles (naut.)	miles (statute)	1.1516
miles (naut.)	yards	2,027.
miles (statute)	centimeters	1.609×10^{5}
miles (statute)	feet	5,280.
miles (statute)	inches	6.336×10^{4}
miles (statute)	kilometers	1.609
miles (statute)	meters	1,609.
miles (statute)	miles (naut.)	0.8684
miles (statute)	yards	1,760.
miles/hr	cms/sec	44.70
miles/hr	feet/min	88.
miles/hr	feet/sec	1.467
miles/hr	kms/hr	1.609
miles/hr	kms/min	0.02682
miles/hr	knots	0.8684
miles/hr	meters/min	26.82
miles/hr	miles/min	0.1667
miles/hr/sec	cms/sec/sec	44.70
miles/hr/sec	feet/sec/sec	1.467
miles/hr/sec	kms/hr/sec	1.609
miles/hr/sec	meters/sec/sec	0.4470
miles/min	cms/sec	2,682.
miles/min	feet/sec	88.
miles/min	kms/min	1.609
miles/min	knots/min	0.8684
miles/min	miles/hr	60.0
mil-feet	cu inches	9.425×10^{-6}
milliers	kilograms	1,000.
Millimicrons	meters	1×10^{-9}
Milligrams	grains	0.01543236
milligrams	grams	0.001
milligrams/liter	parts/million	1.0
millihenries	henries	0.001
milliliters	liters	0.001
millimeters	centimeters	0.1
millimeters	feet	3.281×10^{-3}
millimeters	inches	0.03937
millimeters	kilometers	10^{-6}
millimeters	meters	0.001
millimeters	miles	6.214×10^{-7}
millimeters	mils	39.37
millimeters	yards	1.094×10^{-3}
million gals/day	cu ft/sec	1.54723
mils	centimeters	2.540×10^{-3}
mils	feet	8.333×10^{-5}
mils	inches	0.001
mils	kilometers	2.540×10^{-8}
mils	yards	2.778×10^{-5}
miner's inches	cu ft/min	1.5
Minims (British)	cubic cm.	0.059192
Minims (U.S., fluid)	cubic cm.	0.061612
minutes (angles)	degrees	0.01667
minutes (angles)	quadrants	1.852×10^{-4}
minutes (angles)	radians	2.909×10^{-4}
minutes (angles)	seconds	60.0
myriagrams	kilograms	10.0
myriameters	kilometers	10.0
myriawatts	kilowatts	10.0
N		
nepers	decibels	8.686
Newton	Dynes	1 x 105

Units of measurement from one system to another

TO CONVERT	INTO	MULTIPLY BY
O		
OHM (International)	OHM (absolute)	1.0005
ohms	megohms	10^{-6}
ohms	microhms	10^{6}
ounces	drams	16.0
ounces	grains	437.5
ounces	grams	28.349527
ounces	pounds	0.0625
ounces	ounces (troy)	0.9115
ounces	tons (long)	2.790×10^{-5}
ounces	tons (metric)	2.835×10^{-5}
ounces (fluid)	cu inches	1.805
ounces (fluid)	liters	0.02957
ounces (troy)	grains	480.0
ounces (troy)	grams	31.103481
ounces (troy)	ounces (avdp.)	1.09714
ounces (troy)	pennyweights (troy)	20.0
ounces (troy)	pounds (troy)	0.08333
Ounce/sq. inch	Dynes/sq. cm.	4309
ounces/sq in.	pounds/sq in.	0.0625
P		
Parsec	Miles	19×10^{12}
Parsec	Kilometers	3.084×10^{13}
parts/million	grains/U.S. gal	0.0584
parts/million	grains/Imp. gal	0.07016
parts/million	pounds/million gal	8.345
Pecks (British)	cubic inches	554.6
Pecks (British)	liters	9.091901
Pecks (U.S.)	bushels	0.25
Pecks (U.S.)	cubic inches	537.605
Pecks (U.S.)	liters	8.809582
Pecks (U.S.)	quarts (dry)	8
pennyweights (troy)	grains	24.0
pennyweights (troy)	ounces (troy)	0.05
pennyweights (troy)	grams	1.55517
pennyweights (troy)	pounds (troy)	4.1667×10^{-3}
pints (dry)	cu inches	33.60
pints (liq.)	cu cms.	473.2
pints (liq.)	cu feet	0.01671
pints (liq.)	cu inches	28.87
pints (liq.)	cu meters	4.732×10^{-4}
pints (liq.)	cu yards	6.189×10^{-4}
pints (liq.)	gallons	0.125
pints (liq.)	liters	0.4732
pints (liq.)	quarts (liq.)	0.5
Planck's quantum	Erg – second	6.624×10^{-27}
Poise	Gram/cm. sec.	1.00
Pounds (avoirdupois)	ounces (troy)	14.5833
poundals	dynes	13,826.
poundals	grams	14.10
poundals	joules/cm	1.383×10^{-3}
poundals	joules/meter (newtons)	0.1383
poundals	kilograms	0.01410
poundals	pounds	0.03108
pounds	drams	256.
pounds	dynes	44.4823×10^{4}
pounds	grains	7,000.
pounds	grams	453.5924
pounds	joules/cm	0.04448
pounds	joules/meter (newtons)	4.448
pounds	kilograms	0.4536
pounds	ounces	16.0
pounds	ounces (troy)	14.5833
pounds	poundals	32.17
pounds	pounds (troy)	1.21528
pounds	tons (short)	0.0005
pounds (troy)	grains	5,760.
pounds (troy)	grams	373.24177

TO CONVERT	INTO	MULTIPLY BY
pounds (troy)	ounces (avdp.)	13.1657
pounds (troy)	ounces (troy)	12.0
pounds (troy)	pennyweights (troy)	240.0
pounds (troy)	pounds (avdp.)	0.822857
pounds (troy)	tons (long)	3.6735×10^{-4}
pounds (troy)	tons (metric)	3.7324×10^{-4}
pounds (troy)	tons (short)	4.1143×10^{-4}
pounds of water	cu feet	0.01602
pounds of water	cu inches	27.68
pounds of water	gallons	0.1198
pounds of water/min	cu ft/sec	2.670×10^{-4}
pound-feet	cm-dynes	1.356×10^{7}
pound-feet	cm-grams	13,825.
pound-feet	meter-kgs	0.1383
pounds/cu ft	grams/cu cm	0.01602
pounds/cu ft	kgs/cu meter	16.02
pounds/cu ft	pounds/cu in.	5.787×10^{-4}
pounds/cu ft	pounds/mil-foot	5.456×10^{-9}
pounds/cu in.	gms/cu cm	27.68
pounds/cu in.	kgs/cu meter	2.768×10^{4}
pounds/cu in.	pounds/cu ft	1,728.
pounds/cu in.	pounds/mil-foot	9.425×10^{-6}
pounds/ft	kgs/meter	1.488
pounds/in.	gms/cm	178.6
pounds/mil-foot	gms/cu cm	2.306×10^{6}
pounds/sq ft	atmospheres	4.725×10^{-4}
pounds/sq ft	feet of water	0.01602
pounds/sq ft	inches of mercury	0.01414
pounds/sq ft	kgs/sq meter	4.882
pounds/sq ft	pounds/sq in.	6.944×10^{-3}
pounds/sq in.	atmospheres	0.06804
pounds/sq in.	feet of water	2.307
pounds/sq in.	inches of mercury	2.036
pounds/sq in.	kgs/sq meter	703.1
pounds/sq in.	pounds/sq ft	144.0
Q		
quadrants (angle)	degrees	90.0
quadrants (angle)	minutes	5,400.0
quadrants (angle)	radians	1.571
quadrants (angle)	seconds	3.24×10^{5}
quarts (dry)	cu inches	67.20
quarts (liq.)	cu cms	946.4
quarts (liq.)	cu feet	0.03342
quarts (liq.)	cu inches	57.75
quarts (liq.)	cu meters	9.464×10^{-4}
quarts (liq.)	cu yards	1.238×10^{-3}
quarts (liq.)	gallons	0.25
quarts (liq.)	liters	0.9463
R		
radians	degrees	57.30
radians	minutes	3,438.
radians	quadrants	0.6366
radians	seconds	2.063×10^{5}
radians/sec	degrees/sec	57.30
radians/sec	revolutions/min	9.549
radians/sec	revolutions/sec	0.1592
radians/sec/sec	revs/min/min	573.0
radians/sec/sec	revs/min/sec	9.549
radians/sec/sec	revs/sec/sec	0.1592
revolutions	degrees	360.0
revolutions	quadrants	4.0
revolutions	radians	6.283
revolutions/min	degrees/sec	6.0
revolutions/min	radians/sec	0.1047
revolutions/min	revs/sec	0.01667

Units of measurement from one system to another

TO CONVERT	INTO	MULTIPLY BY
revolutions/min/min	radians/sec/sec	1.745×10^{-3}
revolutions/min/min	revs/min/sec	0.01667
revolutions/min/min	revs/sec/sec	2.778×10^{-4}
revolutions/sec	degrees/sec	360.0
revolutions/sec	radians/sec	6.283
revolutions/sec	revs/min	60.0
revolutions/sec/sec	radians/sec/sec	6.283
revolutions/sec/sec	revs/min/min	3,600.0
revolutions/sec/sec	revs/min/sec	60.0
Rod	Chain (Gunters)	.25
Rod	Meters	5.029
Rods (Surveyors' meas.)	yards	5.5
rods	feet	16.5
	S	
Scruples	grains	20
seconds (angle)	degrees	2.778×10^{-4}
seconds (angle)	minutes	0.01667
seconds (angle)	quadrants	3.087×10^{-6}
seconds (angle)	radians	4.848×10^{-6}
Slug	Kilogram	14.59
Slug	Pounds	32.17
Sphere	Steradians	12.57
square centimeters	circular mils	1.973×10^{5}
square centimeters	sq feet	1.076×10^{-3}
square centimeters	sq inches	0.1550
square centimeters	sq meters	0.0001
square centimeters	sq miles	3.861×10^{-11}
square centimeters	sq millimeters	100.0
square centimeters	sq yards	1.196×10^{-4}
square feet	acres	2.296×10^{-5}
square feet	circular mils	1.833×10^{8}
square feet	sq cms	929.0
square feet	sq inches	144.0
square feet	sq meters	0.09290
square feet	sq miles	3.587×10^{-8}
square feet	sq millimeters	9.290×10^{4}
square feet	sq yards	0.1111
square inches	circular mils	1.273×10^{6}
square inches	sq cms	6.452
square inches	sq feet	6.944×10^{-3}
square inches	sq millimeters	645.2
square inches	sq mils	10^{6}
square inches	sq yards	7.716×10^{-4}
square kilometers	acres	247.1
square kilometers	sq cms	10^{10}
square kilometers	sq ft	10.76×10^{6}
square kilometers	sq inches	1.550×10^{9}
square kilometers	sq meters	10^{6}
square kilometers	sq miles	0.3861
square kilometers	sq yards	1.196×10^{6}
square meters	acres	2.471×10^{-4}
square meters	sq cms	10^{4}
square meters	sq feet	10.76
square meters	sq inches	1,550.
square meters	sq miles	3.861×10^{-7}
square meters	sq millimeters	10^{6}
square meters	sq yards	1.196
square miles	acres	640.0
square miles	sq feet	27.88×10^{6}
square miles	sq kms	2.590
square miles	sq meters	2.590×10^{6}
square miles	sq yards	3.098×10^{6}
square millimeters	circular mils	1,973.
square millimeters	sq cms	0.01
square millimeters	sq feet	1.076×10^{-5}
square millimeters	sq inches	1.550×10^{-3}
square mils	circular mils	1.273

TO CONVERT	INTO	MULTIPLY BY
square mils	sq cms	6.452×10^{-6}
square mils	sq inches	10^{-6}
square yards	acres	2.066×10^{-4}
square yards	sq cms	8,361.
square yards	sq feet	9.0
square yards	sq inches	1,296.
square yards	sq meters	0.8361
square yards	sq miles	3.228×10^{-7}
square yards	sq millimeters	8.361×10^{5}
	T	
temperature (°C) +273	absolute temperature (°C)	1.0
temperature (°C) +17.78	temperature (°F)	1.8
temperature (°F) +460	absolute temperature (°F)	1.0
temperature (°F) −32	temperature (°C)	5/9
tons (long)	kilograms	1,016.
tons (long)	pounds	2,240.
tons (long)	tons (short)	1.120
tons (metric)	kilograms	1,000.
tons (metric)	pounds	2,205.
tons (short)	kilograms	907.1848
tons (short)	ounces	32,000.
tons (short)	ounces (troy)	29,166.66
tons (short)	pounds	2,000.
tons (short)	pounds (troy)	2,430.56
tons (short)	tons (long)	0.89287
tons (short)	tons (metric)	0.9078
tons (short)/sq ft	kgs/sq meter	9,765.
tons (short)/sq ft	pounds/sq in.	2,000.
tons of water/24 hrs	pounds of water/hr	83.333
tons of water/24 hrs	gallons/min	0.16643
tons of water/24 hrs	cu ft/hr	1.3349
	V	
Volt/inch	Volt/cm.	.39370
Volt (absolute)	Statvolts	.003336
	W	
watts	Btu/hr	3.4129
watts	Btu/min	0.05688
watts	ergs/sec	107.
watts	foot-lbs/min	44.27
watts	foot-lbs/sec	0.7378
watts	horsepower	1.341×10^{-3}
watts	horsepower (metric)	1.360×10^{-3}
watts	kg-calories/min	0.01433
watts	kilowatts	0.001
Watts (Abs.)	B.T.U. (mean)/min.	0.056884
Watts (Abs.)	joules/sec.	1
watt-hours	Btu	3.413
watt-hours	ergs	3.60×10^{10}
watt-hours	foot-pounds	2,656.
watt-hours	gram-calories	859.85
watt-hours	horsepower-hrs	1.341×10^{-3}
watt-hours	kilogram-calories	0.8605
watt-hours	kilogram-meters	367.2
watt-hours	kilowatt-hrs	0.001
Watt (International)	Watt (absolute)	1.0002
webers	maxwells	10^{8}
webers	kilolines	10^{5}

Units of measurement from one system to another

TO CONVERT	INTO	MULTIPLY BY
webers/sq in.	gausses	1.550×10^7
webers/sq in.	lines/sq in.	10^8
webers/sq in.	webers/sq cm	0.1550
webers/sq in.	webers/sq meter	1,550.
webers/sq meter	gausses	10^4
webers/sq meter	lines/sq in.	6.452×10^4
webers/sq meter	webers/sq cm	10^{-4}
webers/sq meter	webers/sq in.	6.452×10^{-4}

TO CONVERT	INTO	MULTIPLY BY
	Y	
yards	centimeters	91.44
yards	kilometers	9.144×10^{-4}
yards	meters	0.9144
yards	miles (naut.)	4.934×10^{-4}
yards	miles (stat.)	5.682×10^{-4}
yards	millimeters	914.4

Viscosity—equivalents of absolute viscosity

Absolute or Dynamic Viscosity		Centipoise (μ)	Poise; Gram / Cm Sec; Dyne Sec / Cm^2 ($100\ \mu$)	Slugs / Ft Sec; *$Pound_f$ Sec / Ft^2 (μ'_e)	†$Pound_m$ / Ft Sec; Poundal Sec / Ft^2 (μ_e)
Centipoise	(μ)	1	0.01	$2.09\ (10^{-5})$	$6.72\ (10^{-4})$
Poise; Gram / Cm Sec; Dyne Sec / Cm^2	($100\ \mu$)	100	1	$2.09\ (10^{-3})$	0.0672
Slugs / Ft Sec; *$Pound_f$ Sec / Ft^2	(μ'_e)	47 900	479	1	g or 32.2
†$Pound_m$ / Ft Sec; Poundal Sec / Ft^2	(μ_e)	1487	14.87	$\frac{1}{g}$ or .0311	1

*$Pound_f$ = Pound of Force †$Pound_m$ = Pound of Mass

To convert absolute or dynamic viscosity from one set of units to another, locate the given set of units in the left hand column and multiply the numerical value by the factor shown horizontally to the right under the set of units desired.

As an example, suppose a given absolute viscosity of 2 poise is to be converted to slugs/foot second. By referring to the table, we find the conversion factor to be $2.09\ (10^{-3})$. Then, 2 (poise) times $2.09\ (10^{-3}) = 4.18\ (10^{-3}) = 0.00418$ slugs/foot second.

Courtesy Crane Co. (Technical Paper #410).

Viscosity—equivalents of kinematic viscosity

Kinematic Viscosity		Centistokes (ν)	Stokes $\frac{Cm^2}{Sec}$ ($100\ \nu$)	$\frac{Ft^2}{Sec}$ (ν')
Centistokes	(ν)	1	0.01	1.076 (10^{-5})
Stokes $\frac{Cm^2}{Sec}$	($100\ \nu$)	100	1	1.076 (10^{-3})
$\frac{Ft^2}{Sec}$	(ν')	92 900	929	1

To convert kinematic viscosity from one set of units to another, locate the given set of units in the left hand column and multiply the numerical value by the factor shown horizontally to the right, under the set of units desired.

As an example, suppose a given kinematic viscosity of 0.5 square foot/second is to be converted to centistokes. By referring to the table, we find the conversion factor to be 92,900. Then, 0.5 (sq ft/sec) times 92,900 = 46,450 centistokes.

Courtesy Crane Co. and American Society for Testing and Materials.

Viscosity—equivalents of kinematic and Saybolt Universal Viscosity

Kinematic Viscosity, Centistokes v	Equivalent Saybolt Universal Viscosity, Sec — At 100 F Basic Values	At 210 F
1.83	32.01	32.23
2.0	32.62	32.85
4.0	39.14	39.41
6.0	45.56	45.88
8.0	52.09	52.45
10.0	58.91	59.32
15.0	77.39	77.93
20.0	97.77	98.45
25.0	119.3	120.1
30.0	141.3	142.3
35.0	163.7	164.9
40.0	186.3	187.6
45.0	209.1	210.5
50.0	232.1	233.8
55.0	255.2	257.0
60.0	278.3	280.2
65.0	301.4	303.5
70.0	324.4	326.7
75.0	347.6	350.0
80.0	370.8	373.4
85.0	393.9	396.7
90.0	417.1	420.0
95.0	440.3	443.4
100.0	463.5	466.7
120.0	556.2	560.1
140.0	648.9	653.4
160.0	741.6	
180.0	834.2	Saybolt Seconds equal Centistokes times 4.6673
200.0	926.9	
220.0	1019.6	
240.0	1112.3	
260.0	1205.0	
280.0	1297.7	
300.0	1390.4	
320.0	1483.1	
340.0	1575.8	
360.0	1668.5	
380.0	1761.2	
400.0	1853.9	
420.0	1946.6	
440.0	2039.3	
460.0	2132.0	
480.0	2224.7	
500.0	2317.4	
Over 500	Saybolt Seconds equal Centistokes times 4.6347	

Note: To obtain the Saybolt Universal viscosity equivalent to a kinematic viscosity determined at t, multiply the equivalent Saybolt Universal viscosity at 100 F by $1 + (t - 100)\ 0.000\ 064$.
For example, 10 v at 210 F are equivalent to 58.91 multiplied by 1.0070 or 59.32 sec Saybolt Universal at 210 F.

Courtesy Crane Co. (Technical Paper #410).

Viscosity—equivalents of kinematic and Saybolt Furol Viscosity at 122° F

Kinematic Viscosity, Centistokes υ	Equivalent Saybolt Furol Viscosity, Sec At 122 F	At 210 F
48	25.3	
50	26.1	25.2
60	30.6	29.8
70	35.1	34.4
80	39.6	39.0
90	44.1	43.7
100	48.6	48.3
125	60.1	60.1
150	71.7	71.8
175	83.8	83.7
200	95.0	95.6
225	106.7	107.5
250	118.4	119.4
275	130.1	131.4
300	141.8	143.5
325	153.6	155.5
350	165.3	167.6
375	177.0	179.7
400	188.8	191.8
425	200.6	204.0
450	212.4	216.1
475	224.1	228.3
500	235.9	240.5
525	247.7	252.8
550	259.5	265.0
575	271.3	277.2
600	283.1	289.5
625	294.9	301.8
650	306.7	314.1
675	318.4	326.4
700	330.2	338.7
725	342.0	351.0
750	353.8	363.4
775	365.5	375.7
800	377.4	388.1
825	389.2	400.5
850	400.9	412.9
875	412.7	425.3
900	424.5	437.7
925	436.3	450.1
950	448.1	462.5
975	459.9	474.9
1000	471.7	487.4
1025	483.5	499.8
1050	495.2	512.3
1075	507.0	524.8
1100	518.8	537.2
1125	530.6	549.7
1150	542.4	562.2
1175	554.2	574.7
1200	566.0	587.2
1225	577.8	599.7
1250	589.5	612.2
1275	601.3	624.8
1300	613.1	637.3
Over 1300	*	†

*OVER 1300 CENTISTOKES AT 122 F:
Saybolt Fluid Sec = Centistokes x 0.4717

†OVER 1300 CENTISTOKES AT 210 F:
Log (Saybolt Furol Sec − 2.87) = 1.0276 [Log (Centistokes)] − 0.3975

Courtesy Crane Co. and American Society for Testing and Materials.

Viscosity—general conversions

APPROXIMATE CONVERSION TABLE

SECONDS SAYBOLT UNIVERSAL	SECONDS SAYBOLT FUROL	ENGLER DEGREES	ENGLER TIME	BARBEY	REDWOOD STANDARD	REDWOOD ADMIRALTY	KINEMATIC VISCOSITY STOKES
35		1.18	60	2,800	32		0.026
50		1.60	82	880	44		0.074
75		2.30	102	460	65		0.141
100	15	3.00	153	320	88		0.202
150	19	4.40	230	205	128		0.318
200	23	5.90	305	148	170	18	0.431
250	28	7.60	375	118	212	23	0.543
300	33	8.90	450	98	254	27	0.651
400	42	11.80	550	72	338	36	0.876
500	52	14.50	750	59	423	45	1.10
600	61	17.50	900	48	518	53	1.32
700	71	20.6	1050	41	592	62	1.54
800	81	23	1,200	36.5	677	71	1.76
900	91	27	1,300	32.0	762	78	1.98
1,000	100	29	1,500	29.5	846	87	2.20
1,500	150	42	2,300	19.5	1,270	135	3.30
2,000	200	59	3,000	14.5	1,695	175	4.40
2,500	250	73	3,750	11.5	2,120	230	5.50
3,000	300	87	4,500	9.6	2,540	260	6.60
4,000	400	117	6,000	7.4	3,380	350	8.80
5,000	500	145	7,500	6.0	4,230	435	11.00
6,000	600	175	9,000	5.2	5,080	530	13.20
7,000	700	205	10,500	4.1	5,925	610	15.40
8,000	800	230	12,000	3.7	6,770	700	17.60
9,000	900	260	13,500	3.2	7,620	780	19.8
10,000	1,000	290	15,000	2.9	8 460	870	22
20,000	2,000	590	30,000	1.4	16,920	1,760	44
40,000	4,000	1,170	60,000		33,850	3,600	88
60,000	6,000	1,750	90,000		50,800	5,300	132
80,000	8,000	2,300	120,000		67,700	7,000	176
100,000	10,000	2,900	150,000		84,600	8,700	220
200,000	20,000	5,900	300,000		169,200	17,600	440
400,000	40,000	11,700	600,000		338,500	36,000	880
600,000	60,000	17,500	900,000		508,000	53,000	1,320
800,000	80,000	23,000	1,200,000		677,000	70,000	1,760
1,000,000	100,000	29,000	1,500,000		846,000	87,000	2,200

ABSOLUTE VISCOSITY is the force required to move a unit plane surface over another plane surface at a unit velocity when surfaces are separated by a layer of fluid of unit thickness

KINEMATIC VISCOSITY is the absolute viscosity divided by the density of the fluid.

SAE VISCOSITY NUMBERS are arbitrary numbers which have been assigned to classify lubricating oils and do not have any functional relationship with other systems.

VISCOSITY INDEX is an arbitrary system relating the effect of temperature change to viscosity. It is the slope of a plot of viscosity versus temperature for a given fluid.

1 Stoke = 1 Poise/Density in Grams
1 Centistoke = 1.076×10^{-5} Ft²/Sec
1 Poise = 1 Gram/(Cm)(Sec)
1 Centipoise = 1/100 Poise

1 Centistoke = 1/100 Stoke
1 Centipoise = 0.000672 Lbs/(Ft)(Sec)
1 Centipoise = 0.000672 (Poundals)(Sec)/Ft²
1 Centipoise = 0.0000209 (Lbs)(Sec)/Ft²

1 Centipoise = 0.0000209 Slugs/(Ft)(Sec)
1 Centipoise = 2.42 Lbs/(Ft)(Hr)
1 Slug/(Ft)(Sec) = 1/0.002089 Poise
1 Ft²/Sec = 92903 Centistokes

Courtesy J. H. Jones Company.

Pressure Conversion Chart

This line chart provides an easy method for converting units of pressure. Draw a line perpendicular to the scale lines through a known value of flow and read the equivalent value on any of the other scales. For values off the scale, known values can be read in multiples of 10 or 100.

Example: Known Value —120 psi is equivalent to:

278 ft. of water
1,920 ounces per sq. in.
17,400 pounds per sq. ft.
3,300 inches of water
242 inches of mercury
6,180 mm of mercury

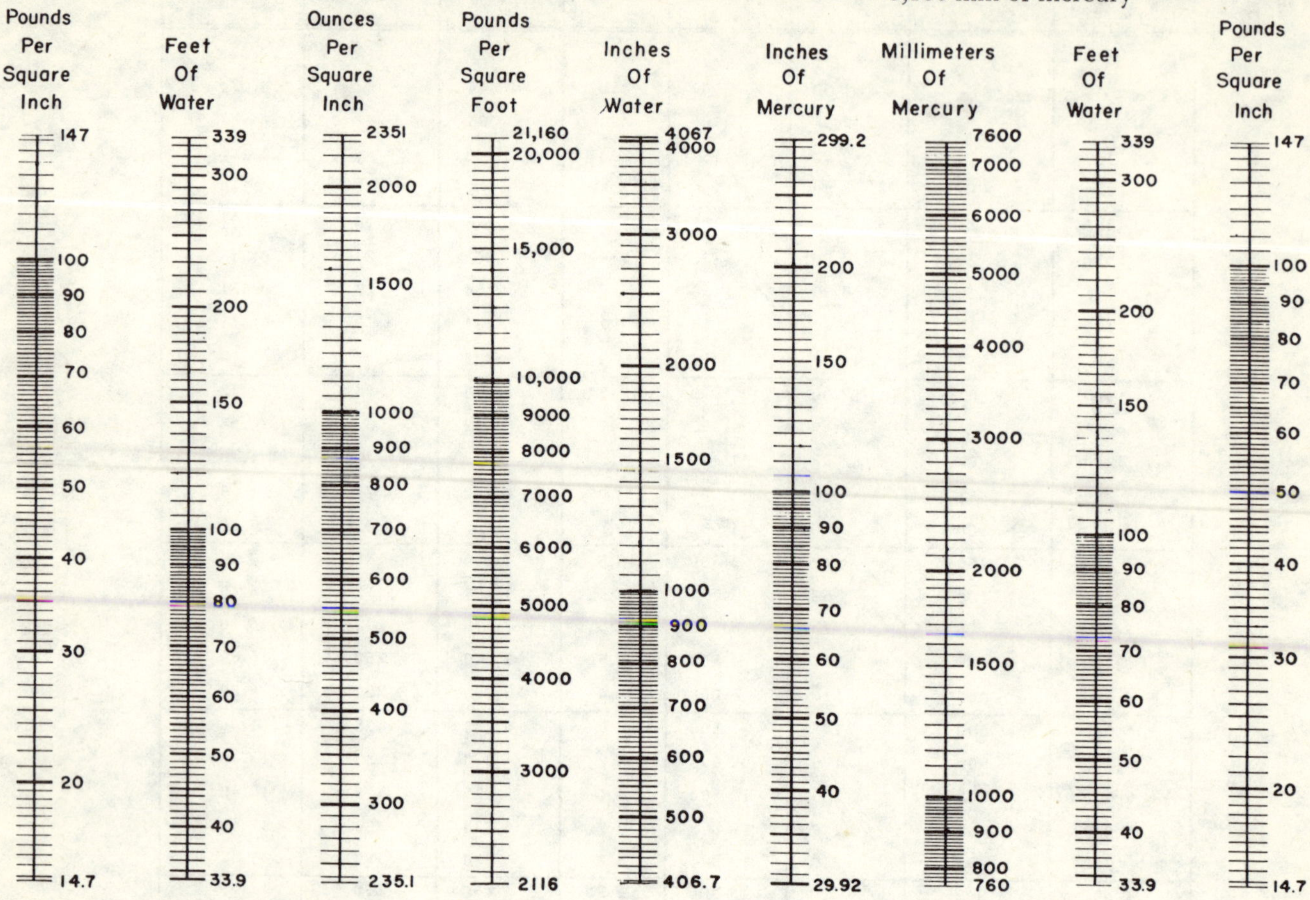

A simple method to determine square root

Problem: determine the square root of the number 6,280

Solution:

1. **Guess the square root to an even number.** For the example, use 80 since $(80)^2$ equals 6400.
2. **Divide the number used into the number of which the square is desired** and carry the quotient to only two figures. Thus $6280 \div 80 = 78$.
3. **Average this quotient with the divisor.** $80 + 78 = 158$, and $158 \div 2 = 79$.
4. **Divide this figure into the number** and carry quotient to four or five figures, if desired. $6280 \div 79 = 79.5$.
5. **Average this quotient with #3** for the square root. $79 + 79.5 = 158.5$, and $158.5 \div 2 = 79.25$.
6. **To check answer,** multiply answer by itself. $79.25 \times 79.25 = 6280.5$.

Multiplying prefixes adopted in 1962 by the International Committee on Weights and Measures

10^{12} = tera = T	10^{-6} = micro = μ
10^{9} = giga = G	10^{-9} = nano = n
10^{6} = mega = M	10^{-12} = pico = p
10^{3} = kilo = k	10^{-15} = femto = f
10^{-3} = milli = m	10^{-18} = atto = a

With one exception the multipliers increasing the unit use capital letters. The exception is k for kilo, and present usage increasingly writes K for this symbol.

With one exception, all multipliers reducing the unit use small letters. The exception is the Greek letter μ for micro. This is firmly established.

Energy Conversion Chart

This nomograph provides an easy method for determining the fuel requirements for various types of fuel. Since the Btu content per unit volume of natural gas and crude oil vary over a limited, but significant range of values, nomographs for the energy value of several gases and oils are included rather than using an average value.

Example: 30 cubic feet of a gas with 1100 Btu per cubic foot yields 33,000 Btu. Conversion of 33,000 Btu to the other units of energy is obtained by drawing a straight line connecting the values 33,000 on two end Btu scales. Using the right-hand nomograph, an oil with 11,000 Btu per gallon (about 4.6 million Btu per barrel) will produce 33,000 Btu from 0.3 gallon (about 0.0072 barrel).

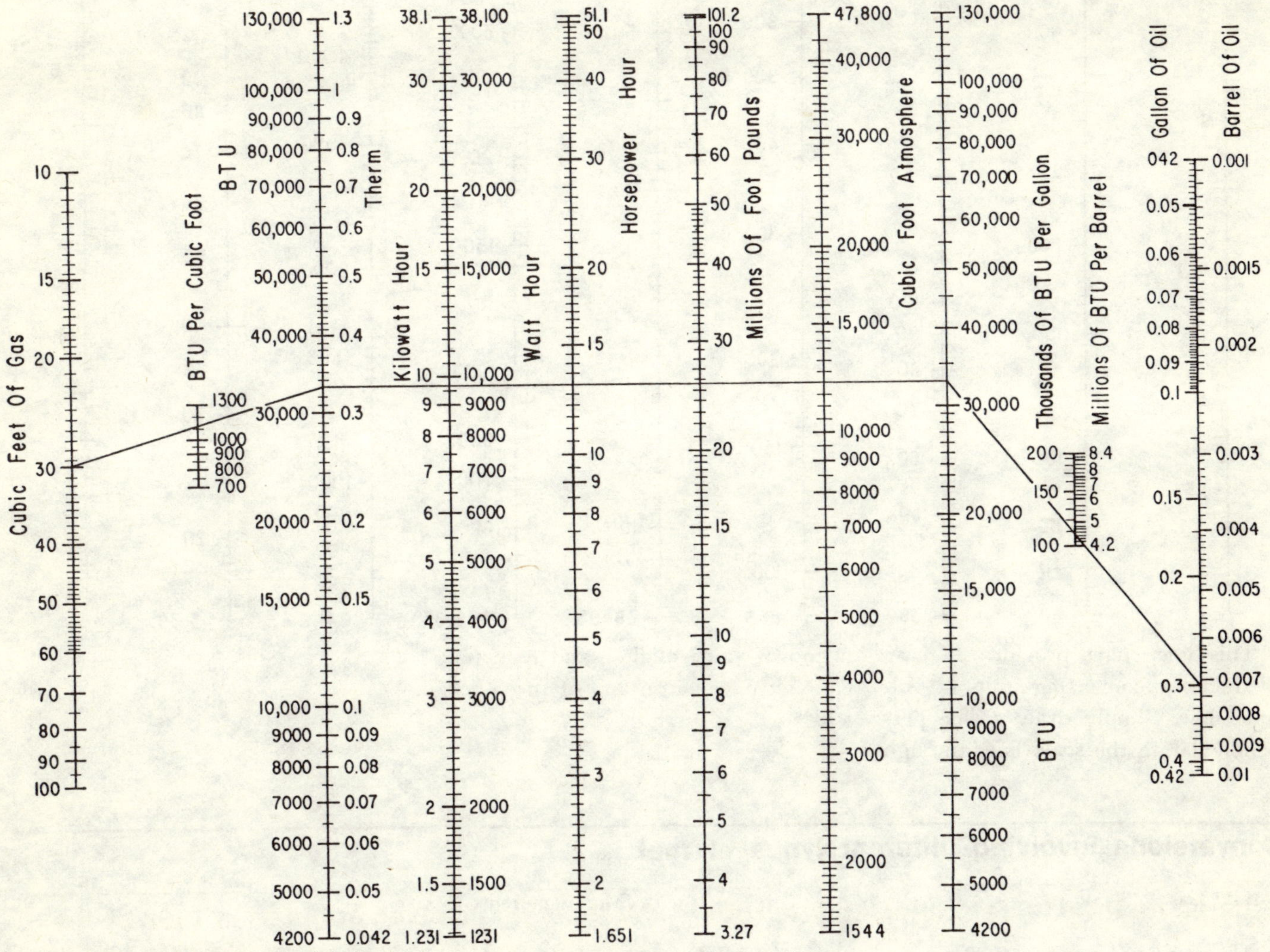

Flow Conversion Chart

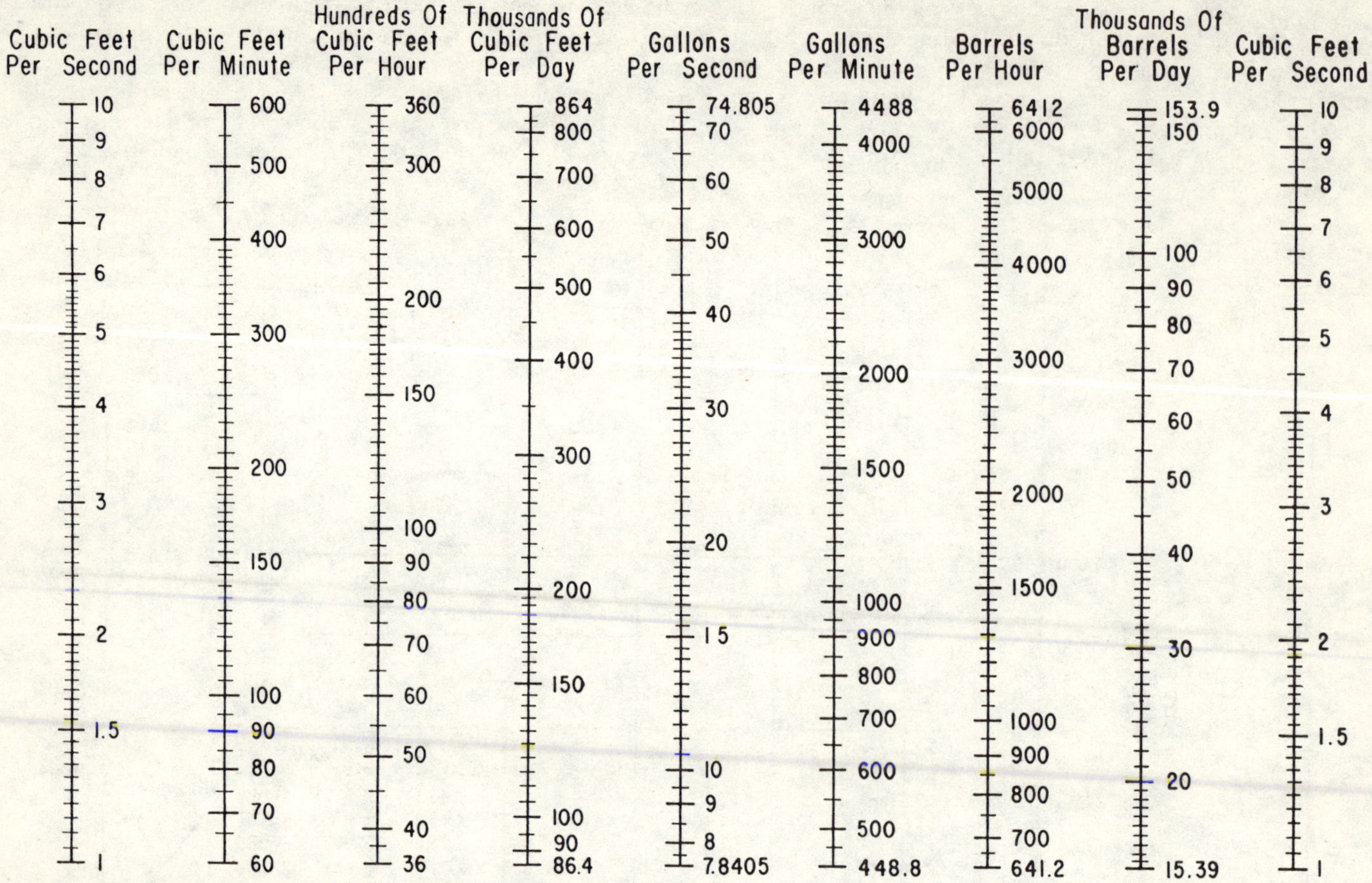

This line chart provides an easy method for converting units of volume flow. Simply draw a line perpendicular to the scale lines through a known value of flow and read the equivalent value on any of the other scales.

Conversions involving different types of fuel

Natural Gas

The calorific value of natural gas varies according to its composition, and virtually no two fields produce gas of identical quality; the composition of gas from any one field can vary over its producing life. Thus there is no such thing as a "standard" natural gas, nor a corresponding "standard" calorific value. However, when referring to natural gas in *hypothetical cases* it is quite common to indicate an "average" gross calorific value of 1,000 Btu/ft^3 (the equivalent of which is taken, for purposes here, to be 9,400 kcal/Nm^3 gross, see note to Table 3). Although in the U.S. most natural gas transmitted is at or slightly above this heat value, this does not generally apply in other countries. As an important example, Groningen (Netherlands) gas, because of a high nitrogen content, has a gross calorific value no greater than 898 Btu/ft^3 measured at 60° F, 30 inHg, dry (7,980 kcal/m^3, or 8,420 kcal/Nm^3).

When precise values are required, reference should be made to the proprietors of the gas field concerned or to published data.

Manufactured Gases

These are derived from primary energy sources by processes involving chemical reaction, e.g. gases produced from coal and naphtha, also propane, butane etc. Gas as piped to consumers can comprise both manufactured gases and natural gas used for enrichment. Calorific values of the component gases can thus vary widely depending on origin, but the resulting blend of what is known as *Town Gas* has a fairly constant heating value which, depending on locality, is usually about half that of natural gas. Exact figures can only come from the distribution company concerned.

Solid Fuels

The calorific value of solid fuels varies widely between countries, grades, qualities and even from seam to seam. It is particularly important therefore to use local data wherever possible. Measurements in Btu/lb or kcal/kg will normally include moisture and ash content as received.

N.B. Btu/lb × 0.5556 = kcal/kg
kcal/kg × 1.8 = Btu/lb

Where actual values are not known, an average gross calorific value of 12,400 Btu/lb (6,890 kcal/kg) is considered reasonable for hard coal grades of good quality. This also serves as the basis of a common unit, "coal equivalent," to which the many grades of solid fuel and other forms of energy may be reduced for the purposes of aggregation. Standard fuel for the purposes of European Economic Community statistics is considered to have a heat value of 7,000 kcal/kg.

Electricity

The basic unit of electric power is the *Watt* (W), with the *Kilowatt-hour* (kWh) as the derivative of most importance.

1 kilowatt-hour (kWh) = 3,412.14 Btu
= 859.845 kcal
= 3,600 kJ

See note to Table 4

Liquid Fuels

In the case of liquid fuels, the actual calorific value of the particular fuel under consideration should be used where possible for conversions to other units, for example when calculating the natural gas equivalent of some specific application. For more general conversions, refer to Table 1.

Table 1
Petroleum fuels volume/weight/calorific value factors

Hydrocarbons	Specific Gravity	Weight/Volume bbl/ton	Gross Calorific Value Btu/Imp. gal	Btu/lb	kcal/kg	10^6 Btu/bbl	10^6 kcal/bbl	b.o.e./bbl
Propane	0.501	12.790	......	21,500	11,943	3.795	0.956	0.654
Butane	0.579	11.060	......	21,140	11,742	4.315	1.088	0.744
Motor Gasoline	0.733	8.740	148,000	20,160	11,200	5.176	1.304	0.892
Kerosene	0.780	8.213	156,000	20,000	11,110	5.456	1.377	0.941
Vaporizing Oil	0.823	7.783	163,000	19,800	11,000	5.701	1.442	0.983
Gas Oil	0.840	7.625	164,000	19,500	10,835	5.736	1.445	0.988
Fuel Oils	0.850	7.536	165,660	19,490	10,830	5.794	1.460	0.999
	0.860	7.448	166,840	19,400	10,780	5.835	1.470	1.006
	0.870	7.362	168,080	19,320	10,730	5.878	1.481	1.013
	0.880	7.278	169,310	19,240	10,690	5.921	1.492	1.021
	0.890	7.196	170,440	19,150	10,640	5.961	1.502	1.028
	0.900	7.116	171,640	19,070	10,590	6.002	1.513	1.035
	0.910	7.038	172,730	18,980	10,540	6.041	1.522	1.042
	0.920	6.961	173,800	18,890	10,490	6.078	1.532	1.048
	0.930	6.887	174,860	18,800	10,440	6.115	1.541	1.054
	0.940	6.813	175,890	18,710	10,390	6.151	1.550	1.061
	0.950	6.741	176,910	18,620	10,340	6.187	1.559	1.067
	0.960	6.671	177,910	18,530	10,290	6.222	1.568	1.073
	0.970	6.602	178,890	18,440	10,240	6.256	1.577	1.079
	0.980	6.535	179,860	18,350	10,190	6.290	1.585	1.084
	0.990	6.469	180,800	18,260	10,140	6.323	1.593	1.090
	1.000	6.404	181,730	18,170	10,090	6.356	1.602	1.096

Sources:
W.P.C. Publication "Technical Data on Fuel" (Editor H. M. Spiers) 6th Edition 1961; all products except fuel oil. Latter from H. Hyams, "Petroleum Cargoes, Measurement and Sampling, 1966."

Notes:
1. Btu/gal × 34.9726 = Btu/bbl
2. Btu/bbl × 0.252 = kcal/bbl
3. b.o.e./bbl = $\frac{10^6 \text{ Btu/bbl}}{5.8}$
4. kcal/kg × 1.8 = Btu/lb
5. To convert 10^6 Btu/bbl to ft^3 natural gas (1000 Btu), multiply by 10^3
6. To convert 10^6 kcal/bbl to m^3 natural gas (9400 kcal), multiply by 106.4

Table 2
Conversion factors for calorific values of gases under different conditions of measurement

A \ B	Btu/ft³ @ 60°F 30 inHg dry	Btu/ft³ @ 60°F 30 inHg saturated	Btu/ft³ @ 60°F 760 mmHg (60°F) dry	kcal/m³ @ 0°C 760 mmHg dry	kcal/m³ @ 0°C 760 mmHg saturated	kcal/m³ @ 15°C 760 mmHg dry	kcal/m³ @ 15°C 760 mmHg saturated
Btu/ft³ @ 60°F 30 inHg dry	1	0.9826	0.9946	9.381	9.325	8.893	8.743
Btu/ft³ @ 60°F 30 inHg saturated	1.0177	1	1.0122	9.547	9.490	9.050	8.898
Btu/ft³ @ 60°F 760 mmHg (60°F) dry	1.0054	0.9879	1	9.431	9.375	8.940	8.790
kcal/m³ @ 0°C 760 mmHg dry	0.10659	0.10474	0.10602	1	0.9940	0.9480	0.9320
kcal/m³ @ 0°C 760 mmHg saturated	0.10724	0.10537	0.10666	1.0060	1	0.9537	0.9376
kcal/m³ @ 15°C 760 mmHg dry	0.11245	0.11049	0.11184	1.0548	1.0486	1	0.9832
kcal/m³ @ 15°C 760 mmHg saturated	0.11437	0.11238	0.11375	1.0729	1.0665	1.0171	1

Notes

1. For equal conditions of temperature, pressure and water vapor content, the following factors apply:

$$1 \text{Btu/ft}^3 = 8.8992 \text{ kcal/m}^3$$
$$1 \text{ kcal/m}^3 = 0.11237 \text{ Btu/ft}^3$$

2. To convert one unit of A to its equivalent heat value in terms of alternative units/conditions B, multiply by the appropriate factor.

Table 3
Heat value conversions and natural gas equivalents of various fuel units

	10^6 Btu gross	10^6 kcal gross	bbl oil equiv.	tons oil equiv.	metric t.c.e.	ft^3 natural gas	m^3 natural gas	10^6 kWh
1 barrel oil equivalent	5.8	1.462	1	0.1349	0.21217	5,800	155.50	0.0017
1 metric tonne coal equivalent	27.337	6.888	4.713	0.6357	1	27,337	732.90	0.0080
1 ton oil equivalent	43	10.836	7.414	1	1.5730	43,000	1152.82	0.0126
1 ton fuel oil equivalent	41.4	10.433	7.138	0.9628	1.5144	41,400	1109.92	0.0121

Notes

The natural gas equivalents above and in Tables 1 (notes) and 4 are based on gas with a gross heat content of 1,000 Btu/ft^3 measured dry at 60° F and 30 inHg, which within 0.2% is equal to 9,400 kcal/m^3 gross measured dry at O°C and 760 mmHg. On this basis, 37.3 cubic feet = 1 cubic meter.

The abbreviation m.t.c.e. is often used for metric tonnes of coal equivalent or million tons coal equivalent. Neither practice is sound as small m is the symbol for meter and not for metric or million. Instead, the ton unit used should be stated in full and the quantity expressed by the use of exponents; for example, 36×10^6 metric t.c.e.

Conversions above and in Table 4 are based on gross calorific values in line with normal statistical practice both inside and outside the gas industry. Net calorific values may however be more appropriate in situations where it is desired to compare or convert fuels in terms of useful heat content, due allowance having been made for changes in end-use efficiency.

Standard Reference Fuels

To convert different types and grades of fuel to a common basis for the purposes of presentation or comparison, it is the practice to express them in terms of some standard unit of a defined calorific value, for example:

Barrel of Oil Equivalent (b.o.e.), with a heat content of 5.8×10^6 Btu gross.

Ton of Oil Equivalent (t.o.e.), simply a hypothetical "ton of oil" with an average heating value of 43×10^6 Btu gross. Oil equivalents may, if required, also be expressed in terms of standard fuel oil equivalent with a calorific value of 41.4×10^6 Btu per ton gross, i.e. 18,500 Btu/lb x 2240.

Metric Tonne Coal Equivalent (t.c.e.), with a heat content of 27.337×10^6 Btu gross, i.e. 12,400 Btu/lb x 2204.62.

The heat values of these different units and conversions between them are given in Table 3, together with natural gas equivalents.

Table 4
Conversion for daily/annual rates of energy consumption (gross heat basis)

	1 x 10^6 metric t.c.e. per year	1 x 10^6 ft^3 nat. gas per day	1 x 10^9 m^3 nat. gas per year	bbl daily oil equiv.	1 x 10^{12} (US trillion) ft^3 nat. gas per year	1 x 10^6 kWh per year	1 x 10^6 thermies per year
1 x 10^6 metric t.c.e. per year =	1	74.896	0.7329	12,913	0.027337	8,012	6,891
100 x 10^6 ft^3 nat. gas per day =	1.3352	100	0.9785	17,241	0.0365	10,697	9,201
1 x 10^9 m^3 nat. gas per year =	1.3645	102.2	1	17,620	0.0373	10,932	9,403
1 x 10^4 bbl daily oil equiv. =	0.7744	58	0.5675	1 x 10^4	0.0212	6,204	5,336
1 x 10^{12} (US trillion) ft^3 nat. gas per year =	36,580	2740	26,808	472,367	1	293,071	252,074
1 x 10^6 kWh per year =	0.1248 x 10^{-3}	0.9348 x 10^{-2}	0.9147 x 10^{-4}	1.6118	0.341 x 10^{-5}	1	0.8601
1 x 10^6 thermies per year =	0.1451 x 10^{-3}	0.1087 x 10^{-1}	0.1064 x 10^{-3}	1.8739	0.397 x 10^{-5}	1.1626	1

Notes

The table is good for natural gas conversions on the basis of a gas of 1,000 Btu/ft^3 gross (9,400 kcal/Nm^3), see note to Table 3 for conditions of measurement. Corrections may be applied as follows:

Gases with other calorific values: apply a correction based on the ratio of the calorific values; e.g. to use for Groningen gas, multiply the horizontal figures shown against 1 x 10^9 m^3 per year by 8420 ÷ 9400 i.e. 0.896 and the corresponding column by 1.116

Gases measured at other conditions of temperature and pressure: apply a correction based on the appropriate factor shown in Table 2; e.g. for gas measured saturated at 60°F and 30_6inHg, multiply the horizontal figures shown against 100 x 10^6 ft^3 per day by 0.9826 and the corresponding column by 1.0177

Calorific value and conditions of measurement different: apply a combined correction; e.g. to use for Gronigen gas measured dry at 15°C and 760 mmHg (m^3_s), multiply the horizontal figures shown against 1 x 10^9 m^3 by 0.948 x 8420 ÷ 9400 i.e. 0.849 and the corresponding column by the reciprocal of this number, i.e. 1.178

To use the table for *1 x 10^6 therms per year,* multiply U.S. trillion row by 10^{-4} and the corresponding column by 10^4

To use the table for *1 x 10^6 tons oil equivalent,* multiply the horizontal figures shown against 1 x 10^6 metric t.c.e. per year by 1.573 and the corresponding column by 0.6357

The factors relating to kilowatt-hours are based on the heat content of the electricity when used and not to the corresponding power station intakes of coal, oil or natural gas, which would need to reflect station generating efficiencies and load factor.

Weight of water per cubic foot at various temperatures

Temperature ° F.	Weight per Cubic Foot, Lbs.	Temperature ° F.	Weight per Cubic Foot, Lbs.	Temperature ° F.	Weight per Cubic Foot, Lbs.	Temperature ° F.	Weight per Cubic Foot, Lbs.	Temperature ° F.	Weight per Cubic Foot, Lbs.
32	62.42	150	61.18	260	58.55	380	54.36	500	48.7
40	62.42	160	60.98	270	58.26	390	53.94	510	48.1
50	62.41	170	60.77	280	57.96	400	53.5	520	47.6
60	62.37	180	60.55	290	57.65	410	53.0	530	47.0
70	62.31	190	60.32	300	57.33	420	52.6	540	46.3
80	62.23	200	60.12	310	57.00	430	52.2	550	45.6
90	62.13	210	59.88	320	56.66	440	51.7	560	44.9
100	62.02	212	59.83	330	56.30	450	51.2	570	44.1
110	61.89	220	59.63	340	55.94	460	50.7	580	43.3
120	61.74	230	59.37	350	55.57	470	50.2	590	42.6
130	61.56	240	59.11	360	55.18	480	49.7	600	41.8
140	61.37	250	58.83	370	54.78	490	49.2	...	.....

From "National Pipe Standards" The National Tube Co.

Engineering constants

Atmosphere:

Theoret. air to burn 1 lb. carbon = 11.52 lbs.
Vol. of 1 lb. air at 32 deg. F and 14.7 lbs. per sq. in. = 12.387 cu. ft.
Vol. of 1 lb. air at 62 deg. F and 14.7 lb. per sq. in. = 13.144 cu. ft.
14.223 lb. per sq. in. = 1 Kg. per sq. cm. = 1 metric atm.
28.8 = equivalent mol. wgt. of air
30 in. Hg. at 62 deg. = atm. press
0.07608 lb. = wgt. 1 cu. ft. air at 62 deg. F and 14.7 lb. per sq. in.
0.08073 lb. = wgt. 1 cu. ft. air at 32 deg. F and 14.7 lb. per sq. in.

Heat:

8 lb. oxygen required to burn 1 lb. hydrogen (H)
777.5 ft. lb. = 1 B.T.U.
62,000 B.T.U. = cal. val. hydrogen (H)
14,600 B.T.U. per lb. cal. val. of carbon (C)
1.8 F deg. = 1 C deg.
2547 B.T.U. per hr. = 1 h.p.
— 270 deg. C = absolute zero
3,415 B.T.U. = 1 kw. hr.
34.56 lb. = wt. air to burn 1 lb. hydrogen (H)
3.9683 B.T.U. = 1 Kg. calorie
4,050 B.T.U. = cal. val. of sulphur (S)
4.32 lb. = wgt. air to burn 1 lb. sulphur (S)

Water:

Freezing point = 0 deg. C, 32 deg. F
Boiling point, atm. press = 100 deg. C, 212 deg. F
Water power available from 1 cu. ft.-sec. falling 1 ft. = 0.1134 h.p.
1.134 ft. at 62 deg. F = 1 in. Hg. at 62 deg F
2.309 ft. at 62 deg. F = 1 lb. per sq. in.
27.71 in. at 62 deg. F = 1 lb. per sq. in.
33.947 ft. at 62 deg. F = atm. press.
5.196 lb. per sq. ft. = 1 in. at 62 deg. F
59.76 lb. = wgt. 1 cu. ft. at 212 deg. F
62.355 lb. = wgt. 1 cu. ft. at 62 deg. F
1 in. at 62 deg. F = 0.0735 in. Hg. at 62 deg. F
8.3356 lb. = wgt. 1 gal. at 62 deg. F
970.4 B.T.U. = latent heat of evap. at 212 deg. F

Steam:

Total heat sat. at atm. press = 1150.4 B.T.U.
0.47 B.T.U. per lb. deg. F = approx. specific heat of superheated steam at atm. press.

Surfaces:

1,273,239 circular mils = 1 sq. in.
1.4142 = square root of 2
144 square ins. = 1sq. ft.
1728 cu. in. = 1 cu. ft.
1.7321 = square root of 3
2.54 cm. = 1 in.
57.296 deg. = 1 radian (angle)
0.062428 lb. per cu. ft. = 1 Kg. per cu. meter
area of circle = 0.7854 (= 3.1416 ÷ 4) x diameter squared

Miscellaneous:

2,116.3 lb. per sq. ft. = atm. press.
2.3026 x log 10a log ea
2.7183 = c = base hyperbolic logs
288,000 B.T.U. per 24 hrs. = 1 ton of refrigeration
3.1416 = π = ratio of circumference of circle to diameter = ratio area of circle to square of radius
550 ft.-lb. per sec. = 1 h.p.
58.349 grains per gal. = 1 gram per liter
61.023 cu. in. = 1 liter
745.7 watts = 1 h.p.
8,760 hr. = 1 year of 365 days
88 ft. per sec. (min.) = 1 mile per min. (hr.)
2.666 lb. = wgt. oxygen required to burn 1 lb. carbon

Mensuration units

Diameter of a circle × 3.1416 = Circumference
Radius of a circle × 6.283185 = Circumference

Square of the radius of a circle × 3.1416 = Area
Square of the diameter of a circle × 0.7854 = Area
Square of the circumference of a circle × 0.07985 = Area
Half the circumference of a circle × half its diameter = Area

Circumference of a circle × 0.159155 = Radius
Square root of the area of a circle × 0.56419 = Radius

Circumference of a circle × 0.31831 = Diameter
Square root of the area of a circle × 1.12838 = Diameter

Diameter of a circle × 0.86 = Side of an inscribed equilateral triangle
Diameter of a circle × 0.7071 = Side of an inscribed square
Circumference of a circle × 0.225 = Side of an inscribed square
Circumference of a circle × 0.282 = Side of an equal square
Diameter of a circle × 0.8862 = Side of an equal square

Base of a triangle × ½ the altitude = Area
Multiplying both diameters and .7854 together = Area of an Elipse

Surface of sphere × 1/6 of its diameter = Solidity

Circumference of a sphere × its diameter = Surface
Square of the diameter of a sphere × 3.1416 = Surface
Square of the circumference of a sphere × 0.3183 = Surface

Cube of the diameter of a sphere × 0.5236 = Solidity
Cube of the radius of a sphere × 4.1888 = Solidity
Cube of the circumference of a sphere × 0.016887 = Solidity

Square root of the surface of a sphere × 0.56419 = Diameter
Square root of the surface of a sphere × 1.772454 = Circumference

Cube root of the solidity of a sphere × 1.2407 = Diameter
Cube root of the solidity of a sphere × 3.8978 = Circumference

Radius of a sphere × 1.1547 = Side of inscribed cube
Square root of (⅓ of the square of) the diameter of a sphere = Side of inscribed cube

Area of its base × ⅓ of its altitude = Solidity of a cone or pyramid, whether round, square or triangular

Area of one of its side × 6 = Surface of cube
Altitude of trapezoid × ½ the sum of its parallel sides = Area

Minutes to decimal hours conversion table

Minutes	Hours	Minutes	Hours
1	.017	31	.517
2	.034	32	.534
3	.050	33	.550
4	.067	34	.567
5	.084	35	.584
6	.100	36	.600
7	.117	37	.617
8	.135	38	.634
9	.150	39	.650
10	.167	40	.667
11	.184	41	.684
12	.200	42	.700
13	.217	43	.717
14	.232	44	.734
15	.250	45	.750
16	.267	46	.767
17	.284	47	.784
18	.300	48	.800
19	.317	49	.817
20	.334	50	.834
21	.350	51	.850
22	.368	52	.867
23	.384	53	.884
24	.400	54	.900
25	.417	55	.917
26	.434	56	.934
27	.450	57	.950
28	.467	58	.967
29	.484	59	.984
30	.500	60	1.000

How to compare costs of gas and alternate fuels

Growing shortage of natural gas prompts consideration of practical replacements, including coal, heavy fuel oil, middle distillates and LPG

Conversion from natural gas to other fuels has progressed from possibility to reality for many companies and will become necessary for many others in months and years ahead.

Fuels considered practical replacements for gas include coal, heavy fuel oils, middle distillates (such as kerosine-type turbo fuel and burner fuel oils) and liquefied petroleum gas (LPG).

To compare relative costs of fuels, the quoted prices normally are converted to the price per million British thermal units (Btu). With HHV being the higher (gross) heating value of the fuel, the following equations may be used:

Coal

$$\text{Cents/million Btu} = \frac{\text{dollars/ton} \times 100 \times 10^6}{\text{lb./ton} \times \text{HHV}}$$

Heavy fuel oils

$$\text{Cents/million Btu} = \frac{\text{dollars/barrel} \times 100 \times 10^6}{\text{gallons/barrel} \times \text{HHV}}$$

Middle distillates and LPG

$$\text{Cents/million Btu} = \frac{\text{cents/gallon} \times 10^6}{\text{HHV}}$$

Natural gas

$$\text{Cents/million Btu} = \frac{\text{cents/Mcf} \times 10^6}{\text{HHV} \times 1{,}000}$$

The approximate heat contents (HHV) of the fuels are as follows:

Coal	
Anthracite	13,900 Btu/lb.
Bituminous	14,000 Btu/lb.
Sub-bituminous	12,600 Btu/lb.
Lignite	11,000 Btu/lb.
Heavy fuel oils and middle distillates	
Kerosine	134,000 Btu/gal.
No. 2 burner fuel oil	140,000 Btu/gal.
No. 4 heavy fuel oil	144,000 Btu/gal.
No. 5 heavy fuel oil	150,000 Btu/gal.
No. 6 heavy fuel oil, 2.7% sulfur	152,000 Btu/gal.
No. 6 heavy fuel oil, 0.3% sulfur	143,800 Btu/gal.
Gas	
Natural	1,000 Btu/cu. ft.
Liquefied butane	103,300 Btu/gal.
Liquefied propane	91,600 Btu/gal.

Typical characteristics of fuel oils

	Kerosine	No. 2 burner fuel
Gravity, °API	41.4	36.3
Pounds per gallon	6.814	7.022
Flash point, °F	133	165
Btu/gal.	134,000	140,000
Viscosity, SSU at 100°F	31.5	35.0
Pour point, °F	−30	0
Sulfur, wt. %	0.04	0.18
Carbon residue, wt. %	0.01	0.05
BS&W, %	—	1.30[2]

[1]Saybolt Furol Viscosity at 122°F
[2]mg/100 ml

LPG

Vapor pressure, psig at 100°F........................188
Distillation, 95% point, °F..........................−40
Residue, vol. %...................................0.05
Oil-ring test..................................none

Notes

Notes

Notes